Advances in Food, Bioproducts and Natural Byproducts for a Sustainable Future: From Conventional to Innovative Processes

Advances in Food, Bioproducts and Natural Byproducts for a Sustainable Future: From Conventional to Innovative Processes

Editors

Isidoro Garcia-Garcia
Jesus Simal-Gandara
Maria Gullo

MDPI • Basel • Beijing • Wuhan • Barcelona • Belgrade • Manchester • Tokyo • Cluj • Tianjin

Editors
Isidoro Garcia-Garcia
University of Cordoba
Spain

Jesus Simal-Gandara
University of Vigo
Spain

Maria Gullo
University of Modena and
Reggio Emilia
Italy

Editorial Office
MDPI
St. Alban-Anlage 66
4052 Basel, Switzerland

This is a reprint of articles from the Special Issue published online in the open access journal *Applied Sciences* (ISSN 2076-3417) (available at: https://www.mdpi.com/journal/applsci/special_issues/Food_Bioproducts_Byproducts).

For citation purposes, cite each article independently as indicated on the article page online and as indicated below:

LastName, A.A.; LastName, B.B.; LastName, C.C. Article Title. *Journal Name* **Year**, *Volume Number*, Page Range.

ISBN 978-3-0365-3957-7 (Hbk)
ISBN 978-3-0365-3958-4 (PDF)

Contents

About the Editors

Isidoro Garcia-Garcia, Full Professor of Chemical Engineering, University of Córdoba, Spain. He works in Biochemical Engineering, especially on bioreactors for acetic acid bacteria (AAB), and is responsible of many research projects as well as contracts with various companies. Despite the limitations on publishing, the transfer of research results to the productive sector has been a fundamental part of his career. A collaboration with international research groups and companies has allowed him to work in different countries (UK, Italy, Germany, El Salvador, Switzerland, Japan and China) and for the coordination of a research network on vinegar and AAB. His contributions to modelling and optimization of biochemical processes, particularly with AAB, could be highlighted with direct applications in some industrial vinegar plants. Additionally, he worked in the development of new fermented products and optimal use of natural resources. Additionally, he has developed an important teaching activity through courses on Chemical and Biochemical Engineering at the University of Cordoba and international universities. https://scholar.google.com/citations?user=iS1vYFwAAAAJ&hl=es.

Jesus Simal-Gandara is Full Professor in Nutrition and Food Science at the Faculty of Food Science and Technology, University of Vigo (Spain). He was the first recipient of the Spanish Award of Completion of Pharmacy and PhD Prize at the Faculty of Pharmacy, University of Santiago de Compostela (Spain). He was Associate Professor in 1991 at the University of Vigo, and has been full Professor since 1999. He is a corresponding member of the Royal Academy of Medicine and Surgery of Galicia (1991) and a member of the Scientific Committee of the Spanish Agency for Consumption, Food Safety and Nutrition (2013–2016). Additional honors include the Research Medal of the Royal Galician Academy of Sciences 2020 Antonio Casares Rodriguez (Chemistry and Geology), President of the International Association of Dietary Nutrition and Safety (2020), Member of the Royal Academy of Pharmacy of Galicia (2021), and Associate Editor in Food Science and Nutrition (Wiley) and Food Chemistry (Elsevier). He leads a research group of excellence, and was leading CIA3 - Environmental, Agricultural and Food Research Center (2008–2018). He was the Head of the Department of Analytical Chemistry and Food Science (2013–2018), and Vice-Chancellor for Internationalization at the University of Vigo (2018). He was nominated by Clarivate Analytics for highly cited research (2018 and 2020) and was invited for research stays at the Université de Paris-Sud (France), University of Delaware (USA), Fraunhofer-Institut für Lebensmitteltechnologie und Verpackung (Germany), Central Science Laboratory (UK), TNO-Voeding (The Netherlands), Packaging Industries Research Association (UK) and The Swedish Institute for Food and Biotechnology (Sweden). He has 19000 citations in 500 papers= 38 per paper; h-index= 73. http://scholar.google.es/citations?user=rmeHFXIAAAAJ&hl=es&oi=ao.

Maria Gullo is Associate Professor in Agricultural Microbiology (AGR/16) at the University of Modena and Reggio Emilia. She is the scientific coordinator of UMMC (Unimore Microbial Culture Collection). Her research activities include the ecophysiology and industrial applications of Acetic Acid Bacteria, biopolymers synthesis and applications, the development of new biotechnological products by selective fermentations and industrial microorganisms managed within UMCC. She has experience in the coordination of national and international projects also related to transfer scientific results to small and medium enterprises. She conceived and organized the "Vinegar and Acetic Acid Bacteria-International Symposium", which was the first meeting devoted to acetic acid bacteria and the products of their metabolism. Currently, she is involved in the scientific committees of the established regular symposia held every 3 years. She has consolidated experience in training PhD, masters and undergraduate students, as well as visiting scientists. She is member of the PhD School Council STEBA (Research Doctorate in Food and Agricultural Science, Technology and Biotechnology). She is Delegate member General Assembly of the Joint Research Unit (JRU) in the frame of Microbial Resource Research Infrastructure MIRRI-IT (MIRRI-ITALIA). https://personale.unimore.it/rubrica/dettaglio/mgullo; https://www.umcc.unimore.it/.

Editorial

Advances in Food, Bioproducts and Natural Byproducts for a Sustainable Future: From Conventional to Innovative Processes

Isidoro Garcia Garcia [1,*], Jesus Simal-Gandara [2] and Maria Gullo [3]

[1] Department of Inorganic Chemistry and Chemical Engineering, Chemical Engineering Area, Faculty of Sciences, Instituto Universitario de Nanoquímica (IUNAN), Universidad de Córdoba, Campus Universitario de Rabanales, Ctra(a). N-IV, km 396, Building Marie Curie (C-3), CP/14071 Cordoba, Spain

[2] Department of Analytical Chemistry and Food Science, Faculty of Food Science and Technology, University of Vigo, Ourense Campus, E-32004 Ourense, Spain; jsimal@uvigo.es

[3] Department of Life Sciences, University of Modena and Reggio Emilia, 42122 Reggio Emilia, Italy; maria.gullo@unimore.it

* Correspondence: isidoro.garcia@uco.es

Citation: Garcia, I.G.; Simal-Gandara, J.; Gullo, M. Advances in Food, Bioproducts and Natural Byproducts for a Sustainable Future: From Conventional to Innovative Processes. *Appl. Sci.* **2022**, *12*, 2893. https://doi.org/10.3390/app12062893

Received: 28 February 2022
Accepted: 7 March 2022
Published: 11 March 2022

Publisher's Note: MDPI stays neutral with regard to jurisdictional claims in published maps and institutional affiliations.

The world population is expected to reach almost 10,000 million in 2050, which entails the need to focus on sustainability and its three pillars: the economy, the environment, and society. Within this context, it is necessary to use our resources efficiently; for instance, we will need to produce much more food using less land and while polluting less to optimize the production of biomass from diversified resources, along with its subsequent conversion, fractionation, and processing. To achieve this, new approaches and processes, with special emphasis from a biotechnological perspective, may need to be implemented to move towards a circular model that will confer environmental sustainability. Global projections of food losses constitute an abundant pool of complex carbohydrates, proteins, lipids, and functional compounds. Hence, the deployment of food waste streams as raw materials will encompass the formulation of added-value products that will be ideally reintroduced in the food supply chain to close the loop.

Therefore, the analysis and optimization of any food and bioproduct process, as well as the development of innovative and emerging food and by-product processing methods, are important as a necessity for the sustainable transition to a bioeconomy era. The valorization, bioprocessing, and biorefining of food-industry-based streams, the role of industrial microorganisms, the isolation of high-added-value compounds, applications of the resulting bio-based chemicals in food manufacturing, novel food formulations, economic policies for food waste management, along with sustainability or techno-economic assessment of processing methods constitute subject areas that need to be addressed. More specifically, bioprocess design to valorize food-industry waste and by-product streams should be initiated by characterizing the composition of the onset raw material with the aim of identifying the target end-products, whereas the generation of multiple high-added-value products is a prerequisite for cost-effective processes to establish economic sustainability. On top of that, the feasibility of innovative processes could be sustained by encompassing food applications, driven by the constantly emerging consumers' demand for functional foods and beverages with enhanced nutritional value. Equally, a growing awareness for bio-based and natural food components is being developed, thereby imposing challenges on the substitution of chemically derived ingredients with their natural counterparts.

In this context, many non-exclusive methods are possible, through which we can approach the general and specific objectives outlined in the previous paragraphs. The papers collected in this book could show some examples that can help to achieve a bioeconomy. The topics they address could be classified under several non-exclusive headings, including the following:

- Use of by-products and/or raw materials that are currently not used sufficiently;
- Waste management;

- Improvement and optimization of existing bioprocesses;
- Development of new processes and products;
- New potentials of traditional products;
- Deepening in the basic aspects behind each product and process.

The interest in the development of molecules with biomedical interest is a topic of great current interest, an example of which is addressed in [1], a publication in which the extraction of Agglutinin is sought from wheat germ, a by-product of the flour-making industry of this cereal. Agglutinin is a molecule with cytotoxic properties that can find very interesting applications in anti-cancer treatments.

The interest in health promotes the search for functional ingredients in different raw materials, in particular, in relatively unknown fruits, many of them tropical, given that these seem to be a very interesting source of phenolic compounds, pectin. and other compounds. An example of this is addressed in the publication [2], in which the possibilities of using the pulp, seed, and peel of the jackfruit are analyzed. In this way, not only would the use of this large fruit, which can weigh up to 40 kg, be optimized or improved, but at the same time, the sources and types of functional ingredients for the pharmaceutical and food industries be increased.

On the other hand, it is not possible to forget because of both resource optimization and environmental reasons the problem of residues. Specifically, those from the agri-food industry have great potential for their use, since they are considered to be a very important source of multiple compounds: food dyes, antioxidants, anti-inflammatory, immunoactive, analgesic, antimicrobial, and other types. In this context, the publications [3,4] address the study of the use of eggplant residues and olive mill wastewater, respectively, as resources that offer many of these possibilities at the same time, as it would reduce its environmental impact.

The interest in a better use of our renewable resources, together with the development of new products, attracts the attention of many researchers. For example, the development of foams, taking into account the multiple applications of these materials, can be an interesting objective, in particular for the food industry, since they can be used as matrices for food or for the development of materials intended for to food packaging. In the publication [5], the possibilities of using starch, cellulose, and proteins from various plants for this purpose are analyzed. On the other hand, the publication [6], also addresses the possibility of developing new products, in this case, new beverages fermented from renewable resources, specifically, molasses mixed with certain medicinal plants. The effect of the culture medium and two lactic acid bacteria has been studied, both separately and together, on some of the properties of the product obtained, e.g., the content of polyphenols, sugars, aromatic compounds and antioxidant capacity.

The search for new products and applications for existing ones is not limited to the study of the use of by-products, residues, or other renewable resources. The possibilities offered by consolidated products are also considered, such as in the case of sourdoughs, see [7]. This is a complex system in which, from the microbial point of view, fundamentally yeasts and lactic acid bacteria can be found. Although they have traditionally been used to improve the sensory, textural, and other properties of bread and other baked products, new potentials linked to the production of products with higher added-value that are healthier and more natural are being studied, for example, related to the formation of exopolysaccharides and antifungal compounds, all this while using new raw materials that allow optimal flours to be obtained without the problems of allergies and intolerances that are occurring at the moment.

The improvement and optimization of existing processes, as well as the development of new ones, is of great importance in the context of the general objectives indicated earlier. The optimization of any process implies an adequate use of all the material and energy resources that it consumes; therefore, it is possible to increase productivity and reduce the environmental footprint that it supposes. When dealing with an increase in the world population while trying to reduce the impact that population has on the planet, this is of great importance. Examples of studies that we could classify in this context can be found in other publications of this book. For example, [8,9] deal with the study and modeling of an alternative way of working in the acetification stage of the wine vinegar bioprocess, a product with multiple uses and great importance: For instance, it is a high-quality condiment. For another example, see the publication [10], which deals with the study of the influence of operational variables, temperature, and time on the properties and durability of the processing of another food, sous vide chicken breast fillets; this case is an example of the need to optimize processing conditions in order to reduce the generation of waste from food that has already been prepared and not consumed. Another example that shows the interest in optimizing the use of raw materials could be the one presented in the publication [11], in which the resistance of paper against moisture and oil is analyzed after coating with Poly (-3-hydroxybutyrate-co-3-hydroxyvalerate) (PHBV) and Polycaprolactone (PCL). The work shows the possibilities and interest in the development of recyclable cellulosic materials that are in contact with food, helping to maintain it.

On the other hand, another topic of interest is related to the development of new processes for obtaining products that are already on the market, or products similar to some already existing ones, but with various advantages in terms of new uses, the use of other raw materials to obtain them or increase the productivity and/or efficiency of the new process. For example, in the publication [12], an interesting topic related to the production of bacterial cellulose is addressed; unlike the cellulose of vegetal origin, this product has a series of properties, among which its high purity, its water-retention capacity, its mechanical resistance, and its biocompatibility can be highlighted, which make it very attractive for various uses. However, the main problem for its development is related to the high production costs. The selection of new producing species could contribute to reduce these costs and the development of specific processes; for this reason, a search for new bacterial species with high cellulose production capability is carried out, Kombucha being one of the most interesting niches where this type of bacteria could be found. Although Kombucha is the result of the activity of a complex microbiota, its bacterial fraction is mainly made up of bacteria of the *Komagataeibacter* genus. Another matter of interest is related to enzymatic processes in which, instead of working with microorganisms, enzymes are previously obtained that, after purification, are used to carry out very specific reactions; in this way, the selectivity of the biotransformations that are intended is much higher, not as it happens in the work with microorganisms. In this context, in order to reuse enzymes as much as possible, a recurring theme is the possibility of immobilizing them on supports and, in this way, being able to retain them. Publication [13] analyzes the behavior of pectinolytic enzymes immobilized on Nylon.

On the other hand, every day, more work is conducted on the development of crops when, for whatever reason, there is no adequate natural environment (terrain, lighting, etc.). In this sense, the use of LED lighting to replace natural light is very interesting. The publications [14,15] address an example of this topic, specifically, the germination and growth of *Ocimum Basilicum* is analyzed, in the absence of natural light, depending on the type of LED light, the fertilizers that are used, the distance between plants, and other aspects. The optimization of the operating conditions of plant growth can be of great interest for various applications.

Finally, the publication [16], in the context of winemaking, deals with an example of the studies that are increasingly important for food production, knowing the processes at their most basic, molecular scale. The current availability of powerful analysis techniques makes it possible to identify, in complex samples, the basic molecules involved in the metabolisms responsible for the biotransformations behind the bioprocesses with which we produce many of our foods or with which we carry out other types of biotransformations. This knowledge is essential to optimize the transformations that are carried out and to improve the quality of the products obtained.

Appl. Sci. **2022**, *12*, 2893

Author Contributions: All authors have contributed equally to conceptualization, methodology, writing—original draft preparation, writing—review and editing. All authors have read and agreed to the published version of the manuscript.

Funding: This research received no external funding.

Acknowledgments: We would like to thank all the authors and peer reviewers for their valuable contributions to this Special Issue. This issue would not be possible without their valuable and professional work. In addition, we would also like to take the opportunity to show our gratitude to the Journal's editorial team for its work.

Conflicts of Interest: The authors declare no conflict of interest.

References

1. Balčiūnaitė-Murzienė, G.; Dzikaras, M. Wheat Germ Agglutinin—From Toxicity to Biomedical Applications. *Appl. Sci.* **2021**, *11*, 884. [CrossRef]
2. Cruz-Casillas, F.; García-Cayuela, T.; Rodriguez-Martinez, V. Application of Conventional and Non-Conventional Extraction Methods to Obtain Functional Ingredients from Jackfruit (*Artocarpus heterophyllus* Lam.) Tissues and By-Products. *Appl. Sci.* **2021**, *11*, 7303. [CrossRef]
3. Silva, G.; Pereira, E.; Melgar, B.; Stojković, D.; Sokovic, M.; Calhelha, R.; Pereira, C.; Abreu, R.; Ferreira, I.; Barros, L. Eggplant Fruit (*Solanum melongena* L.) and Bio-Residues as a Source of Nutrients, Bioactive Compounds, and Food Colorants, Using Innovative Food Technologies. *Appl. Sci.* **2021**, *11*, 151. [CrossRef]
4. Foti, P.; Romeo, F.; Russo, N.; Pino, A.; Vaccalluzzo, A.; Caggia, C.; Randazzo, C. Olive Mill Wastewater as Renewable Raw Materials to Generate High Added-Value Ingredients for Agro-Food Industries. *Appl. Sci.* **2021**, *11*, 7511. [CrossRef]
5. Jarpa-Parra, M.; Chen, L. Applications of Plant Polymer-Based Solid Foams: Current Trends in the Food Industry. *Appl. Sci.* **2021**, *11*, 9605. [CrossRef]
6. Gadhoumi, H.; Gullo, M.; De Vero, L.; Martinez-Rojas, E.; Saidani Tounsi, M.; Hayouni, E. Design of a New Fermented Beverage from Medicinal Plants and Organic Sugarcane Molasses via Lactic Fermentation. *Appl. Sci.* **2021**, *11*, 6089. [CrossRef]
7. De Vero, L.; Iosca, G.; Gullo, M.; Pulvirenti, A. Functional and Healthy Features of Conventional and Non-Conventional Sourdoughs. *Appl. Sci.* **2021**, *11*, 3694. [CrossRef]
8. Álvarez-Cáliz, C.; Santos-Dueñas, I.; Jiménez-Hornero, J.; García-García, I. Modelling of the Acetification Stage in the Production of Wine Vinegar by Use of Two Serial Bioreactors. *Appl. Sci.* **2020**, *10*, 9064. [CrossRef]
9. Álvarez-Cáliz, C.; Santos-Dueñas, I.; Jiménez-Hornero, J.; García-García, I. Optimization of the Acetification Stage in the Production of Wine Vinegar by Use of Two Serial Bioreactors. *Appl. Sci.* **2021**, *11*, 1217. [CrossRef]
10. Haghighi, H.; Belmonte, A.; Masino, F.; Minelli, G.; Lo Fiego, D.; Pulvirenti, A. Effect of Time and Temperature on Physicochemical and Microbiological Properties of Sous Vide Chicken Breast Fillets. *Appl. Sci.* **2021**, *11*, 3189. [CrossRef]
11. Lo Faro, E.; Menozzi, C.; Licciardello, F.; Fava, P. Improvement of Paper Resistance against Moisture and Oil by Coating with Poly(-3-hydroxybutyrate-co-3-hydroxyvalerate) (PHBV) and Polycaprolactone (PCL). *Appl. Sci.* **2021**, *11*, 8058. [CrossRef]
12. La China, S.; De Vero, L.; Anguluri, K.; Brugnoli, M.; Mamlouk, D.; Gullo, M. Kombucha Tea as a Reservoir of Cellulose Producing Bacteria: Assessing Diversity among *Komagataeibacter* Isolates. *Appl. Sci.* **2021**, *11*, 1595. [CrossRef]
13. Ben-Othman, S.; Rinken, T. Immobilization of Pectinolytic Enzymes on Nylon 6/6 Carriers. *Appl. Sci.* **2021**, *11*, 4591. [CrossRef]
14. Barbi, S.; Barbieri, F.; Bertacchini, A.; Barbieri, L.; Montorsi, M. Effects of Different LED Light Recipes and NPK Fertilizers on Basil Cultivation for Automated and Integrated Horticulture Methods. *Appl. Sci.* **2021**, *11*, 2497. [CrossRef]
15. Barbi, S.; Barbieri, F.; Bertacchini, A.; Montorsi, M. Statistical Optimization of a Hyper Red, Deep Blue, and White LEDs Light Combination for Controlled Basil Horticulture. *Appl. Sci.* **2021**, *11*, 9279. [CrossRef]
16. González-Jiménez, M.; Mauricio, J.; Moreno-García, J.; Puig-Pujol, A.; Moreno, J.; García-Martínez, T. Differential Response of the Proteins Involved in Amino Acid Metabolism in Two *Saccharomyces cerevisiae* Strains during the Second Fermentation in a Sealed Bottle. *Appl. Sci.* **2021**, *11*, 12165. [CrossRef]

Review

Wheat Germ Agglutinin—From Toxicity to Biomedical Applications

Gabrielė Balčiūnaitė-Murzienė [1,2,*] **and Mindaugas Dzikaras** [2]

[1] Institute of Pharmaceutical Technologies, Faculty of Pharmacy, Academy of Medicine, Lithuanian University of Health Sciences, 44307 Kaunas, Lithuania
[2] Panevėžys Institute of Technologies and Business, Kaunas University of Technology, 44249 Kaunas, Lithuania; Mindaugas.Dzikaras@ktu.edu
* Correspondence: gabriele.balciunaite-murziene@lsmuni.lt; Tel.: +370-638-85187

Featured Application: Wheat germ agglutinin has the potential for enabling and improving targeted drug delivery systems, anticancer drugs, and antibacterial and antifungal therapeutics due to its cytotoxic mechanisms and specific carbohydrate binding.

Abstract: Wheat germ agglutinin is a hevein class N-Acetylglucosamine–binding protein with specific toxicity and biomedical potential. It is extractable from wheat germ—a low-value byproduct of the wheat industry—using well–established extraction methods based on salt precipitation and affinity chromatography. Due to its N-Acetylglucosamine affinity, wheat germ agglutinin exhibits antifungal properties as well as cytotoxic properties. Its anticancer properties have been demonstrated for various cancer cells, and toxicity mechanisms are well described. Wheat germ agglutinin has been demonstrated as a viable solution for various biomedical and therapeutic applications, such as chemotherapy, targeted drug delivery, antibiotic-resistant bacteria monitoring and elimination. This is performed mostly in conjunction with nanoparticles, liposomes, and other carrier mechanisms via surface functionalization. Combined with abundant wheat byproduct sources, wheat germ agglutinin has the potential to improve the biomedical field considerably.

Keywords: wheat; germ; wheat byproducts; agglutinin; WGA; toxicity; glycosylation; N-Acetylglucosamine; GlcNAc; carbohydrate

Citation: Balčiūnaitė-Murzienė, G.; Dzikaras, M. Wheat Germ Agglutinin—From Toxicity to Biomedical Applications. *Appl. Sci.* **2021**, *11*, 884. https://doi.org/10.3390/app11020884

Received: 2 December 2020
Accepted: 15 January 2021
Published: 19 January 2021

Publisher's Note: MDPI stays neutral with regard to jurisdictional claims in published maps and institutional affiliations.

1. Introduction

Wheat (*Triticum aestivum* L.) is one of the most essential agricultural staple foods used for human consumption and animal feed. Approximately 21% of the world's food supplies depend on annual wheat crop harvest [1], which causes the production of byproducts that are not always used efficiently or, in some cases, entirely discarded as waste [2]. Increasing trends of consumption-based economy restructuring to more ecologically-minded circular economy models encourage the scientific development of new technologies for byproduct valorization to create products with high nutritional value. One of the wheat processing byproducts which takes 2–3% of whole wheat grain [3] is wheat germ. They are a source of oils [4], tocopherols [5], various polyphenols [6], and specific proteins, such as agglutinins (lectins) [7]. Even though agglutinins are considered to be antinutrients [8], they might have various prospective applications in biomedicine, biotechnology, and agriculture itself. Considering the high quantities of wheat germ available as low-value byproducts throughout the world, any valorization attempts by extracting specific proteins may lead to high impact worldwide. This review is focused on wheat germ agglutinin, its structure and specificity, extraction and purification methods, biological activity, and possible applications.

2. Wheat Germ Agglutinin Structure

Wheat germ agglutinin (WGA) is one of the first purified lectins extracted at the very beginning of the lectinomics field. WGA structure was described during the 1970s. WGA is a mixture of three closely related major isoforms, named WGA1, WGA2, and WGA3 [9], which are a 36.0 kDa size stable 18.0 kDa polypeptide chain homodimer [10] with twofold axis symmetry [11]. Variability of these three isolectins is observed at 10 sequence positions, with WGA3 being the most distinct form, differing from WGA1 by 8 positions and from WGA2 by 7 positions [12]. The polypeptide chain is stable under high-temperature exposure [13]. Moreover, WGA monomers are highly resistant to acidity, and conformational changes can be reversed by increasing pH [14].

Each polypeptide chain is composed of four hevein domains, named A, B, C, and D [15,16] (Figure 1). WGA3 exhibits higher interdomain similarity than WGA1 or WGA2, suggesting closer relatedness to the common ancestral molecule [12]. Amino acid analysis shows that WGA contains high amounts of glycine and half-cystine, features not typical to most of the lectins [17]. Additionally, the protein is rich in disulfide bridges, with each hevein domain containing eight disulfide-forming cysteines [18]. These disulfides have been shown to play an important role in hevein stability since heveins lack a hydrophobic core [19]. Additionally, the disulfides seem to explain low pH value stability [14]. WGA has been shown to undergo cotranslational processing of glycan addition to the C-terminus, which is later post-translationally removed before WGA reaches mature form [20]. Affinity studies show that subunit specificity to oligosaccharides is much better than to monosaccharides. A, B, and C subunits bind *N*-Acetylglucosamine (GlcNAc) residues, such as chitin, whereas subunit D accommodates the glycoside aglycones. Monosaccharides are only bound to subunit C [21]. However, low affinity to other carbohydrates, such as N-Acetylneuraminic (sialic) acid, plays an important role in WGA activity [22]. Based on the structure and carbohydrate specificity, WGA is classified as chitin-binding lectin composed of hevein domains [23]. Lectin structure and carbohydrate specificity enable its broad spectrum of biological activities.

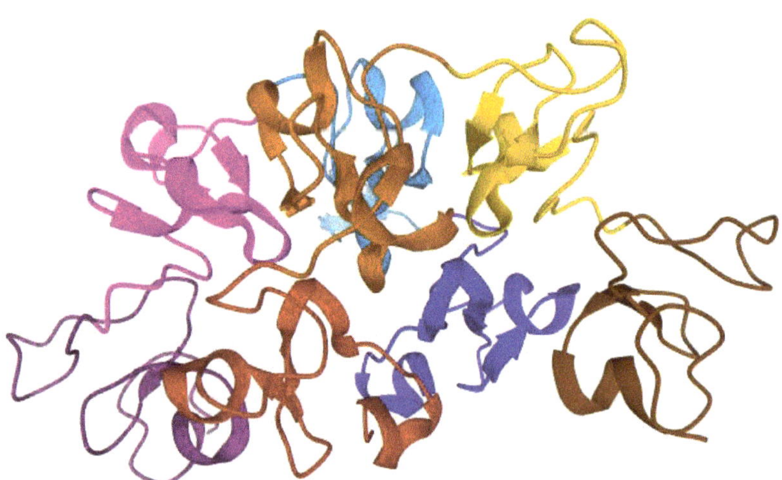

Figure 1. Dimeric wheat germ agglutinin 3 (WGA3) structure by [24]. Red, orange, yellow, and brown colors represent the subunits A, B, C, and D of the first protein, respectively. Blue, light blue, pink, and purple represent the subunits A, B, C, and D of the second protein, respectively. The sugar-binding site is located at the interface of both WGA monomers.

3. WGA Extraction and Purification

The most widely used natural WGA purification from wheat germ strategies include a series of protein precipitation and chromatographic purification steps. Untreated wheat germ material goes through a defatting step followed by material disruption with a laboratory mill. The material preparation is followed by extraction in water or aqueous buffers. Crude wheat extracts are used for protein precipitation with salts, mostly ammonium sulfate. Precipitated wheat proteins are resuspended and dialyzed. These steps are followed by chromatographic purification steps, including ion exchange and affinity chromatography. Table 1 represents summarized WGA purification steps and WGA yield (mg per 1 g of raw germ material).

Table 1. Wheat germ agglutinin (WGA) purification strategies and yields.

Defatting Agent	Purification Strategy	Number of Steps	Yield of WGA (mg Per 1 g Raw Wheat Germ)	Reference
Light petroleum	Extraction in water Protein precipitation 55% $(NH_4)_2SO_4$ Dialysis in water Ion exchange chromatography using a DEAE [1]-cellulose column, two SE [2]-Sephadex columns, and QAE [3]-Sephadex ($\times 4$)	7	0.61	[22]
Light petroleum	Extraction in water Protein precipitation 55% $(NH_4)_2SO_4$ Dialysis in water Affinity chromatography CNAG [4]-Sepharose ($\times 1$)	4	0.3	[25]
Acetone	Extraction in 0.05 N HC1 buffer Protein precipitation 35% $(NH_4)_2SO_4$ n-Butanol treatment Dialysis in 0.05 N HC1 and second protein precipitation 35% $(NH_4)_2SO_4$ Ion exchange chromatography on DEAE-cellulose column ($\times 1$)	6	0.48	[17]
n-Hexane	Extraction in 0.05 N HC1 buffer Tangential flow filtration MWCO [5] 10 kDa Batch affinity purification on chitosan matrices ($\times 1$)	3		[26]

[1]–DEAE–diethylaminoethyl; [2]–sulphoethyl; [3]–Diethyl-(2-hydroxypropyl)aminoethyl; [4]–CNAG-2-acetamido-N-(ε-aminocaproyl)-2-deoxy-β-D-glucopyranosylamine; [5]–MWCO–Molecular weight cut-off.

4. WGA Biological Activity

4.1. Antifungal Activity

The main fungal cell wall component is chitin, which is a $\beta(1,4)$-homopolymer of GlcNAc. WGA specifically binds to fungal cell walls and disrupts the cell's structural integrity [27]. It demonstrates WGA function in plant immunity against pathogenic fungi. However, there are not many results published regarding native WGA antifungal activity to human pathogenic fungi [28,29]. Recombinant WGA linked to the effector Fc region of murine IgG2a was designed for mycosis treatment. WGA-Fc antibodies were successfully bound to chitin standard fungal cultures and inhibited the growth of *Histoplasma capsulatum*, *Cryptococcus neoformans*, *Candida albicans*, and *Saccharomyces cerevisiae*. It also augmented antifungal macrophage functions and fully protected mice against *H. capsulatum* infections [29]. Other hevein class lectins demonstrate strong antifungal activity against *Candida albicans* (URM 5901), minimal inhibitory concentration (MIC) 25 μg/mL, against *Candida krusei* (URM 6391)–MIC 12.5 μg/mL [30] and *Candida tropicalis* (ATCC 750)–MIC 11.9 μg/mL [31].

4.2. Cytotoxicity

Since the early discovery of lectins, they were considered toxic due to the obvious toxicity of first-identified lectins, such as abrin [32] or ricin [33]. Even though only a few lectins are highly toxic in low doses, cytotoxic activity is typically dose-dependent. Lectin cytotoxicity can occur in a lectin-dependent cell-mediated manner or as direct toxicity to cells [34]. Lectins interact with immune cells as antigens and exhibit their cytotoxic activities and induce a humoral response. Concentration-dependent increases with IFN-γ, and TNF-α release were observed after whole blood samples were treated with WGA [35]. Another study has found that WGA binds to glycosylated T3 receptor, which is followed by the formation of inositol phosphate—an intracellular secondary messenger that takes part in the regulation of cell functions, such as apoptosis—in human blood T cells [36]. On the other hand, WGA inhibits T lymphocyte proliferation by inhibiting the responsiveness of the lymphocytes to interleukin 2 (IL-2) rather than interfering with IL-2 production and IL-2 receptor expression [37,38].

Direct lectin cytotoxicity depends on interaction time and affinity to glycoconjugates on the cell surface. WGA demonstrates high specificity to GlcNAc, but interaction with sialic acid is important for WGA cell-killing activity. WGA-induced acute myeloid leukemia cell death is dependent on both GlcNAc binding and interaction with sialic acids. The authors noticed a reduction in WGA cell binding and killing after sialic acid was removed by neuraminidase treatment [39]. Sialic acid specificity plays a crucial role because most of the cancer cells are hyper sialylated [40,41]. Experiments show that WGA is endocytosed into the Chinese hamster ovary (CHO) cells after 30 min of treatment [42]. Later research confirmed the same time-dependent endocytosis with leukemic Jurkat cell line [43]. WGA is endocytosed into cells by clathrin-independent endocytosis mechanism depending on interactions with N-acetylneuraminic acid or N-acetylglucosamine (GlcNAc) on the cell surface of normal and malignant cells [44,45]. WGA binds high molecular weight glycoproteins (HMWAG) expressed on the cell surface. WGA-HMWAG complexes rapidly become localized within endosomes and then move more slowly to a tubular endocytic network within the perinuclear region of the cell [42]. Endocytosed WGA interacts with the cell nuclear envelope pore membrane protein POM 121 and blocks RNA export [46]. This interaction with nuclear pore proteins depends on GlcNAc glycosylation moieties in other cell models of the nuclear transport blockade [47,48]. Endocytosed WGA can suppress rat pancreatic acinar tumor cell line proliferation by blocking auto stimulation and prevention of the trophic effect of glycosylated cholecystokinin or gastrin–like peptides [49].

WGA is shown to exert cytotoxic effects to differing extents with a final outcome (apoptosis or necrosis) depending on the cell line used in toxicity assays. WGA induced paraptosis in L929 murine fibroblast cell line. Vacuolation, loss of cell architecture, and upregulation of the apoptosis-related proteins Bax and caspase-3 were observed. However, internucleosomal DNA fragmentation, apoptotic bodies, and related apoptotic phenotypes were not detected [50]. Meanwhile, WGA intake into pancreatic carcinoma cell lines caused chromatin condensation, nuclear fragmentation, and DNA release consistent with apoptosis [51]. Interaction with the leukemic Jurkat cell line induced a loss of transmembrane potential, disruption of the inner mitochondrial membrane, and release of cytochrome c and caspase-9 activation of apoptosis [43]. These variances of outcomes might also be caused by differences between cell glycosylation patterns [52].

5. WGA Biomedical Perspectives

5.1. Advanced Drug Delivery

Despite the WGA toxicity, glycosylation-targeted reactivity with biological molecules might have great potential for WGA application in modern biomedicine. Recent studies focus on the improvement of drug delivery systems. WGA–anchored nanoparticles improved nanoparticle adhesion onto cells [53]. Paclitaxel-loaded nanoparticles with immobilized WGA molecules demonstrated higher cytotoxicity to cancer cell lines than the paclitaxel alone [54,55]. This efficacy improvement could be attributed to a more efficient cellular uptake via WGA-receptor-mediated endocytosis and isopropyl myristate-facilitated release of paclitaxel from the nanoparticles. This efficacy enhancement is observed with other chemotherapeutic agents, too, such as. docetaxel [56] or doxorubicin [57].

Biotechnologically synthesized macromolecular drugs are considered to be the future of the pharmacy. Good response to these drugs is limited due to poor absorption and high susceptibility to the loss of their activity. WGA as a polypeptide delivery mediator was tested with a fluorescein-labeled (F) bull serum albumin (BSA) complex on the Caco-2 cell line. Results indicated up to 8.7 times higher F-BSA/WGA complex binding to Caco-2 cells as compared with glycyl-F-BSA complex alone. Moreover, about 75% of F-BSA-WGA were bound specifically to Caco-2 cells [58]. More complex nanoparticles (liposomes) with a polymer–WGA surface conjugate led to an increase in the association between liposomes and lung epithelial cells enhanced systemic absorption of calcitonin model peptide improved bio adhesion to lung epithelia also significantly enhanced and prolonged the therapeutic efficacy of calcitonin [59]. WGA-functionalized carboxymethylated kappa–carrageenan microparticles with entrapped insulin molecules were protected from hydrolysis and proteolysis by stomach acids and enzymes. Microparticle interactions with intestinal wall cells and insulin absorption were improved by WGA grafts [60]. Technologically advanced systems with functionalized lectins demonstrate higher lectin efficacy than standalone lectins. Magnetic nanoparticles with immobilized WGA increased apoptosis in prostate cancer cells in vitro and in vivo in the presence of a magnetic field [61]. Lectin interaction with cell glycosylation points facilitated the internalization of chemotherapeutic compounds and magnetic particles, which results in higher targeted efficacy in cancer treatment. WGA was also used to develop a method for photodynamic inactivation of methicillin-resistant *S. aureus* and *Pseudomonas aeruginosa* [62]. WGA-modified liposomes loaded with generation II photosensitizers exhibited considerably higher photosensitizers deliverance to the bacteria than the unmodified liposomes, demonstrating WGA–liposome complex use not only for antibiotic-resistant bacteria detection but also elimination. Table 2 summarizes various developed drug delivery systems utilizing WGA.

5.2. Overcoming Physiological Barriers

Drug delivery usually involves crossing specific biological barriers. Nonspecific distribution and inadequate accumulation of therapeutics remain formidable challenges for drug developers. The blood-brain barrier (BBB) might also be overcome with WGA incorporation on nanoparticles [63–65]. PEGylated (Polyethylene glycol modified) fourth-generation poly(amidoamine) (PAMAM) dendrimer-based nanocarrier containing doxorubicin (DOX) inside and WGA with transferrin (Tf) as targeting molecules were developed for BBB crossing. A PA-MAM–WGA–Tf nanocarrier increased DOX delivery through BBB by 13.5% compared to free doxorubicin. Experiments also showed increased DOX accumulation in the tumor site due to dual targeting mechanisms. Accumulation led to the complete breakage of the avascular C6 glioma spheroids in vitro [66]. Recent studies suggest a novel intranasal nanoparticle delivery strategy to overcome BBB. WGA-modified PEG-poly(lactic acid) (PLA) nanoparticles with miR132 were administered to rodents. Transmission electron microscopy results demonstrated that the accumulation of WGA–modified nanoparticles in the brain was significantly higher than that of unmodified nanoparticles. Moreover, the results of Morris water maze analysis showed that the intranasal delivery of WGA–PLA–miR132 improved the learning and memory function of APP/PS1 mice [67]. These results suggest novel drug delivery through BBB strategies, which might overcome treatment-resistant intracranial tumors and neurodegenerative diseases, such as Alzheimer's disease.

Table 2. Summarized studies on WGA grafted drug delivery systems.

Carrier Particle	Drug	Excipients	Animal/Cell Line	Outcome	Reference
Polymeric nanoparticles	Paclitaxel	Poly (lactic-co-glycolic acid) Isopropyl myristate	A549 H1299 CCL-186 cell lines	Demonstrated higher cytotoxicity	[54]
Polymeric nanoparticles	Paclitaxel	Poly (lactic-co-glycolic acid)	Caco-2 HT-29 cell lines	Increased intracellular retention in the Caco-2 and HT-29 cells Endocytosed nanoparticles could successfully escape from the endo-lysosome compartment and release into the cytosol with increasing incubation time	[55]
Polymeric nanoparticles	Docetaxel	Thiolated sodium alginate	HT-29 L929 cell lines	IC_{50} values of 52.9 µg/mL for HT-29 cells and 201.6 µg/mL for L929 cells Selectivity towards HT-29 cells over L929 cells	[56]
Polymeric nanoparticles	Etoposide Carmustine Doxorubicin	Methoxy poly(ethylene glycol)-poly(ε-caprolactone) Folic acid	U87MG cell line	WGA-modified surface promoted BBB permeation, and folic acid facilitated target site on U87MG cells Anti-proliferation against U87MG cells	[57]
Liposomes	Calcitonin	Carbopol	A549 cell line	Carbopol-WGA modification enhanced interaction with A549 lung epithelial cells compared with unmodified or CP-modified liposomes Enhanced and prolonged efficacy of calcitonin.	[59]
Microparticles	Insulin	Carboxymethylated kappa–carrageenan microparticles	Caco-2 cell line Sprague–Dawley rats	WGA-functionalized microparticles at 20 mg/mL showed a reduction in cell viability upon exposure The oral administration of insulin entrapped in the microparticles led to a prolonged duration of the hypoglycemic effect, up to 12–24 h, in diabetic rats.	[60]
Metal-oxide based nanoparticles	-	Iron oxide Fe3O4	PZ-HPV-7 DU-145 PC-3 LNCaP cell lines DU-145 e BALB/c-nu/nu mice	2.46 nM nanomagnet lectin exposure for 15 min induce apoptosis of cancer cells; Xenografted (DU-145) e BALB/c-nu/nu mice, where the tumor was not only completely arrested but also reduced.	[61]
Liposomes	Temoporfin	1,2-dipalmitoyl-sn-glycero-3-phosphocholin N-[1-(2,3-dioleoyloxy)propyl]-N,N,N-trimethylammonium methyl sulfate 1,2-distearoyl-sn-glycero-3-phosphoethanolamine-N-[3-(N-succinimidyloxyglutaryl) aminopropyl (polyethyleneglycol)-2000-carbamyl]	Methicillin-resistant *Staphylococcus aureus* (MRSA) and *Pseudomonas aeruginosa* cultures	The WGA-modified liposomes eradicated all MRSA and significantly enhanced the photodynamic inactivation of *P. aeruginosa*.	[62]

WGA-incorporated nanoparticles also demonstrate promising features for targeting drugs to bladder cancer cells. Due to the urothelial barrier, constant active drug substance washout, and short dwelling time, bladder cancers and urinary tract infections are another difficulty for efficient drug delivery. Surgical tumor resection is usually followed by repeated intravesical chemotherapy. WGA alone demonstrates high binding to 5637 bladder cancer cells [68]. Later investigations in flow-cytometric and fluorometric experiments with single cells and cell monolayers showed fluorescein (Fc)-labeled poly-lglutamic acid (PGA)–WGA drug delivery system binding and the internalization of high molecular weight (>100 kDa) conjugates. Results of specificity studies showed that the interaction between the Fc–PGA–WGA conjugates and the cell surface depended solely on the WGA component [69]. A recent study by Brauner and colleagues demonstrated poly(d,l-lactic-co-glycolic acid) (PLGA)–WGA nanoparticles' inherent adhesion capability to immortalized human uroepithelial cells. A slight adhesion capacity increase was WGA biospecificity-dependant [70].

WGA–liposome–cyclodextrin complexes were bioengineered for oral cell therapeutics [71]. The designed carrier showed fast attachment to oral cells and sustained co-drug release of ciprofloxacin and betamethasone in saliva in vitro. The complex significantly increased oral cell survival against *Aggregatibacter actinomycetemcomitans*, a bacterial pathogen responsible for chronic periodontal disease, and simultaneously reduced the inflammation. Crucially, WGA-facilitated cell-binding prevented complex degradation and loss due to salivary flushing and/or mucus turnover and allowed for sustained 24 h drug release.

These promising results might eventually bring some breakthroughs in medicine. However, in vitro and some in vivo results with animal models still have to go through in-depth clinical trials before advanced WGA-facilitated drug delivery systems can be applied in clinics.

6. Non-Medical WGA Applications

Outside of medical applications, WGA potential has been demonstrated as an option for basic research and biotechnological use. Quantum dot–WGA complexes have been used for elucidating the endocytic and exocytic lectin processes [72]. Single-particle tracking techniques of these complexes showed that WGA processes are both actin- and microtubule-dependent, and each contains a five-stage transport route that is transferable to design processes of any lectins as drug carriers or antineoplastic drugs.

Anisotropic silver nanoparticles functionalized with WGA or *Lens Culinaris* agglutinin have been demonstrated to detect specific Gram-positive and Gram-negative bacteria [73]. The designed photometric assay could detect *S. aureus* down to 10^3 cells/mL and *E. Coli* down to 3×10^3 cells/mL by having the bacteria-bound nanoparticle–lectin compounds resilient to NaCl-induced agglomeration. The assay was demonstrated on urine samples as well.

Extraction of two recombinant human glycoproteins—erythropoietin and Darbepoetin—from equine plasma by immunoaffinity chromatography has been shown to be improved by 15% when pretreating the samples with wheat germ agglutinin immobilized on Sepharose gel [74]. The technique is easily scalable and indicates the WGA potential for naturally or synthetically glycosylated recombinant protein industry.

The various biotechnological applications are already demonstrating the WGA value.

7. Conclusions

A 36 kDa size, hevein class, GlcNAc-specific lectin WGA is considered an antinutrient compound. Carbohydrate specificity enables its biological activities, such as fungistatic, cytotoxic, pro-apoptotic, and pro-necrotic. WGA might modulate immune cell response to antigens and cause changes in immune cell signaling. With regard to biotechnological applications, WGA is a highly promising tool in nanotechnology and advanced drug delivery systems with examples WGA-based nanoparticle, liposome, and other complexes for production, research, bacteria monitoring, and elimination. Considering the large quantities of wheat germ produced worldwide as a waste or low value byproduct, WGA extraction and utilization might be a viable solution for relatively cheap biomedical improvements in the future.

Author Contributions: Conceptualization, G.B.-M.; investigation, G.B.-M. and M.D.; writing—original draft preparation, G.B.-M. and M.D.; writing—review and editing, G.B.-M. and M.D. All authors have read and agreed to the published version of the manuscript.

Funding: This research was funded by the European Regional Development Fund according to the supported activity 'Attracting scientists from abroad to carry out research' under Measure No. 01.2.2-LMT-K-718 (project No. 01.2.2-LMT-K-718-02-0012).

Acknowledgments: Authors also acknowledge Panevėžys Mechatronics Center for providing workspace, literature, and the technical base.

Conflicts of Interest: The authors declare no conflict of interest.

References

1. Enghiad, A.; Ufer, D.; Countryman, A.M.; Thilmany, D.D. An Overview of Global Wheat Market Fundamentals in an Era of Climate Concerns. Available online: https://www.hindawi.com/journals/ija/2017/3931897/ (accessed on 1 April 2020).
2. Bacenetti, J.; Duca, D.; Negri, M.; Fusi, A.; Fiala, M. Mitigation Strategies in the Agro-Food Sector: The Anaerobic Digestion of Tomato Purée by-Products. An Italian Case Study. *Sci. Total Environ.* **2015**, *526*, 88–97. [CrossRef] [PubMed]
3. Šramková, Z.; Gregová, E.; Šturdík, E. Chemical Composition and Nutritional Quality of Wheat Grain. *Acta Chim. Slovaca* **2009**, *2*, 115–138.
4. Edwin Geo, V.; Prabhu, C.; Thiyagarajan, S.; Maiyalagan, T.; Aloui, F. Comparative Analysis of Various Techniques to Improve the Performance of Novel Wheat Germ Oil—An Experimental Study. *Int. J. Hydrog. Energy* **2020**, *45*, 5745–5756. [CrossRef]
5. Boukid, F.; Folloni, S.; Ranieri, R.; Vittadini, E. A Compendium of Wheat Germ: Separation, Stabilization and Food Applications. *Trends Food Sci. Technol.* **2018**, *78*, 120–133. [CrossRef]
6. Teslić, N.; Bojanić, N.; Rakić, D.; Takači, A.; Zeković, Z.; Fišteš, A.; Bodroža-Solarov, M.; Pavlić, B. Defatted Wheat Germ as Source of Polyphenols—Optimization of Microwave-Assisted Extraction by RSM and ANN Approach. *Chem. Eng. Process.-Process Intensif.* **2019**, *143*, 107634. [CrossRef]
7. Schwefel, D.; Maierhofer, C.; Beck, J.G.; Seeberger, S.; Diederichs, K.; Möller, H.M.; Welte, W.; Wittmann, V. Structural Basis of Multivalent Binding to Wheat Germ Agglutinin. *J. Am. Chem. Soc.* **2010**, *132*, 8704–8719. [CrossRef]
8. Shi, L.; Arntfield, S.D.; Nickerson, M. Changes in Levels of Phytic Acid, Lectins and Oxalates during Soaking and Cooking of Canadian Pulses. *Food Res. Int.* **2018**, *107*, 660–668. [CrossRef]
9. Smith, J.J.; Raikhel, N.V. Nucleotide Sequences of CDNA Clones Encoding Wheat Germ Agglutinin Isolectins A and D. *Plant Mol. Biol.* **1989**, *13*, 601–603. [CrossRef]
10. Peumans, W.J.; Stinissen, H.M.; Carlier, A.R. Isolation and Partial Characterization of Wheat-Germ-Agglutinin-like Lectins from Rye (Secale Cereale) and Barley (Hordeum Vulgare) Embryos. *Biochem. J.* **1982**, *203*, 239–243. [CrossRef]
11. Rice, R.H.; Etzler, M.E. Subunit Structure of Wheat Germ Agglutinin. *Biochem. Biophys. Res. Commun.* **1974**, *59*, 414–419. [CrossRef]
12. Wright, C.S.; Raikhel, N. Sequence Variability in Three Wheat Germ Agglutinin Isolectins: Products of Multiple Genes in Polyploid Wheat. *J. Mol. Evol.* **1989**, *28*, 327–336. [CrossRef] [PubMed]
13. Matucci, A.; Veneri, G.; Dalla Pellegrina, C.; Zoccatelli, G.; Vincenzi, S.; Chignola, R.; Peruffo, A.D.B.; Rizzi, C. Temperature-Dependent Decay of Wheat Germ Agglutinin Activity and Its Implications for Food Processing and Analysis. *Food Control* **2004**, *15*, 391–395. [CrossRef]
14. Portillo-Téllez, M.D.C.; Bello, M.; Salcedo, G.; Gutiérrez, G.; Gómez-Vidales, V.; García-Hernández, E. Folding and Homodimerization of Wheat Germ Agglutinin. *Biophys. J.* **2011**, *101*, 1423–1431. [CrossRef] [PubMed]
15. Rodríguez-Romero, A.; Ravichandran, K.G.; Soriano-García, M. Crystal Structure of Hevein at 2.8 Å Resolution. *Febs. Lett.* **1991**, *291*, 307–309. [CrossRef]
16. Leyva, E.; Medrano-Cerano, J.L.; Cano-Sánchez, P.; López-González, I.; Gómez-Velasco, H.; Del Río-Portilla, F.; García-Hernández, E. Bacterial Expression, Purification and Biophysical Characterization of Wheat Germ Agglutinin and Its Four Hevein-like Domains. *Biopolymers* **2019**, *110*, e23242. [CrossRef]
17. Nagata, Y.; Goldberg, A.R.; Burger, M.M. The isolation and purification of wheat germ and other agglutinins. In *Methods in Enzymology*; Elsevier: Amsterdam, The Netherlands, 1974; Volume 32, pp. 611–615. ISBN 978-0-12-181895-1.
18. Wright, H.T.; Sandrasegaram, G.; Wright, C.S. Evolution of a Family of N-Acetylglucosamine Binding Proteins Containing the Disulfide-Rich Domain of Wheat Germ Agglutinin. *J. Mol. Evol.* **1991**, *33*, 283–294. [CrossRef]
19. Hernández-Arana, A.; Rojo-Domínguez, A.; Soriano-García, M.; Rodríguez-Romero, A. The Thermal Unfolding of Hevein, a Small Disulfide-Rich Protein. *Eur. J. Biochem.* **1995**, *228*, 649–652. [CrossRef]
20. Mansfield, M.A.; Peumans, W.J.; Raikhel, N.V. Wheat-Germ Agglutinin Is Synthesized as a Glycosylated Precursor. *Planta* **1988**, *173*, 482–489. [CrossRef]
21. Allen, A.K.; Neuberger, A.; Sharon, N. The Purification, Composition and Specificity of Wheat-Germ Agglutinin. *Biochem. J.* **1973**, *131*, 155–162. [CrossRef]
22. Gallagher, J.T.; Morris, A.; Dexter, T.M. Identification of Two Binding Sites for Wheat-Germ Agglutinin on Polylactosamine-Type Oligosaccharides. *Biochem. J.* **1985**, *231*, 115–122. [CrossRef]
23. Peumans, W.J.; van Damme, E.J.M.; Barre, A.; Rougé, P. Classification of Plant Lectins in Families Of Structurally and Evolutionary Related Proteins. *Adv. Exp. Med. Biol.* **2001**, *491*, 27–54. [CrossRef] [PubMed]

24. Harata, K.; Nagahora, H.; Jigami, Y. X-ray Structure of Wheat Germ Agglutinin Isolectin 3. *Acta Cryst. D* **1995**, *51*, 1013–1019. [CrossRef] [PubMed]
25. Lotan, R.; Gussin, A.E.S.; Lis, H.; Sharon, N. Purification of Wheat Germ Agglutinin by Affinity Chromatography on a Sepharose-Bound N-Acetylglucosamine Derivative. *Biochem. Biophys. Res. Commun.* **1973**, *52*, 656–662. [CrossRef]
26. Baieli, M.F.; Urtasun, N.; Miranda, M.V.; Cascone, O.; Wolman, F.J. Efficient Wheat Germ Agglutinin Purification with a Chitosan-Based Affinity Chromatographic Matrix. *J. Sep. Sci.* **2012**, *35*, 231–238. [CrossRef]
27. Ciopraga, J.; Gozia, O.; Tudor, R.; Brezuica, L.; Doyle, R.J. Fusarium Sp. Growth Inhibition by Wheat Germ Agglutinin. *Biochim. Biophys. Acta* **1999**, *1428*, 424–432. [CrossRef]
28. Tonkal, A.M. In Vitro Antitrichomonal Effect of Nigella Sativa Aqueous Extract and Wheat Germ Agglutinin. *J. King Abdulaziz Univ. Med. Sci.* **2009**, *16*, 17–34. [CrossRef]
29. Liedke, S.C.; Miranda, D.Z.; Gomes, K.X.; Gonçalves, J.L.S.; Frases, S.; Nosanchuk, J.D.; Rodrigues, M.L.; Nimrichter, L.; Peralta, J.M.; Guimarães, A.J. Characterization of the Antifungal Functions of a WGA-Fc (IgG2a) Fusion Protein Binding to Cell Wall Chitin Oligomers. *Sci. Rep.* **2017**, *7*, 12187. [CrossRef]
30. da Silva, P.M.; de Moura, M.C.; Gomes, F.S.; da Silva Trentin, D.; Silva de Oliveira, A.P.; de Mello, G.S.V.; da Rocha Pitta, M.G.; de Melo Rego, M.J.B.; Coelho, L.C.B.B.; Macedo, A.J.; et al. PgTeL, the Lectin Found in Punica Granatum Juice, Is an Antifungal Agent against Candida Albicans and Candida Krusei. *Int. J. Biol. Macromol.* **2018**, *108*, 391–400. [CrossRef]
31. Kanokwiroon, K.; Teanpaisan, R.; Wititsuwannakul, D.; Hooper, A.B.; Wititsuwannakul, R. Antimicrobial Activity of a Protein Purified from the Latex of Hevea Brasiliensis on Oral Microorganisms. *Mycoses* **2008**, *51*, 301–307. [CrossRef]
32. Dickers, K.J.; Bradberry, S.M.; Rice, P.; Griffiths, G.D.; Vale, J.A. Abrin Poisoning. *Toxicol. Rev.* **2003**, *22*, 137–142. [CrossRef]
33. Lopez Nunez, O.F.; Pizon, A.F.; Tamama, K. Ricin Poisoning after Oral Ingestion of Castor Beans: A Case Report and Review of the Literature and Laboratory Testing. *J. Emerg. Med.* **2017**, *53*, e67–e71. [CrossRef]
34. Gorelik, E. Mechanisms of Cytotoxic Activity of Lectins. *Trends Glycosci. Glycotechnol.* **1994**, *6*, 435–445. [CrossRef]
35. Fraser, S.; Sadofsky, L.; Hart, S. Peripheral Blood Leukocyte Immune Responses Are Distinctly Altered in Sarcoidosis. *Eur. Respir. J.* **2015**, *46*. [CrossRef]
36. Clevers, H.C.; de Bresser, A.; Kleinveld, H.; Gmelig-Meyling, F.H.; Ballieux, R.E. Wheat Germ Agglutinin Activates Human T Lymphocytes by Stimulation of Phosphoinositide Hydrolysis. *J. Immunol.* **1986**, *136*, 3180–3183. [PubMed]
37. Kawakami, K.; Yamamoto, Y.; Onoue, K. Effect of Wheat Germ Agglutinin on T Lymphocyte Activation. *Microbiol. Immunol.* **1988**, *32*, 413–422. [CrossRef]
38. Reed, J.C.; Robb, R.J.; Greene, W.C.; Nowell, P.C. Effect of Wheat Germ Agglutinin on the Interleukin Pathway of Human T Lymphocyte Activation. *J. Immunol.* **1985**, *134*, 314–323.
39. Ryva, B.; Zhang, K.; Asthana, A.; Wong, D.; Vicioso, Y.; Parameswaran, R. Wheat Germ Agglutinin as a Potential Therapeutic Agent for Leukemia. *Front. Oncol.* **2019**, *9*. [CrossRef]
40. Boligan, K.F.; Mesa, C.; Fernandez, L.E.; von Gunten, S. Cancer Intelligence Acquired (CIA): Tumor Glycosylation and Sialylation Codes Dismantling Antitumor Defense. *Cell. Mol. Life Sci.* **2015**, *72*, 1231–1248. [CrossRef]
41. Pearce, O.M.T.; Läubli, H. Sialic Acids in Cancer Biology and Immunity. *Glycobiology* **2016**, *26*, 111–128. [CrossRef]
42. Raub, T.J.; Koroly, M.J.; Roberts, R.M. Endocytosis of Wheat Germ Agglutinin Binding Sites from the Cell Surface into a Tubular Endosomal Network. *J. Cell. Physiol.* **1990**, *143*, 1–12. [CrossRef]
43. Gastman, B.; Wang, K.; Han, J.; Zhu, Z.; Huang, X.; Wang, G.-Q.; Rabinowich, H.; Gorelik, E. A Novel Apoptotic Pathway as Defined by Lectin Cellular Initiation. *Biochem. Biophys. Res. Commun.* **2004**, *316*, 263–271. [CrossRef] [PubMed]
44. Wirth, C.; Schwuchow, J.; Jonas, L. Internalization of Wheat Germ Agglutinin (WGA) by Rat Pancreatic Cells in Vivo and in Vitro. *Acta Histochem.* **1996**, *98*, 165–172. [CrossRef]
45. Pellegrina, C.D.; Matucci, A.; Zoccatelli, G.; Rizzi, C.; Vincenzi, S.; Veneri, G.; Andrighetto, G.; Peruffo, A.D.B.; Chignola, R. Studies on the Joint Cytotoxicity of Wheat Germ Agglutinin and Monensin. *Toxicol. Vitr.* **2004**, *18*, 821–827. [CrossRef] [PubMed]
46. Hallberg, E.; Wozniak, R.; Blobel, G. An Integral Membrane Protein of the Pore Membrane Domain of the Nuclear Envelope Contains a Nucleoporin-like Region. *J. Cell. Biol.* **1993**, *122*, 513–521. [CrossRef] [PubMed]
47. Miller, M.W.; Hanover, J.A. Functional Nuclear Pores Reconstituted with Beta 1-4 Galactose-Modified O-Linked N-Acetylglucosamine Glycoproteins. *J. Biol. Chem.* **1994**, *269*, 9289–9297. [CrossRef]
48. Heese-Peck, A.; Cole, R.N.; Borkhsenious, O.N.; Hart, G.W.; Raikhel, N.V. Plant Nuclear Pore Complex Proteins Are Modified by Novel Oligosaccharides with Terminal N-Acetylglucosamine. *Plant Cell* **1995**, *7*, 1459–1471. [CrossRef] [PubMed]
49. Ebert, C.; Nebe, B.; Walzel, H.; Weber, H.; Jonas, L. Inhibitory Effect of the Lectin Wheat Germ Agglutinin (WGA) on the Proliferation of AR42J Cells. *Acta Histochem.* **2009**, *111*, 336–343. [CrossRef]
50. Tsai, T.L.; Wang, H.C.; Hung, C.H.; Lin, P.C.; Lee, Y.S.; Chen, H.H.W.; Su, W.C. Wheat Germ Agglutinin-Induced Paraptosis-like Cell Death and Protective Autophagy Is Mediated by Autophagy-Linked FYVE Inhibition. *Oncotarget* **2017**, *8*, 91209–91222. [CrossRef]
51. Schwarz, R.E.; Wojciechowicz, D.C.; Picon, A.I.; Schwarz, M.A.; Paty, P.B. Wheatgerm Agglutinin-Mediated Toxicity in Pancreatic Cancer Cells. *Br. J. Cancer* **1999**, *80*, 1754–1762. [CrossRef]
52. Garcia, M.; Seigner, C.; Bastid, C.; Choux, R.; Payan, M.J.; Reggio, H. Carcinoembryonic Antigen Has a Different Molecular Weight in Normal Colon and in Cancer Cells Due to N-Glycosylation Differences. *Cancer Res.* **1991**, *51*, 5679–5686.

53. Weissenböck, A.; Wirth, M.; Gabor, F. WGA-Grafted PLGA-Nanospheres: Preparation and Association with Caco-2 Single Cells. *J. Control. Release* **2004**, *99*, 383–392. [CrossRef] [PubMed]

54. Mo, Y.; Lim, L.-Y. Preparation and in Vitro Anticancer Activity of Wheat Germ Agglutinin (WGA)-Conjugated PLGA Nanoparticles Loaded with Paclitaxel and Isopropyl Myristate. *J. Control. Release* **2005**, *107*, 30–42. [CrossRef] [PubMed]

55. Wang, C.; Ho, P.C.; Lim, L.Y. Wheat Germ Agglutinin-Conjugated PLGA Nanoparticles for Enhanced Intracellular Delivery of Paclitaxel to Colon Cancer Cells. *Int. J. Pharm.* **2010**, *400*, 201–210. [CrossRef] [PubMed]

56. Chiu, H.I.; Ayub, A.D.; Mat Yusuf, S.N.A.; Yahaya, N.; Abd Kadir, E.; Lim, V. Docetaxel-Loaded Disulfide Cross-Linked Nanoparticles Derived from Thiolated Sodium Alginate for Colon Cancer Drug Delivery. *Pharmaceutics* **2020**, *12*, 38. [CrossRef] [PubMed]

57. Kuo, Y.-C.; Chang, Y.-H.; Rajesh, R. Targeted Delivery of Etoposide, Carmustine and Doxorubicin to Human Glioblastoma Cells Using Methoxy Poly(Ethylene Glycol)-poly(E-caprolactone) Nanoparticles Conjugated with Wheat Germ Agglutinin and Folic Acid. *Mater. Sci. Eng. C* **2019**, *96*, 114–128. [CrossRef] [PubMed]

58. Gabor, F.; Schwarzbauer, A.; Wirth, M. Lectin-Mediated Drug Delivery: Binding and Uptake of BSA-WGA Conjugates Using the Caco-2 Model. *Int. J. Pharm.* **2002**, *237*, 227–239. [CrossRef]

59. Murata, M.; Yonamine, T.; Tanaka, S.; Tahara, K.; Tozuka, Y.; Takeuchi, H. Surface Modification of Liposomes Using Polymer-Wheat Germ Agglutinin Conjugates to Improve the Absorption of Peptide Drugs by Pulmonary Administration. *J. Pharm. Sci.* **2013**, *102*, 1281–1289. [CrossRef]

60. Leong, K.H.; Chung, L.Y.; Noordin, M.I.; Onuki, Y.; Morishita, M.; Takayama, K. Lectin-Functionalized Carboxymethylated Kappa-Carrageenan Microparticles for Oral Insulin Delivery. *Carbohydr. Polym.* **2011**, *86*, 555–565. [CrossRef]

61. AlSadek, D.M.M.; Badr, H.A.; Al-Shafie, T.A.; El-Bahr, S.M.; El-Houseini, M.E.; Djansugurova, L.B.; Li, C.-Z.; Ahmed, H. Cancer Cell Death Induced by Nanomagnetolectin. *Eur. J. Cell Biol.* **2017**, *96*, 600–611. [CrossRef]

62. Yang, K.; Gitter, B.; Rüger, R.; Albrecht, V.; Wieland, G.D.; Fahr, A. Wheat Germ Agglutinin Modified Liposomes for the Photodynamic Inactivation of Bacteria†. *Photochem. Photobiol.* **2012**, *88*, 548–556. [CrossRef]

63. Gao, X.; Wu, B.; Zhang, Q.; Chen, J.; Zhu, J.; Zhang, W.; Rong, Z.; Chen, H.; Jiang, X. Brain Delivery of Vasoactive Intestinal Peptide Enhanced with the Nanoparticles Conjugated with Wheat Germ Agglutinin Following Intranasal Administration. *J. Control. Release* **2007**, *121*, 156–167. [CrossRef] [PubMed]

64. Chauhan, N.; Xiao, C.; Davis, F.; Lambert, M.; Viola, K.; Lacor, P.; Dravid, V.; Klein, W. Intranasal Passive Immunization Using WGA-Modified Oligomer Antibodies Greatly Improves Learning and Memory in Alzheimer's 5XFAD Mice. *Alzheimer's Dement.* **2012**, *8*, P197. [CrossRef]

65. Zhang, Y.; Walker, J.B.; Minic, Z.; Liu, F.; Goshgarian, H.; Mao, G. Transporter Protein and Drug-Conjugated Gold Nanoparticles Capable of Bypassing the Blood-Brain Barrier. *Sci. Rep.* **2016**, *6*, 25794. [CrossRef] [PubMed]

66. He, H.; Li, Y.; Jia, X.-R.; Du, J.; Ying, X.; Lu, W.-L.; Lou, J.-N.; Wei, Y. PEGylated Poly(Amidoamine) Dendrimer-Based Dual-Targeting Carrier for Treating Brain Tumors. *Biomaterials* **2011**, *32*, 478–487. [CrossRef] [PubMed]

67. Su, Y.; Sun, B.; Gao, X.; Dong, X.; Fu, L.; Zhang, Y.; Li, Z.; Wang, Y.; Jiang, H.; Han, B. Intranasal Delivery of Targeted Nanoparticles Loaded With MiR-132 to Brain for the Treatment of Neurodegenerative Diseases. *Front. Pharmacol.* **2020**, *11*. [CrossRef]

68. Plattner, V.E.; Wagner, M.; Ratzinger, G.; Gabor, F.; Wirth, M. Targeted Drug Delivery: Binding and Uptake of Plant Lectins Using Human 5637 Bladder Cancer Cells. *Eur. J. Pharm. Biopharm.* **2008**, *70*, 572–576. [CrossRef]

69. Apfelthaler, C.; Anzengruber, M.; Gabor, F.; Wirth, M. Poly-(l)-Glutamic Acid Drug Delivery System for the Intravesical Therapy of Bladder Cancer Using WGA as Targeting Moiety. *Eur. J. Pharm. Biopharm.* **2017**, *115*, 131–139. [CrossRef]

70. Brauner, B.; Semmler, J.; Rauch, D.; Nokaj, M.; Haiss, P.; Schwarz, P.; Wirth, M.; Gabor, F. Trimethoprim-Loaded PLGA Nanoparticles Grafted with WGA as Potential Intravesical Therapy of Urinary Tract Infections—Studies on Adhesion to SV-HUCs Under Varying Time, PH, and Drug-Loading Conditions. *Acs Omega* **2020**, *5*, 17377–17384. [CrossRef]

71. Wijetunge, S.S.; Wen, J.; Yeh, C.-K.; Sun, Y. Wheat Germ Agglutinin Liposomes with Surface Grafted Cyclodextrins as Bioadhesive Dual-Drug Delivery Nanocarriers to Treat Oral Cells. *Colloids Surf. B Biointerfaces* **2020**, *185*, 110572. [CrossRef]

72. Liu, S.-L.; Zhang, Z.-L.; Sun, E.-Z.; Peng, J.; Xie, M.; Tian, Z.-Q.; Lin, Y.; Pang, D.-W. Visualizing the Endocytic and Exocytic Processes of Wheat Germ Agglutinin by Quantum Dot-Based Single-Particle Tracking. *Biomaterials* **2011**, *32*, 7616–7624. [CrossRef]

73. Mikaelyan, M.V.; Poghosyan, G.G.; Hendrickson, O.D.; Dzantiev, B.B.; Gasparyan, V.K. Wheat Germ Agglutinin and Lens Culinaris Agglutinin Sensitized Anisotropic Silver Nanoparticles in Detection of Bacteria: A Simple Photometric Assay. *Anal. Chim. Acta* **2017**, *981*, 80–85. [CrossRef] [PubMed]

74. Stanley, S.M.R.; Chua, D. Improved Recovery of Erythropoietin and Darbepoetin from Equine Plasma by the Application of a Wheat Germ Agglutinin Mediated Pre-Extraction Prior to Immunoaffinity Chromatography. *Adv. Biosci. Biotechnol.* **2014**. [CrossRef]

Review

Application of Conventional and Non-Conventional Extraction Methods to Obtain Functional Ingredients from Jackfruit (*Artocarpus heterophyllus* Lam.) Tissues and By-Products

Frida Camila Cruz-Casillas, Tomás García-Cayuela and Veronica Rodriguez-Martinez *

Tecnologico de Monterrey, Escuela de Ingenieria y Ciencias, Av. General Ramon Corona No. 2514, Zapopan 45201, Jalisco, Mexico; a01630860@itesm.mx (F.C.C.-C.); tomasgc@tec.mx (T.G.-C.)
* Correspondence: veronica.rodriguezmz@tec.mx

Abstract: In recent years, researchers in the pharmaceutical and food areas focused on finding the best ways to take advantage of functional ingredients present in jackfruit tissues and by-products such as phenolics and pectin. Many of these studies focused on adding value to the by-products and decreasing their negative environmental impact. However, the type, quantity, and characteristics of jackfruit functional ingredients are highly dependent on the extraction method, either through conventional or non-conventional technologies, and the jackfruit tissue used, with peel and seeds being the most studied. The reported studies suggest that extractions and pre-treatments with emerging technologies such as ultrasounds, microwaves, radio frequency, or supercritical fluids can facilitate the release of functional ingredients of jackfruit; reduce the time and energy consumption required; and, in some cases, improve extraction yields. Therefore, emerging technologies could increase the functional potential of jackfruit and its by-products, with promising applications in the pharmaceutical and nutraceutical industries.

Keywords: jackfruit; jackfruit processing; by-products; extraction methods; phenolic compounds; pectin; emerging technologies; innovative technologies; functional ingredients; bioactive compounds

Citation: Cruz-Casillas, F.C.; García-Cayuela, T.; Rodriguez-Martinez, V. Application of Conventional and Non-Conventional Extraction Methods to Obtain Functional Ingredients from Jackfruit (*Artocarpus heterophyllus* Lam.) Tissues and By-Products. *Appl. Sci.* **2021**, *11*, 7303. https://doi.org/10.3390/app11167303

Academic Editors: Isidoro Garcia-Garcia and David Ian Ellis

Received: 10 June 2021
Accepted: 27 July 2021
Published: 9 August 2021

Publisher's Note: MDPI stays neutral with regard to jurisdictional claims in published maps and institutional affiliations.

1. Introduction

The tropical fruit known as Jackfruit (*Artocarpus heterophyllus* Lam.) is native to Thailand, India, and Indonesia, and it is also found in other Asian countries, northern Australia, parts of Africa [1], and Latin America. Jackfruit trees belong to the Moraceae family [2]. They are medium-sized, characterized by reaching between 8.5 m to 24 m in height, and their fruit is born from its main and lateral branches [3]. Fruits can weigh up to 45 kg and measure up to a length of 91 cm and a diameter of 51 cm [4]. Ranasinghe et al. [5] described the outer part of the jackfruit shell (commonly known as rind or peel) to have conical carpel apices that cover a thick rubbery wall (Figure 1A). Inside the fruit, a non-edible core forms a longitudinal axis that is fused with the rags, which in turn are connected to the rind of the syncarp (Figure 1B) [5]. The bulbs are the fleshy edible region found between the rags (Figure 1B); each one is comprised by the pulp surrounding a seed (Figure 1C) [5]. Jackfruit seeds (Figure 1D), which vary in number between 100 and 500, represent 18–25% of the fruit's total weight [1]. The kernels of the seeds represent 90–95% of their weight; on the other hand, the pulp represents 30% of the fruit weight [1]. According to Akter & Haque [6], around 70 and 80% of the jackfruit components are non-edible; from them, about 60% correspond to the outer rind, perianth, and central core and are considered waste and usually discarded.

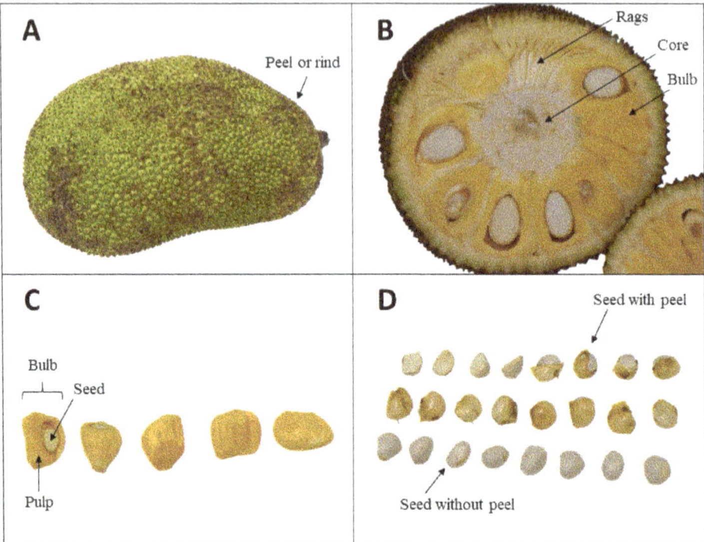

Figure 1. Jackfruit parts: (**A**) peel or rind with conical carpel apices; (**B**) core forming the central longitudinal axis connected to the syncarp's rind by the rags in between which the bulbs are found; (**C**) bulbs formed by pulp and seeds; (**D**) seeds with and without the peel.

One of the jackfruit's main attributes is its high nutritional content. In a review article, Swami et al. [3] presented data from Samaddar regarding the composition of 100 g of ripe jackfruit pulp with a caloric content of 84. The results are as follows: 77% moisture, 18.9% of carbohydrates, 1.9% protein, 1.1% fiber, 0.8% total minerals (30 mg phosphorus, 20 mg calcium, 500 mg iron), 0.1% fat, 30 mg thiamine, and 540 UI vitamin A [3]. In addition to its nutritional content, jackfruit has specific bioactive compounds, which demonstrated positive effects in helping prevent certain chronic diseases, including cancer, cardiovascular diseases, and diseases related to aging [7]. For this reason, jackfruit is considered a medicinal plant in some countries [8].

The most reported bioactive compounds in jackfruit are phenolic compounds [1] and carotenoids [3]. However, there is evidence of other functional ingredients such as prebiotics including indigestible polysaccharides and oligosaccharides [9,10]; pectin [11]; minerals such as Ca, Fe, K, Mg, and Na [12]; essential fatty acids (EFAs) such as alpha-linoleic and linoleic [13]; and other fatty acids such as palmitic, oleic, stearic, myristic, lauric, capric, and arachidic acids [14], among others. Jackfruit functional ingredients are of great interest owing to their possible applications in the food and pharmaceutical industries, specifically for their health-related benefits. However, these components are not necessarily present in a high amount in jackfruit [15]. For those reasons, there has been an increasing interest in finding strategies for their extraction, especially those focusing on using innovative and emerging technologies.

Various studies in the literature showed the effect of conventional and non-conventional extraction methods to obtain jackfruit functional ingredients; among the non-conventional methods, some use emerging technologies to extract or as a pre-treatment. Among the studies using conventional extraction, applying solvents such as methanol [16] and oxalic acid [17] stands out, as well as the extraction with hot water [18]. Regarding extraction by emerging technologies, radio frequency-assisted [19] and supercritical fluid (SFE) with CO_2 [20] extractions have attracted attention owing to their less negative impact on the environment and safety of the final product obtained in comparison with other methods using non-conventional technologies. Regardless of the extraction method, the yields of a particular functional compound will vary depending on the jackfruit tissue utilized for the extraction; the most common selected tissues for this process are peel and seeds, by-products generated from the fruit consumption.

The limited utilization of pericarp and other inedible parts of the jackfruit makes them a source of contamination [21]. Agro-industrial by-products such as peels, seeds, and residual fruit pulp can be a good natural source of antioxidants such as phenolic compounds [22]. Antioxidants include bioactive compounds capable of counteracting the damage of oxidation processes in the body by assisting chemical reactions [23]. Hence, applying emerging technologies in new processes to transform the jackfruit by-products is useful to obtain functional ingredients or components and may help stop the negative impact on the environment.

The purpose of this publication is to review the following: (1) the most relevant functional ingredients in the jackfruit tissues and by-products; (2) the impact of conventional extraction and assisted extraction by emerging technologies in the functional ingredient yield; and, finally, (3) the functionality and applications of those functional ingredients.

2. Bioactive Compounds of Jackfruit

This study will focus on the species of jackfruit *Artocarpus heterophyllus* Lam., the one with the most information reported. Furthermore, information on the other jackfruit species is scarce.

2.1. Pulp

Jackfruit pulp, the edible part of the fruit, has several functional ingredients and bioactive compounds. One of its most notable properties is its vitamin C content, with a reported value of 8.16 mg/100 g [24]. On the other hand, phenolic acids, flavonoids, flavanones, and their derivatives were found in extracts of dietary fiber from jackfruit pulp when characterizing their composition by chromatographic analysis and a quadrupole time-of-flight mass spectrometer in negative ion mode used for detection (UPLC-ESI-QTOF-MS/MS) [25]. Additionally, Singh et al. [26] reported the presence of three phenolic acids in the unripe pulp extracts, gallic (9.70 µg/g), ferulic (8.04 µg/g), and tannic (4.87 µg/g) acids, when analyzed by high-performance liquid chromatography (HPLC). These values differed from those obtained from the ripe pulp, where the amount of gallic acid was almost double (19.31 µg/g), the ferulic acid decreased (2.66 µg/g), while the tannic acid slightly increased (5.24 µg/g) [26]. Moreover, Galvez & Dizon [27] reported values for the fresh pulp's total phenolic content between 125.91 and 127.73 mg CE (catechin equivalent)/100 g. In contrast, the tannin content was found to be between 127.73 and 208.72 mg VE (vanillin equivalent)/100 g depending on the physical, chemical, and functional properties [27]. Regarding the carotenoid presence in the pulp, de Faria et al. [28] found five major types of carotenoids, including 24–44% all-trans-lutein, 24–30% all-trans-β-carotene, 4–19% all-trans-neoxanthin, 4–9% 9-cis-neoxanthin, and 4–10% 9-cis-violaxanthin using high-performance liquid chromatography coupled with photodiode array and mass spectrometry detectors (HPLC-PDA-MS/MS). Their study showed a maximum total value of carotenoids of 150.3 µg/100 g [28].

Regarding fatty acids, Chowdhury et al. [14] reported a total amount of fatty acids of 31.9 mg per 100 g of petroleum ether extract (105 mg per 100 g of whole fruit) in the edible part (fertilized fleshy perianths) of jackfruit.

2.2. Seed

Jackfruit seeds have many bioactive compounds such as polyphenols, and within these, mainly phenolic acids, flavonoids, and stilbenes [29]. Depending on the extraction technique used, those compounds might present variations in their chemical composition or bioactivity [20]. Kumoro et al. [1] reported values for polyphenols (243 ± 27.0 mg of gallic acid equivalent (GAE)/100 g dry seeds), flavonoids (2.03 ± 0.06 mg EC/100 g dry seeds), tannins (0.06–0.229 mg/100 g of seed), ferulic acid (0.216 mg/100 g of seed), and gallic acid (1.105 ± 0.1 2 mg/100 g of seed).

Some studies reported the oils present in jackfruit seeds. The main fatty acids present are linoleic and linolenic acid [1]. Nagala et al. [30] evaluated the oils' fatty acids composition and antioxidant potential in five different jackfruit species. Their study found jackfruit to be a good source of essential fatty acids (EFAs) with a notable antioxidant activity; higher percentages were observed in the DPPH test for *Artocarpus integer* (98.4 ± 0.2% of inhibition/50 µL), followed by *Artocarpus integrifolia* (98.2 ± 0.3% of inhibition/50 µL) and *Artocarpus heterophyllus* (87.4 ± 0.2% of inhibition/50 µL). Additionally, the seeds are good sources of minerals, such as potassium, which has the highest concentration (2470 ppm), followed by calcium, magnesium, and sodium; the seed's kernel contains 148.50 ppm of iron [1]. Jackfruit seeds also contain 400 mg/g extract of non-reducing sugar, a potential prebiotic ingredient [10].

There are reports that seed powder nanoparticles showed antimicrobial activity against *Escherichia coli* and *Bacillus megaterium* [3]. The jackfruit seed crude extracts contain jacalin protein [31]. Regarding the antioxidant activity of peptides from jackfruit seeds, Chai et al. [32] found, by subjecting proteins to trypsin digestion, that the peptide JFS-2 has antioxidant activity, which can be affected by pH, temperature, and gastrointestinal digestion.

2.3. Peel

Approximately half of the ripe jackfruit corresponds to the outer peel [33], a sizeable unused portion of fruit because it is not edible. The peel is rich in calcium [33]. It also contains between 8.94 and 15.14% of pectin in dry matter (DM) [34], making it an important source of this polysaccharide [21]. Pectin is an acid hetero-polysaccharide present in the cell walls and middle lamellae of plants grown in the soil [35]. It consists of α-1,4-linked D-galacturonic acid partially esterified with methyl groups and several neutral sugars, such as L-rhamnose, L-arabinose, and D-galactose [17].

According to Zhang et al. [16], jackfruit peel extracts have a higher total phenolic content and total flavonoids than extracts from pulp, seeds, and flakes. These authors reported 48.04 mg GAE/g DM and 2.79 mg of quercetin equivalents (QE)/g DM, respectively [16]. Sharma et al. [36] reported that the jackfruit shell is a potential source of β-carotene, ascorbic acid, and polyphenols such as catechin and chlorogenic acid; the amounts found of phenolic and flavonoids were 158 ± 0.34 mg GAE and 10.0 ± 0.64 mg CE, respectively. Finally, Chowdhury et al. [14] reported a total amount of fatty acids of 34.3 mg/100 g of petroleum ether extract (12.6 mg/100 g of whole fruit) in the outer rind of jackfruit.

3. Methods Used in the Extraction of Functional Ingredients from Jackfruit

Conventional technologies used to extract functional ingredients from food generally consist of liquid–liquid or solid–liquid extraction [37]. These techniques are efficient in yield, but present some challenges regarding the general cost of the method and the safety of the final product [38]. Consequently, modification and development of the techniques have occurred, including adding pre-treatment steps to help release the compounds from the matrix [37]; this is where emerging technologies come into play. Emerging or modern technologies sometimes overcome the challenges from conventional technologies such as high energy consumption; food product matrix overheating; and loss of stability, functionality, or safety of the final product [38]. Finally, both conventional and emerging technologies have advantages and disadvantages in terms of cost and safety of the final product; however, emerging technologies provide better quality products and reduce the impact on the environment [39]. A summary of the most widely used extraction methods for functional ingredients present in jackfruit is presented in Figure 2.

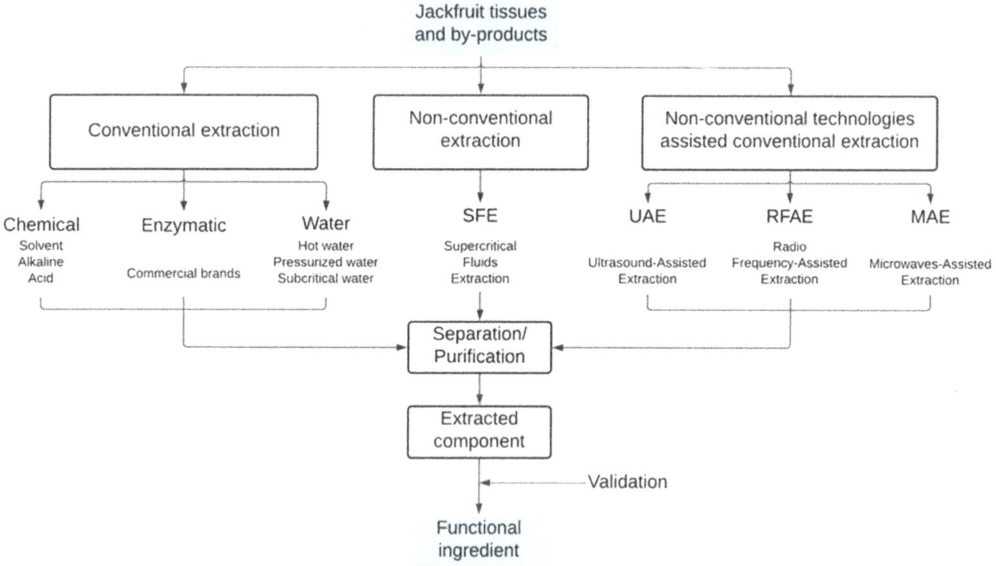

Figure 2. Conventional and non-conventional extraction methods and conventional extraction assisted by non-conventional technologies for functional ingredients from jackfruit tissues and by-products.

3.1. Conventional Technologies

Research performed using conventional technologies to extract jackfruit functional ingredients focused on phenolic compounds, pectin, jacalin, and prebiotics. The majority of the studies in the literature utilized jackfruit by-products instead of the edible part, mostly to give them added value and reduce their environmental impact [40]. Additionally, this approach allows using all parts of the fruit, which means a lower economic cost to obtain the matrix or raw material for the extraction [17]. Table 1 presents some of the reported works comparing the yields obtained from different extraction methods, solvents, and processing conditions.

When comparing reported yields of non-extractable polyphenol (NEPP) extraction processes through alkaline, acidic, or enzymatic hydrolysis from the dietary fiber of jackfruit pulp, those extracted from alkaline hydrolysis had the highest content (64.9 mg GAE/10 g DM) and antioxidant properties, having an ABTS$^{·+}$ elimination capacity nine times greater than the acid extract and four times greater than the enzymatic extract [25]. Islam et al. [40] compared the yields of polyphenol content obtained from pressurized hot-water extraction (PHWE), enzyme-assisted extraction (EnE), and organic solvent extraction (OSE) systems using the peel, seeds, and rags of jackfruit. They observed higher yields using the combination PHWE and peel by obtaining 47.22 ± 2.31 (mg GAE/g DM) [40]. Besides, the PHWE of polyphenols from jackfruit seeds was reported by Shakthi Deve et al. [18], in which jackfruit obtained the highest yield (87.52%) compared with other fruits such as sweet orange (*Citrus sinensis*) (74.79%) of antiglycation activity, concluding that flavonoid compounds like rutin, quercetin, and quercetin-3-O-xyloside could be responsible for that. Another study compared the effect of maceration (MAC) and Soxhlet (SOX) extraction techniques with ethanol and hexane on the composition and biological activity of the corresponding extracts [20]. Their results from extracts of jackfruit seeds showed that the best antioxidant activity and recovery yields correspond to SOX extraction with ethanol (12.08 ± 2.13 mg GAE/g) compared with the 1.89 ± 0.01 mg GAE/g recovered by MAC with ethanol [20]. Subcritical water-assisted extraction (SCWE) is another method that has been used to extract compounds from jackfruit, which Li et al. [21] compared based on pectin yields with a citric acid extraction method. Despite SCWE having a lower yield, 149.6 g/kg (14.96%) against 16.83% from the citric acid method, it showed a reduction in extraction time and energy consumption when operating at 138 °C for 9.15 min and a ratio of 17.03 mL/g liquid/solid (L/S) [21].

Furthermore, Shanooba et al. [29] tested the combination of different solid–solvent ratios, solvent concentrations, and times to extract polyphenols from jackfruit seeds through the conventional solvent extraction (CSE) method. They concluded that the best extraction performance was with 95% ethanol for 30 min at a solvent ratio of 2.5:25 g of jackfruit seed powder/mL of ethanol [29]. Likewise, Zhang et al. [16] reported that, when using 90% methanol for 6 h to extract phenolics, jackfruit peel yields were 4.95, 4.65, and 4.12 times higher than other parts of the fruit such as seeds, pulp, and flake, respectively. However, when comparing the phenolic content obtained from the peel to that from the fiber or core, the one from the fiber is higher (23.28 ± 4.73 mg/g), followed by the peel (17.07 ± 5.16 mg/g), and finally the core (15.68 ± 3.74 mg/g), using a CSE with methanol as solvent followed by 48 h incubation [12].

Sundarraj et al. [17] conducted experiments to determine the best temperature and time conditions for pectin extraction with oxalic acid from jackfruit peel, where they found that the best performance results corresponded to 90 °C and 60 min. Meanwhile Bhornsmithikun et al. [10] studied the effects of temperature, extraction times, and liquid–solid ratios' variations for the extraction of prebiotics, finding that a 50% ethanol extraction for 15 min at 60 °C at a ratio of 10:1 (*v/w*) had the best yields of non-reducing sugar content. More recently, Mohamad et al. [31] reported the extraction of jacalin from jackfruit peel using a sodium bis (2-ethylhexyl) sulfosuccinate/isooctane reverse micellar system.

3.2. Emerging Technologies

Some studies on the extraction of functional ingredients present in jackfruit have applied emerging technologies to assist in the process, such as ultrasounds, microwaves, and radio frequency, as a combined treatment or as a pre-treatment to the samples to facilitate further extraction. The results varied according to the technology used, the matrix, and the compounds of interest extracted, as shown in Table 2. Among those studied, some resulted in higher performance than conventional methods alone when using techniques including ultrasonic-microwave assisted extraction (UMAE) and ultrasound-assisted extraction (UAE) [5]. The implementation of emerging technologies to extract functional ingredients from jackfruit has not been fully explored, which presents a window of opportunity for new research.

Few extraction methodologies for jackfruit pulp compounds reported in the literature use non-conventional technologies compared with those for seeds and peel. One of those is a study by Singh et al. [26], who used UAE (15 min at 4 °C) with ethanol as a solvent to recover phenolic acids from ripe and unripe jackfruit pulp. This study concluded that jackfruit has a good content of phenolic acids in its raw and ripe edible parts, with 19.31 ± 1.8 µg/g of gallic acid being the maximum detected amount in raw fruit pulp [26].

Table 1. Conventional extraction of functional ingredients from jackfruit parts.

Part	Technique	Extraction Conditions	Functional Ingredients and Yields	Ref.
Pulp	Alkaline hydrolysis	2.00 g of sample mixed with 8.00 mol/L of NaOH, liquid–solid ratio of 20 mL/g, and extraction time of 4 h.	Non-extractable polyphenols64.90 mg GAE/10 g DM	[25]
	Acid hydrolysis	2.00 g of sample mixed with 9.00% of H_2SO_4, liquid–solid ratio of 20 mL/mg, and extraction time of 4 h.	Non-extractable polyphenols21.80 mg GAE/10 g DM	
	Enzymatic hydrolysis	2.00 g of sample mixed with 2.40 mg/mL of cellulase and pectinase at mass ratio 2:1, liquid–solid ratio of 20 mL/g, and extraction time of 90 min.For all hydrolysis in this section, pH 2.00 and centrifuged at $2795 \times g$ for 15 min. Ethyl ether/ethyl acetate (1:1, v/v, 30 mL) was used to extract supernatant three times.	Non-extractable polyphenols25.00 mg GAE/10 g DM	
Seed	Conventional solvent	(1) Phenolics were extracted with a solid–solvent ratio of 2.50:25.00 g/mL at 30 min in 95.00% ethanol concentration.	(1) Total phenolic content: 122.17 ± 0.41 µg GAE/mL Total flavonoid content: 0.47 ± 0.02 mg CE/mLCondensed tannin: 0.34 ± 0.08 mg CE/mL	[29]
		(2) Prebiotics were extracted at 60 °C, extraction time of 15 min, and L/S ratio at 10:1(v/w) using 50.00% ethanol as a solvent.	(2) Average extraction yield: 20.25%Average non-reducing sugar: 400 mg/g extract	[10]
	Reverse micellar extraction	Sodium bis(2-ethylhexyl) sulfosuccinate (AOT)-based reverse micellar system. Aqueous phase pH 4.58, 125.00 mM NaCl and 40.00 mM AOT, 300 rpm for 20 min at room temperature.	Jacalin yield: $88.04 \pm 1.30\%$	[31]
	Enzymes	Phenolics were extracted by dissolving 2.50 g of dry powder sample in 50 mL of 0.10 M phosphate buffer (pH 4.00) according to a solid/liquid ratio of 1:20 (g/mL) and adding an aqueous enzyme solution (Viscozyme® L).	Total phenolic content: 10.54 ± 1.41 mg GAE/g DM Total flavonoid content: 0.19 ± 0.03 mg QE/g DM	[40]
	Hot water	Phenolics were extracted by weighing out 0.50 g of powdered material and extract with 25 mL of distilled water by placing it in a boiling water bath at a temperature of 90 °C for 5 min.	Total phenolic content: $41.65 \pm 13.95\%$	[18]
	Pressurized hot water	Phenolics were extracted by mixing 10.00 g of dry powder samples with 200 mL of distilled water in a 1000 mL pressure cooker according to a solid/liquid ratio of 1:20 (g/mL).	Total phenolic content: 7.02 ± 0.39 mg GAE/g DMTotal flavonoid content: 0.48 ± 0.13 mg QE/g DM	[40]
	Maceration	Phenolics were extracted by placing 50.00 g of sample in 200 mL of ethanol for 196 h with daily manual stirring at room temperature (25 °C).	Total phenolic content: 1.892 ± 0.009 mg GAE/g	[20]
	Soxhlet	Recycle 150 mL of ethanol over 5.00 g of sample in a Soxhlet apparatus for 6 h.	Total phenolic content: 12.075 ± 2.131 mg GAE/g	[20]
Peel	Conventional solvent	(1) Phenolics were extracted by dispersing 2.00 g of samples in 60 mL of 90.00% methanol at a solid/liquid ratio of 1:30 (g/mL) and extracted at room temperature for 6 h in a shaker at 100 rpm.	Phenolics (1) Total phenolic content: 48.04 mg GAE/g DM Total flavonoid content: 2.79 mg QE/g DM	[16]
	Conventional solvent	(2) Phenolics were extracted with methanol following a 48 h incubation at 24 °C. (3) Pectin were extracted with oxalic acid, 90 °C and 60 min.	(2) Phenolics: 17.07 ± 5.16 mg/g Total flavonoid content: 28.55 ± 12.42 mg/g (3) PectinPectin yield: 38% 39.05 ± 0.59 g/g of pectin	[12]
	Pressurized hot water	Phenolics were extracted by mixing 10.00 g of dry powder samples with 200 mL of distilled water in a 1000 mL pressure cooker according to a solid/liquid ratio of 1:20 (g/mL).	Total phenolic content: 47.22 ± 2.31 mg GAE/g DM Total flavonoid content: 11.52 ± 0.81 mg QE/g DM	[40]
	Extraction assisted with subcritical water	Extraction temperature 138 °C, extraction time 9.15 min, L/S ratio 17.03 mL/g.	Pectin yield: 149.60 g/kg (dry matter).	[21]
	Enzymes	Phenolics were extracted by dissolving 2.50 g of dry powder sample in 50 mL of 0.10 M phosphate buffer (pH 4) according to a solid/liquid ratio of 1:20 (g/mL) and adding an aqueous enzyme solution (Viscozyme® L).	Total phenolic content: 14.19 ± 0.85 mg GAE/g DM Total flavonoid content: 1.25 ± 0.07 mg QE/g DM	[40]

DM: dry matter, GAE: gallic acid equivalent, QE: quercetin equivalents, CE: catechin equivalent.

Table 2. Emerging technologies of assisted extraction of functional ingredients from jackfruit parts.

Parts	Technique	Extraction Conditions	Functional Ingredients and Yields	Ref.
Pulp	Ultrasound-assisted extraction	1.00 g of sample finely crushed with 5–10 mL of ethanol water (80–20; v/v), ultrasonicated (15 min at 40 °C) and centrifuged (15 min at 7500 rpm).	Phenolic acids Raw pulp Gallic acid: 9.70 µg/g Ferulic acid: 8.04 µg/g Tannic acid: 4.87 µg/g Ripe pulp Gallic acid: 19.31 µg/g Ferulic acid: 2.66 µg/g Tannic acid: 5.24 µg/g	[26]
Seed	Ultrasound-assisted extraction	7.00 g of sample in 210 mL of solvent (ethanol) and mix. Ultrasound equipment with a probe is used for 4 min at 70% power (maximum power-500 W) and a frequency of 20 kHz.	Total phenolic content: 0.841 ± 0.067 mg GAE/g	[20]
	Microwave-assisted extraction	Immerse 1.00 g of the powdered sample in 100 mL of ethanol using an ETHOS-Milestone extractor, 5 min extraction time, 450 W microwave power, and 50 °C of extraction temperature.	Phenolic compounds yield: 17.34 mg/g%	[41]
	Supercritical fluid extraction with CO_2	7.00 g of sample is placed into a column (127.5 mm length × 10 mm diameter and internal volume of 10 cm^3) to form the bed of fixed particles. Conditions: temperature of 50 °C, pressure of 12 MPa, CO_2 flow rate of 4.0 mL min^{-1}, and extraction time of 150 min.	Total phenolic content: 0.937 ± 0.004 mg GAE/g	[20]
Peel	Radio frequency-assisted extraction	Radio frequency time of 61.50 min, the ratio of liquid to solid 20.63:1 mL/g, and pH 2.61.	Pectin yield: 29.40%	[19]
	Ultrasonic microwave-assisted extraction	Ultrasound time: 29 min, microwave time: 10 min, power 50 of W, 86 °C, and solid to liquid ratio of 1:48 g/mL.	Pectin yield: 21.50%	[34]
	Ultrasound-assisted extraction	Liquid–solid ratio of 15:1 mL/g, pH of 1.60, sonication time of 24 min, and temperature of 60 °C.	Pectin yield: 14.50%	

GAE: gallic acid equivalent.

For the extraction of bioactive compounds from the seeds, Tramontin et al. [20] used UAE (4 min, 70% power, and 20 kHz) and supercritical fluids extraction (SFE) with CO_2 (50 °C, 12 MPa, 4.0 mL min^{-1} CO_2, and 150 min). In this study, the authors compared the performance of MAC, SOX, UAE, and SFE methods and concluded that the lowest yields corresponded to SFE at 50 °C/20 MPa/3 mL min^{-1}, obtaining 0.341 ± 0.008 mg GAE/g [20]. However, changing the processing conditions (50 °C/12 MPa/4 mL min^{-1}) resulted in the third-highest value in the study, 0.937 ± 0.004 mg GAE/g. All the extraction methods used in this study had better yields when using ethanol as a solvent; additionally, ethanol extracts had better antioxidant capacity [20]. More recently, Olalere et al. [41] used microwaves-assisted extraction (MAE) in jackfruit seeds (5 min of irradiation, 450 W, and 50 °C) and concluded that MAE produces high-quality extracts with a remarkable impact on properties like texture, phenolic exudation, and lower degradation of thermolabile compounds.

Jackfruit peel is considered an important source of pectin, thus some studies reported in the literature have focused on its extraction using emerging technologies. A couple of examples are the study performed by Xu et al. [34] using ultrasonic microwave-assisted extraction (1:48 *w/v*, 86 °C, and 29 min) and the one by Moorthy et al. [33] utilizing UAE (15:1 mL/g, pH 1.6, 24 min, and 60 °C). When comparing the results from these investigations, the UMAE showed better performance than the UAE regarding pectin extraction yield. Naik et al. [19] applied radio frequency-assisted extraction (RFAE), which, compared with all methods previously described for pectin extraction, showed the highest pectin yield with 29.40%.

4. Functionality and Application of Functional Ingredients from Jackfruit

4.1. Health Benefits

Several studies have highlighted the health-related benefits of functional ingredients from jackfruit; some of them are related to helping mitigate diseases associated with oxidative stress [29]. For example, peel extracts have shown a high radical scavenging capacity; additionally, they inhibit α-glucosidase, a potential antidiabetic property that helps control blood glucose levels [16].

Reports on jackfruit pulp showed it has selective antimutagenic and antiproliferative activities, in addition to some compounds, such as carotenoids, reducing the risk of lymphoma cancer [42]. Similarly, Kumoro et al. [1] reported that phenolic compounds present in seeds, such as polyphenols, flavonoids, tannins, and phenolic acids, have antiproliferative capabilities in cancer cell lines such as lung (A549), breast (MCF-7), liver (HepG2), and colon (HT-29), obtaining similar result to those reported by Li et al. [43]. Swami et al. [3] showed jackfruit seeds contain a non-reducing sugar that has a prebiotic effect, conferring benefits to the bacteria in the digestive system by creating a better environment for their growth and activity. The same study revealed that lignans, isoflavones, and saponins, also present in jackfruit, have anti-cancer, anti-hypertensive, anti-aging, antioxidant, and antiulcer properties [3].

Furthermore, other phenolic compounds present in the jackfruit, such as artocarpesin, norartocarpetin, and oxyresveratrol, have remarkable anti-inflammatory activity by suppressing nitric oxide (NO) and prostaglandin E2 (PGE2) production [44]. Similarly, pharmacological studies on pectin present in jackfruit showed its antioxidant, anti-tumor, and anti-inflammatory biological properties [34]. Jackfruit is also a good source of vitamin C and vitamin A, for which it has shown antiviral, antibacterial, and a good aiding in preventing blindness caused by macular degeneration [45]. Additionally, as Cardona [45] reported, its potassium content may help prevent cardiovascular accidents and reduce the heart attack risk by lowering blood pressure. Swami et al. [3] showed that jackfruit could help strengthen bones owing to its high content of magnesium, an essential nutrient in the calcium absorption process that works synergistically to strengthen and prevent bone-related disorders. Thanks to the positive health effects offered by the properties of jackfruit described above, different industries have focused on taking advantage of them.

Babu et al. [13] researched oil extracted from jackfruit seeds, showing that it is rich in EFA (1.35 g/100 g), including alpha-linoleic acid and linoleic acid, which stood out for its antioxidant properties. EFA help the proper functioning of the body, especially the brain, nervous system, adequate production of hormones, and other regulatory processes such as blood circulation, which helps prevent chronic diseases [13].

4.2. Applications

4.2.1. Pharmaceutical Industry

Some of the most relevant applications for the functional ingredients present in jackfruit are related to the pharmaceutical industry. The antioxidant properties of phenolic and carotenoid compounds, in addition to the anti-inflammatory properties of flavonoids and the antibacterial effects of extracts obtained from different parts of jackfruit, are of great interest owing to their potential incorporation into [4], for example, therapeutic agents for treating infectious diseases and their potential health benefits by positively affecting heart's function, conditions of the skin, and prevention of ulcers and cancer development [3].

The jackfruit seed crude extracts contain jacalin, a type of lectin that presents a high affinity to the Thomsen–Friedenreich disaccharide antigen (Galβ1-3GalNAc), specifically its O-glycoside [3]. This protein also recognizes and reversibly binds to galactose [31]. Jacalin is suitable for studying O-linked glycoproteins and evaluating the immune system of patients infected with certain viruses, especially the human immunodeficiency virus 1 (HIV-1) [3]. Based on jacalin's ability to target overexpressed disaccharides in tumor cells, Arya et al. [46] reported the development of surface-modified gold nanoparticles containing inositol hexaphosphate (IP6) and jacalin to maximize the apoptotic effect of IP6 against lines colon cancer cells (HCT-15). Their results showed significant apoptotic effects, concluding that the findings may raise the hope for a new drug delivery strategy to attack colon cancer [46].

In the case of jackfruit pectin, a multifunctional polysaccharide, its applications include usage as a binder for tablet formulations and as a versatile delivery agent for encapsulating drugs [47]. Govindaraj et al. [48] generated pectin (P)/hydroxyapatite (P/HA) bionanocomposites hybrids. These authors tested their cytocompatibility, alkaline phosphatase (ALP), anti-inflammatory activity, cell adhesion, and effects on fibroblast stem cells to establish their biocompatibility and feasibility as a bone graft [48]. On the other hand, Nayak & Pal [49] developed calcium pectinate-jackfruit seed starch mucoadhesive beads containing metformin HCl to evaluate the encapsulating effects of the drug and its cumulative release after some time. This study found positive effects on the yields, mucoadhesivity, and a significant hypoglycemic effect in diabetic rats after oral administration [49].

4.2.2. Food Industry

Regarding its applications in the food industry, some reports showed jackfruit and its derivatives (peel, seed, flour, chips, and wafers, among others) as functional foods thanks to their antioxidant proper, anti-cancer agents, and effect against the skin and vascular diseases [3]. Thanks to their properties, jackfruit bulbs can be used as ingredients for many foods and beverages such as jams, juices, concentrates, wines, and functional drinks, among others [50]. Further, their compounds can act as food flavoring or coloring agents [51]. Other studies report the combination of jackfruit pulp with other fruits like palmyra palm (*Borassus flabellifer* L) and passion fruit (*Passiflora edulis* Sims) to make functional beverages with a high content of bioactive compounds [52]. In a study regarding the optimization of jackfruit nectar's processing conditions with a thermo-ultrasound, researchers found final nectar properties to be as follows: ascorbic acid content of 420 mg/L, total phenolics of 134 mg GAE/L, antiradical activity assayed with ABTS of 3.51 μmol Trolox Equivalents (TE)/L and with DPPH of 75 μmol TE/L, and residual activity of pectin methylesterase of 60% [53]. In addition, thermo-ultrasound proved to be effective as an alternative to pasteurization and generate good product quality [53].

Odoemelam [54] reported that jackfruit seed flour could be used as a functional ingredient as it helps to reduce the concentration of gluten in baked goods. Thanks to their antimicrobial properties, nanoparticles generated from jackfruit seeds were proposed as agents against foodborne pathogens [3]. Reported research results on improving chocolate cream properties by incorporating jackfruit flour showed a significant increase in the content of polyphenols (127.00 mg/g), carotenoids (160.16 mg/g), and antioxidant activity IC50 (42.75 μg/mL), while positively affecting the product's sensory properties such as viscosity and color [55].

Regarding jackfruit pectin, some possible uses in the food industry are as a gelling agent, emulsifier, stabilizer, and thickener agent in preparing jams, marmalades, and substitute for fats in various food formulations [19]. Meethal et al. [56] prepared snack bars using jackfruit seed flour and tested different formulations to compare their physicochemical, sensory, and nutritional changes during storage. Their results showed that the highest phenolic content in their formulations was 44.94 ± 0.21 mg GAE/g, while the highest antioxidant activity was 59.34 ± 0.26%, concluding that a jackfruit seed meal is a low-cost option for the preparation of value-added nutritional products [56].

4.2.3. Other Applications

Other applications mentioned for the bioactive compounds present in jackfruit include the cosmetic industry, such as using pectin to fabricate nanoparticles with emulsifying capabilities [57]. However, there is almost no research in other applications areas apart from the food and pharmaceutical industry, representing an area of opportunity for future investigations research.

5. Conclusions and Future Perspectives

According to the reviewed scientific research, it can be concluded that the quantity and quality of the compounds extracted from jackfruit depend on several factors. The main factors to consider are the extraction method, whether assisted by any emerging technology or not, the conditions of extraction, and the properties of the matrix used.

Regarding the performance of the technologies reviewed, the conventional technology with the best performance for recovering of phenolic compounds was alkaline hydrolysis extraction using NaOH in pulp samples. Additionally, extraction with methanol at 90% using peel as a matrix had the second higher total phenolic content. From the emerging technologies extraction reports, the best results were for SFE with CO_2 using jackfruit seeds. Concerning the application of emerging technologies as a pre-treatment, even though they may have not necessarily improved yields of extraction, their use can be beneficial for other aspects, such as process sustainability, by decreasing the use of organic solvents, shortening extraction times, and reducing the amount of energy consumed. However, more research is required to understand the effect of other emerging technologies, alone and in combination, when extracting these compounds.

Other research opportunity areas include the following: (1) determining the most convenient technology for carotenoid extraction from jackfruit, (2) applying non-conventional assisted extraction for functional ingredients present in jackfruit pulp, and (3) evaluating conventional and emerging technologies extraction methods for essential fatty acids in jackfruit need. Moreover, future studies should consider other jackfruit varieties/species and compare results based on their place of origin. The application of jackfruit functional ingredients also requires more investigation, not only for the food and pharmaceutical industries, but also for other areas such as the cosmetic area.

Author Contributions: Conceptualization, T.G.-C. and V.R.-M.; methodology, F.C.C.-C. and V.R.-M.; investigation, F.C.C.-C. and V.R.-M.; writing—original draft preparation, F.C.C.-C.; writing—review and editing, T.G.-C. and V.R.-M.; visualization, F.C.C.-C., T.G.-C., and V.R.-M.; supervision, T.G.-C. and V.R.-M.; project administration, V.R.-M. All authors have read and agreed to the published version of the manuscript.

Funding: The authors acknowledge the financial support by Tecnologico de Monterrey within the funding program "Fund for financing the publication of Scientific Articles". Frida Camila Cruz-Casillas thanks Consejo Nacional de Ciencia y Tecnología (CONACyT; CVU: 1078840) Mexico for scholarship funding and Tecnologico de Monterrey for academic support.

Institutional Review Board Statement: Not applicable.

Informed Consent Statement: Not applicable.

Data Availability Statement: Data sharing not applicable to this article.

Conflicts of Interest: The authors declare no conflict of interest.

References

1. Kumoro, A.C.; Alhanif, M.; Wardhani, D.H. A Critical Review on Tropical Fruits Seeds as Prospective Sources of Nutritional and Bioactive Compounds for Functional Foods Development: A Case of Indonesian Exotic Fruits. *Int. J. Food Sci.* **2020**, *2020*, 4051475. [CrossRef] [PubMed]
2. Maia, J.G.S.; Andrade, E.H.A.; Zoghbi, M.D.G.B. Aroma volatiles from two fruit varieties of jackfruit (*Artocarpus heterophyllus* Lam.). *Food Chem.* **2004**, *85*, 195–197. [CrossRef]
3. Swami, S.B.; Thakor, N.J.; Haldankar, P.M.; Kalse, S.B. Jackfruit and Its Many Functional Components as Related to Human Health: A Review. *Compr. Rev. Food Sci. Food Saf.* **2012**, *11*, 565–576. [CrossRef]
4. Baliga, M.S.; Shivashankara, A.R.; Haniadka, R.; Dsouza, J.; Bhat, H.P. Phytochemistry, nutritional and pharmacological properties of *Artocarpus heterophyllus* Lam (jackfruit): A review. *Food Res. Int.* **2011**, *44*, 1800–1811. [CrossRef]
5. Ranasinghe, R.A.S.N.; Maduwanthi, S.D.T.; Marapana, R.A.U.J. Nutritional and Health Benefits of Jackfruit (*Artocarpus heterophyllus* Lam.): A Review. *Int. J. Food Sci.* **2019**, *2019*, 4327183. [CrossRef]
6. Akter, F.; Haque, M.A. Jackfruit Waste: A Promising Source of Food And Feed. *Ann. Bangladesh Agric.* **2019**, *23*, 91–102. [CrossRef]
7. Abuajah, C.I.; Ogbonna, A.C.; Osuji, C.M. Functional components and medicinal properties of food: A review. *J. Food Sci. Technol.* **2015**, *52*, 2522–2529. [CrossRef] [PubMed]
8. Vazhacharickal, P.J.; Sajeshkumar, N.K.; Mathew, J.J.; Kuriakose, A.C.; Abraham, B.; Mathew, R.J.; Albin, A.N.; Thomson, D.; Thomas, R.S.; Varghese, N.; et al. Chemistry and Medicinal Properties of Jackfruit (*Artocarpus heterophyllus*): A Review on Current Status of Knowledge. *Int. J. Innov. Res. Rev.* **2015**, *3*, 2347–4424.
9. Wichienchot, S.; Thammarutwasik, P.; Jongjareonrak, A.; Chansuwan, W.; Hmadhlu, P.; Hongpattarakere, T.; Itharat, A.; Ooraikul, B. Extraction and analysis of prebiotics from selected plants from southern Thailand. *Songklanakarin J. Sci. Technol.* **2011**, *33*, 517–523.
10. Bhornsmithikun, V.; Chetpattananondh, P.; Yamsaengsung, R.; Prasertsit, K. Continuous extraction of prebiotics from jackfruit seeds. *J. Sci. Technol.* **2010**, *32*, 635–642.
11. Begum, R.; Aziz, M.G.; Uddin, M.B.; Yusof, Y.A. Characterization of Jackfruit (*Artocarpus heterophyllus*) Waste Pectin as Influenced by Various Extraction Conditions. *Agric. Agric. Sci. Procedia* **2014**, *2*, 244–251. [CrossRef]
12. Adan, A.A.; Ojwang, R.A.; Muge, E.K.; Mwanza, B.K.; Nyaboga, E.N. Phytochemical composition and essential mineral profile, antioxidant and antimicrobial potential of unutilized parts of jackfruit. *Food Res.* **2020**, *4*, 1125–1134. [CrossRef]
13. Babu, N.G.; Kumar, S.; Sundar, S. Extraction and comparison of properties of jackfruit seed oil and sunflower seed oil. *Int. J. Sci. Eng. Res.* **2017**, *8*, 635–639.
14. Chowdhury, F.A.; Azizur Raman, M.; Jabbar Mian, A. Distribution of free sugars and fatty acids in jackfruit (*Artocarpus heterophyllus*). *Food Chem.* **1997**, *60*, 25–28. [CrossRef]
15. Hamzalioğlu, A.; Gökmen, V. Interaction between Bioactive Carbonyl Compounds and Asparagine and Impact on Acrylamide. In *Acrylamide in Food: Analysis, Content and Potential Health Effects*; Elsevier: Amsterdam, The Netherlands, 2016; ISBN 9780128028759.
16. Zhang, L.; Tu, Z.C.; Xie, X.; Wang, H.; Wang, H.; Wang, Z.X.; Sha, X.M.; Lu, Y. Jackfruit (*Artocarpus heterophyllus* Lam.) peel: A better source of antioxidants and a-glucosidase inhibitors than pulp, flake and seed, and phytochemical profile by HPLC-QTOF-MS/MS. *Food Chem.* **2017**, *234*, 303–313. [CrossRef]
17. Sundarraj, A.A.; Thottiam Vasudevan, R.; Sriramulu, G. Optimized extraction and characterization of pectin from jackfruit (*Artocarpus integer*) wastes using response surface methodology. *Int. J. Biol. Macromol.* **2018**, *106*, 698–703. [CrossRef] [PubMed]
18. Shakthi Deve, A.; Sathish Kumar, T.K.; Kumaresan, K.; Rapheal, V.S. Extraction process optimization of polyphenols from Indian Citrus sinensis—As novel antiglycative agents in the management of diabetes mellitus. *J. Diabetes Metab. Disord.* **2014**, *13*, 1–10. [CrossRef]
19. Naik, M.; Rawson, A.; Rangarajan, J.M. Radio frequency-assisted extraction of pectin from jackfruit (*Artocarpus heterophyllus*) peel and its characterization. *J. Food Process Eng.* **2020**, *43*, e13389. [CrossRef]
20. Tramontin, D.P.; Cadena-Carrera, S.E.; Bella-Cruz, A.; Bella Cruz, C.C.; Bolzan, A.; Quadri, M.B. Biological activity and chemical profile of Brazilian jackfruit seed extracts obtained by supercritical CO_2 and low pressure techniques. *J. Supercrit. Fluids* **2019**, *152*, 104551. [CrossRef]
21. Li, W.J.; Fan, Z.G.; Wu, Y.Y.; Jiang, Z.G.; Shi, R.C. Eco-friendly extraction and physicochemical properties of pectin from jackfruit peel waste with subcritical water. *J. Sci. Food Agric.* **2019**, *99*, 5283–5292. [CrossRef]

22. Campos, D.A.; Gómez-García, R.; Vilas-Boas, A.A.; Madureira, A.R.; Pintado, M.M. Management of fruit industrial by-products—A case study on circular economy approach. *Molecules* **2020**, *25*, 320. [CrossRef] [PubMed]

.   Kumar, S.; Sharma, S.; Vasudeva, N. Review on antioxidants and evaluation procedures. *Chin. J. Integr. Med.* **2017**, *2017*, 1–12. [CrossRef] [PubMed]

24. Xu, F.; He, S.Z.; Chu, Z.; Zhang, Y.J.; Tan, L.H. Effects of Heat Treatment on Polyphenol Oxidase Activity and Textural Properties of Jackfruit Bulb. *J. Food Process. Preserv.* **2016**, *40*, 943–949. [CrossRef]

25. Zhang, X.; Zhu, K.; Xie, J.; Chen, Y.; Tan, L.; Liu, S.; Dong, R.; Zheng, Y.; Yu, Q. Optimization and identification of non-extractable polyphenols in the dietary fiber of jackfruit (*Artocarpus heterophyllus* Lam.) pulp released by alkaline, acid and enzymatic hydrolysis: Content, composition and antioxidant activities. *LWT* **2020**, *138*, 110400. [CrossRef]

26. Singh, A.; Maurya, S.; Singh, M.; Singh, U.P. *Studies on the Phenolic Acid Contents in Different Parts of Raw and Ripe Jackfruit and Their Importance in Human Health*; iMedPub: London, UK, 2015; Volume 2.

27. Galvez, L.; Dizon, E. Physico-chemical and Functional Properties of Two Jackfruit (*Artocarpus heterophyllus* Lam.) Varieties in Eastern Visayas, Philippines. *Ann. Trop. Res.* **2017**, *392*, 100–106. [CrossRef]

28. de Faria, A.F.; de Rosso, V.V.; Mercadante, A.Z. Carotenoid composition of jackfruit (*Artocarpus heterophyllus*), determined by HPLC-PDA-MS/MS. *Plant Foods Hum. Nutr.* **2009**, *64*, 108–115. [CrossRef]

29. Shanooba, P.M.; Tungare, K.; Sunariwal, S.; Sonawane, S. Extraction and Characterization of Polyphenols from *Artocarpus heterophyllus* and Its Effect on Oxidative Stability of Peanut Oil. *Int. J. Fruit Sci.* **2020**, *20*, S1134–S1155. [CrossRef]

30. Nagala, S.; Yekula, M.; Tamanam, R.R. Antioxidant and gas chromatographic analysis of five varieties of jackfruit (Artocarpus) seed oils. *Drug Invent. Today* **2013**, *5*, 315–320. [CrossRef]

31. Mohamad, S.F.S.; Said, F.M.; Munaim, M.S.A.; Mohamad, S.; Sulaiman, W.M.A.W. Response surface optimization of the forward extraction of jacalin from jackfruit seeds using AOT/isooctane reverse micellar system. *IOP Conf. Ser. Mater. Sci. Eng.* **2020**, *716*, 012019. [CrossRef]

32. Chai, T.T.; Xiao, J.; Mohana Dass, S.; Teoh, J.Y.; Ee, K.Y.; Ng, W.J.; Wong, F.C. Identification of antioxidant peptides derived from tropical jackfruit seed and investigation of the stability profiles. *Food Chem.* **2021**, *20*, 59–70. [CrossRef]

33. Moorthy, I.G.; Maran, J.P.; Ilakya, S.; Anitha, S.L.; Sabarima, S.P.; Priya, B. Ultrasound assisted extraction of pectin from waste *Artocarpus heterophyllus* fruit peel. *Ultrason. Sonochem.* **2017**, *34*, 525–530. [CrossRef]

34. Xu, S.Y.; Liu, J.P.; Huang, X.; Du, L.P.; Shi, F.L.; Dong, R.; Huang, X.T.; Zheng, K.; Liu, Y.; Cheong, K.L. Ultrasonic-microwave assisted extraction, characterization and biological activity of pectin from jackfruit peel. *LWT Food Sci. Technol.* **2018**, *90*, 577–582. [CrossRef]

35. Adetunji, L.R.; Adekunle, A.; Orsat, V.; Raghavan, V. Advances in the pectin production process using novel extraction techniques: A review. *Food Hydrocoll.* **2017**, *62*, 239–250. [CrossRef]

36. Sharma, A.; Gupta, P.; Verma, A.K. Preliminary nutritional and biological potential of *Artocarpus heterophyllus* L. shell powder. *J. Food Sci. Technol.* **2015**, *52*, 1339–1349. [CrossRef] [PubMed]

37. Gil-Chávez, G.J.; Villa, J.A.; Ayala-Zavala, J.F.; Heredia, J.B.; Sepulveda, D.; Yahia, E.M.; González-Aguilar, G.A. Technologies for Extraction and Production of Bioactive Compounds to be Used as Nutraceuticals and Food Ingredients: An Overview. *Compr. Rev. Food Sci. Food Saf.* **2013**, *12*, 5–23. [CrossRef]

38. Galanakis, C.M. Recovery of high added-value components from food wastes: Conventional, emerging technologies and commercialized applications. *Trends Food Sci. Technol.* **2012**, *26*, 68–87. [CrossRef]

39. Galanakis, C.M.; Barba, F.J.; Prasad, K.N. Cost and safety issues of emerging technologies against conventional techniques. In *Food Waste Recovery: Processing Technologies and Industrial Techniques*; Elsevier Inc.: Amsterdam, The Netherlands, 2015; pp. 321–336, ISBN 9780128004197.

40. Islam, M.R.; Haque, A.R.; Kabir, M.R.; Hasan, M.M.; Khushe, K.J.; Hasan, S.M.K. Fruit by-products: The potential natural sources of antioxidants and α-glucosidase inhibitors. *J. Food Sci. Technol.* **2020**, *58*, 1715–1726. [CrossRef]

41. Olalere, O.A.; Gan, C.Y.; Abdurahman, H.N.; Adeyi, O.; Ahmad, M.M. Holistic approach to microwave-reflux extraction and thermo-analytical fingerprints of under-utilized *Artocarpus heterophyllus* seed wastes. *Heliyon* **2020**, *6*, e04770. [CrossRef]

42. Ruiz-Montañez, G.; Burgos-Hernández, A.; Calderón-Santoyo, M.; López-Saiz, C.M.; Velázquez-Contreras, C.A.; Navarro-Ocaña, A.; Ragazzo-Sánchez, J.A. Screening antimutagenic and antiproliferative properties of extracts isolated from Jackfruit pulp (*Artocarpus heterophyllus* Lam). *Food Chem.* **2015**, *175*, 409–416. [CrossRef]

43. Li, F.; Li, S.; Li, H.B.; Deng, G.F.; Ling, W.H.; Wu, S.; Xu, X.R.; Chen, F. Antiproliferative activity of peels, pulps and seeds of 61 fruits. *J. Funct. Foods* **2013**, *5*, 1298–1309. [CrossRef]

44. Fang, S.C.; Hsu, C.L.; Yen, G.C. Anti-inflammatory effects of phenolic compounds isolated from the fruits of *Artocarpus heterophyllus*. *J. Agric. Food Chem.* **2008**, *56*, 4463–4468. [CrossRef]

45. Cardona, D. La Yaca (*Artocarpus heterophyllus*) y sus beneficios. *Ecuad. Sci. J.* **2017**, *1*, 12–13. [CrossRef]

46. Arya, M.; Mishra, N.; Singh, P.; Tripathi, C.B.; Parashar, P.; Singh, M.; Gupta, K.P.; Saraf, S.A. In vitro and in silico molecular interaction of multiphase nanoparticles containing inositol hexaphosphate and jacalin: Therapeutic potential against colon cancer cells (HCT-15). *J. Cell. Physiol.* **2019**, *134*, 15527–15536. [CrossRef] [PubMed]

47. Khedmat, L.; Izadi, A.; Mofid, V.; Mojtahedi, S.Y. Recent advances in extracting pectin by single and combined ultrasound techniques: A review of techno-functional and bioactive health-promoting aspects. *Carbohydr. Polym.* **2020**, *229*, 115474. [CrossRef] [PubMed]

48. Govindaraj, D.; Rajan, M.; Hatamleh, A.A.; Munusamy, M.A. From waste to high-value product: Jackfruit peel derived pectin/apatite bionanocomposites for bone healing applications. *Int. J. Biol. Macromol.* **2018**, *106*, 293–301. [CrossRef] [PubMed]
49. Nayak, A.K.; Pal, D. Blends of jackfruit seed starch-pectin in the development of mucoadhesive beads containing metformin HCl. *Int. J. Biol. Macromol.* **2013**, *62*, 137–145. [CrossRef]
50. Anaya-Esparza, L.M.; González-Aguilar, G.A.; Domínguez-Ávila, J.A.; Olmos-Cornejo, J.E.; Pérez-Larios, A.; Montalvo-González, E. Effects of Minimal Processing Technologies on Jackfruit (*Artocarpus heterophyllus* Lam.) Quality Parameters. *Food Bioprocess Technol.* **2018**, *11*, 1761–1774. [CrossRef]
51. Bapat, V.A.; Jagtap, U.B.; Ghag, S.B.; Ganapathi, T.R. Molecular Approaches for the Improvement of Under-Researched Tropical Fruit Trees: Jackfruit, Guava, and Custard Apple. *Int. J. Fruit Sci.* **2020**, *20*, 233–281. [CrossRef]
52. Chakraborty, I.; Chaurasiya, A.K.; Saha, J. Quality of diversified value addition from some minor fruits. *J. Food Sci. Technol.* **2011**, *48*, 750–754. [CrossRef] [PubMed]
53. Cruz-Cansino, N.d.S.; Ariza-Ortega, J.A.; Alanís-García, E.; Ramírez-Moreno, E.; Velázquez-Estrada, R.M.; Zafra-Rojas, Q.Y.; Cervantes-Elizarrarás, A.; Suárez-Jacobo, Á.; Delgado-Olivares, L. Optimal thermoultrasound processing of jackfruit (*Artocarpus heterophyllus* lam.) nectar: Physicochemical characteristics, antioxidant properties, microbial quality, and fatty acid profile comparison with pasteurized nectar. *J. Food Process. Preserv.* **2020**, *45*, e15029. [CrossRef]
54. Odoemelam, S.A. Functional Properties of Raw and Heat Processed Jackfruit (*Artocarpus heterophyllus*) Flour. *Pak. J. Nutr.* **2005**, *4*, 366–370. [CrossRef]
55. Fitriani Nur, U.A.; Yusuf, M.; Pirman; Syahriati; Rahmiah, S. Physicochemical, antioxidant and sensory properties of chocolate spread fortified with jackfruit (*Artocarpus heterophyllus*) flour. *Food Res.* **2020**, *4*, 2147–2155. [CrossRef]
56. Meethal, S.M.; Kaur, N.; Singh, J.; Gat, Y. Effect of addition of jackfruit seed flour on nutrimental, phytochemical and sensory properties of snack bar. *Curr. Res. Nutr. Food Sci.* **2017**, *5*, 154–158. [CrossRef]
57. Jin, B.; Zhou, X.; Guan, J.; Yan, S.; Xu, J.; Chen, J. Elucidation of stabilizing pickering emulsion with jackfruit filum pectin-soy protein nanoparticles obtained by photocatalysis. *J. Dispers. Sci. Technol.* **2019**, *40*, 909–917. [CrossRef]

Article

Eggplant Fruit (*Solanum melongena* L.) and Bio-Residues as a Source of Nutrients, Bioactive Compounds, and Food Colorants, Using Innovative Food Technologies

Gabriel F. Pantuzza Silva [1], Eliana Pereira [1,*], Bruno Melgar [1], Dejan Stojković [2], Marina Sokovic [2], Ricardo C. Calhelha [1], Carla Pereira [1], Rui M. V. Abreu [1], Isabel C. F. R. Ferreira [1] and Lillian Barros [1,*]

[1] Centro de Investigação de Montanha (CIMO), Instituto Politécnico de Bragança, Campus de Santa Apolónia, 5300-253 Bragança, Portugal; gabriel.pantuzza@hotmail.com (G.F.P.S.); bruno.melgarc@gmail.com (B.M.); calhelha@ipb.pt (R.C.C.); carlap@ipb.pt (C.P.); ruiabreu@ipb.pt (R.M.V.A.); iferreira@ipb.pt (I.C.F.R.F.)

[2] Institute for Biological Research "Siniša Stanković"-National Institute of Republic of Serbia, University of Belgrade Bulevar Despota Stefana 142, 11000 Belgrade, Serbia; dejanbio@ibiss.bg.ac.rs (D.S.); mris@ibiss.bg.ac.rs (M.S.)

* Correspondence: eliana@ipb.pt (E.P.); lillian@ipb.pt (L.B.)

Abstract: Consumers are very concerned with following a healthy diet, along with some precautions that may influence environmental impact. *Solanum melongena* L. is one of the most consumed vegetables due to its excellent nutritional value and antioxidant action. Associated with its high consumption, considerable amounts of agro-food wastes are produced. This work targets the valorization of this matrix, through the use of its bio-residues to study the obtention of coloring pigments, applying innovative technologies. Its nutritional value, chemical composition, and bioactive potential were evaluated, and the ultrasound-assisted extraction to obtain coloring pigments of high industrial interest was optimized. Considering the results, low contents of fat and carbohydrates and energy value were evident, as well as the presence of compounds of interest (free sugars, organic acids, unsaturated fatty acids, and phenolic acids). In addition, the antioxidant and antimicrobial potential was detected. Response surface methodology was performed to optimize the extraction of natural pigments, showing a concentration of 11.9 mg/g of anthocyanins/g of extract, applying optimal conditions of time, solvent, and solid/liquid ratio of 0.5 min, 68.2% (v/v) and 5 g/L, respectively. *S. melongena* proved to be a good source of bioactive compounds and natural pigments, which can generate great interest in the food industry.

Keywords: eggplant; anthocyanins; natural colorants; bioactivity

Citation: Silva, G.F.P.; Pereira, E.; Melgar, B.; Stojković, D.; Sokovic, M.; Calhelha, R.C.; Pereira, C.; Abreu, R.M.V.; Ferreira, I.C.F.R.; Barros, L. Eggplant Fruit (*Solanum melongena* L.) and Bio-Residues as a Source of Nutrients, Bioactive Compounds, and Food Colorants, Using Innovative Food Technologies. *Appl. Sci.* **2021**, *11*, 151. https://doi.org/10.3390/app11010151

Received: 30 November 2020
Accepted: 22 December 2020
Published: 25 December 2020

Publisher's Note: MDPI stays neutral with regard to jurisdictional claims in published maps and institutional affiliations.

1. Introduction

The food industry is one of the sectors in Europe that generates great amounts of bio-waste. Annually, in Europe, food waste is around 1.3 billion tons, of which approximately 700 million tons are from agriculture. These residues result essentially from the processing of grains, fruits, and vegetables and have a high potential for reuse [1,2]. These residues can be transformed into resources, using intensified conversion processes that can produce potentially sustainable bio-products, such as energy, fertilizers, materials, and molecules. In this sense, the valorization of residues and agri-food by-products is currently presented not only as a necessity, but also as an opportunity to obtain new products of high value and of great impact on the economy of the industrial sector [3].

Solanum melongena L. (Solanaceae), known by its fruit, eggplant, is a non-tuberous species of great agronomic and economic value [4]; however, mature eggplant fruits are considered unmarketable due to their unpleasant fruit color, texture, pith, and bitter taste as well as the large amount of mature seeds, thus becoming a bio-waste of the food industry [5].

Appl. Sci. **2021**, *11*, 151. https://dx.doi.org/10.3390/app11010151   https://www.mdpi.com/journal/applsci

This matrix is considered a versatile food, as it is part of the gastronomy of many countries and different sorts of diets, namely Mediterranean and vegetarian [6]. In addition to their culinary interest, several authors describe the use of eggplants in traditional medicine, acting therapeutically on several health problems. This curative potential is manifested by its rich composition of ascorbic acid and phenolic compounds, both with high antioxidant potential [4,7].

Some of the main phenolic compounds found in eggplant are anthocyanins, responsible for the majority of red, purple, and blue colors of flowers and fruits [8,9]. Anthocyanins are increasingly valued by the industrial sector, due to their application as natural colorant ingredients. However, in addition to their coloring potential, they also exhibit bioactive properties related to their high antioxidant activity [10]. The potential use of anthocyanins as food colorants has attracted the attention of the scientific community, for combining the pigmentation capacity with health benefits [11,12]. In the case of eggplant, most of these anthocyanic compounds are concentrated in the epicarp [13].

Considering the nutritional and bioactive potential of *Solanum melongena* L. fruit pulp and the coloring compounds present in the epicarp, this study aims to optimize ultrasound-assisted extraction methodologies to obtain an anthocyanin-rich extract from *S. melongena* epicarp, to be used as a natural coloring ingredient, and thus, contributing to the sector sustainability and improvement of environmental impact. It also targets the nutritional and chemical characterization of the fruit, as well as at its bioactive properties' assessment.

In this sense, this work addresses a subject with high relevance for several sectors, such as industrial, environmental, scientific, economic, and social. This wide relevance is justified by the proposal to develop strategies that reduce food waste, improving the quality of some commercial products. Furthermore, it leads to the application of innovative extraction technologies, an increase in the quality of products available on the market, an expansion in consumers' preference for these products and, consequently, to an improvement in the environmental impact, economic growth, and the development of a sustainable economy.

2. Materials and Methods

2.1. Samples

The *Solanum melongena* L. fruits (eggplant) were purchased in April 2019, in a local market in Bragança (northeast Portugal). After the acquisition of the samples (15 fruits; approximately 3 kg in fresh weight), the epicarp and the pulp were separated by mechanical methods, and the different parts of samples were selected (whole fruit, pulp, and peel) in order to perform the different analyses. Subsequently, the samples were frozen, dehydrated by lyophilization (FreeZone 4.5, Labconco, Kansas City, MO, USA), and reduced to a fine dried powder. The samples were then stored in a dry place away from light and heat, until further analysis.

2.2. Determination of Color in Eggplant Epicarp

Color evaluation was performed in the epicarp of the fruit, in its fresh and lyophilized state, using a colorimeter (model CR-400, Konica Minolta Sensing, Inc., Osaka, Japan) (model CR-A50). For that, methodology previously described by Roriz et al. [14] was used. The values of the CIE (international lighting commission) $L^* a^* b^*$ coordinates were obtained on a computer system using the illuminant C and an 8 mm diameter diaphragm. For data processing, the "Spectra Magic Nx" software (version CM-S100W 2.03.0006, Konica Minolta, Japan) was used.

2.3. Determination of Nutritional Composition of Eggplant (Pulp and Whole Fruit)

The nutritional composition was determined in pulp and whole fruits, separately, considering their form of consumption. Approximately 10 g of dry sample were used for these analyses. For this purpose, the official analysis methodologies AOAC (association of official analytical chemists) [15] were used. The crude protein (N × 6.25) was evaluated using the macro-Kjeldahl method (AOAC 991.02); fat content was determined applying a soxhlet extraction of the dry sample with petroleum ether (AOAC 989.05); the ash was estimated by incineration at $550 \pm 15\ ^\circ$C (AOAC 935.42); total carbohydrates were calculated by difference, and the energy was estimated using the following equation: Energy (kcal) = 4 × (g protein + g carbohydrates) + 9 × (g fat).

2.4. Free Sugars, Organic Acids, and Fatty Acids Profile of Eggplant (Pulp and Whole Fruit)

The chemical composition was also determined in pulp and whole fruits, separately, also according to their form of consumption. Approximately 5 g of dry sample were used for these analyses. Free sugars were analyzed by high performance liquid chromatography coupled to a refraction index detector (HPLC-RI, Knauer, Smartline system 1000), according to a practice earlier explained by other authors [16]. The compounds were identified by chromatographic comparison with authentic standards (D(−)-fructose, D(+)-sucrose, D(+)-glucose, D(+)-trehalose and D(+)-raffinose pentahydrate), purchased from Sigma-Aldrich (St. Louis, MO, USA). Melezitose was used as the internal standard and the results of sugars were expressed in g/100 g of fresh weight (fw).

Organic acids were evaluated following a procedure previously described by Barros et al. [16]. Approximately 5 g of dry sample were used for these analyses. Ultra-Fast Liquid Chromatography (UFLC, Shimadzu 20A series, Kyoto, Japan) coupled to a diode array detector was used, and the quantification was performed through a calibration curve obtained from commercial standards (L(+)-ascorbic acid, citric acid, malic acid, oxalic acid, shikimic acid, succinic acid, fumaric acid, and quinic acid; purchased from Sigma-Aldrich; St. Louis, MO, USA). The results were expressed in g/100 g of fresh weight (fw).

The fatty acids profile was determined using chromatographic methods, namely using gas chromatography coupled with a flame ionization detector (GC-FID, DANI model GC 1000, Contone, Switzerland) and the separation was achieved with a Macherey–Nagel (Düren, Germany) column (50% cyanopropyl-methyl-50% phenylmethylpolysiloxane, 30 m × 0.32 mm i.d. × 0.25 μm df). The compounds were identified by comparing the relative retention times of FAME (Fatty Acid Methyl Esters) peaks from samples with commercial standards (FAME reference standard mixture, standard 47885-U, Sigma-Aldrich, St. Louis, MO, USA) [16]. Approximately 9 g of dry sample were used for these analyses. The results were expressed in relative percentages (%).

2.5. Determination of Phenolic Compounds of Eggplant

The analysis of non-anthocyanin phenolic compounds was carried out on the epicarp, pulp, and whole fruit of *S. melongena*, whereas the determination of anthocyanin phenolic compounds was performed only in the epicarp, due to its evident purple color. For that purpose, methodology described in previous works was applied [17–20]. Approximately 3 g of dry sample were used for these analyses.

2.5.1. Non-Anthocyanin Compounds

Extraction procedure. An extract was prepared from the lyophilized samples (fruits of *S. melongena*; 1 g) and a maceration with ethanol/water solution (80:20, v/v; 30 mL) at room temperature was used. The alcoholic fraction of the extracts was evaporated under reduced pressure (Büchi R-210, Flawil, Switzerland) and the aqueous fraction was lyophilized (47 $^\circ$C, 0.045 bar; FreeZone 4.5, Labconco, Kansas City, MO, USA) for further analysis. A quantity of the obtained dry extract (10 mg) was subsequently re-dissolved in an ethanol/water solution for further HPLC analysis [18].

Analytical method. The chromatographic data were acquired using a Dionex Ultimate 3000 UPLC (Thermo Scientific, San Jose, CA, USA), coupled to a diode array detector (280, 330, and 370 nm) and an electrospray ionization mass detector (Linear Ion Trap LTQ XL, Thermo Finnigan, San Jose, CA, USA), working in the negative mode. The chromatographic separation was performed using a Waters Spherisorb S3 ODS-2 C18 (3 μm, 4.6 mm × 150 mm, Waters, Milford, MA, USA) column at 35 $^\circ$C. The compounds were identified considering the retention time, UV-Vis, and mass spectra in comparison with available standards (caffeic acid (y = 388,345x + 406,369), protocatequic acid (y = 214,168x + 27,102), and chlorogenic acid (y = 168,823x − 161,172)) and with literature data. Calibration curves of the available phenolic standards were constructed based on the UV-Vis signal to obtain quantitative analysis. In the case of unavailable commercial standards, the compounds were quantified via a calibration curve of the most similar standard available. The results were expressed as mg/g of extract [17].

2.5.2. Anthocyanin Compounds

Extraction procedure. The lyophilized samples (1 g) were extracted using the conventional methodology applied by this research group [20]; ethanol/water (80:20, v/v) acidified with 0.5% of citric acid was used as extraction solvent.

Analytical method. The chromatographic analysis was made using a system previously described in Section 2.5.1 and the separation was achieved using an AQUA® (Phenomenex) reverse phase C18 column (5 μm, 150 × 4.6 mm i.d.) thermostatted at 35 °C, using a UPLC-DAD-ESI/MSn system (Ultra-Performance Liquid Chromatography coupled with Diode Array Detection and an Electrospray Ionization Mass Spectrometry) previously mentioned. Detection was carried out using a DAD (520 nm) and a mass spectrometer (Linear Ion Trap LTQ XL Thermo Finnigan) equipped with an ESI source, operating in positive mode. Compounds identification was performed using the retention time, UV-Vis, and mass spectra data in comparison with available standards and literature review. For quantitative analysis, calibration curves were obtained by injection of standard solutions with known concentrations (0.25–50 μg/mL): cyanidin-3-O-glucoside (y = 97,787x − 743,469; R2 = 0.99993), peonidine (y = 110,391x − 1 × 10⁶; R2 = 0.9999), and pelargonidin-3-O-glucoside (y = 43,781x − 275,315; R2 = 0.9989). The results were expressed in mg of anthocyanin/g of extract [19,20].

2.6. Evaluation of the Bioactive Properties of Solanum melongena L. Fruits: Epicarp, Pulp, and Whole Fruit, through In Vitro Tests

The lyophilized extract, provided by the extraction procedures (Section 2.5.1), was re-dissolved: (i) in culture medium (10 mg/mL) for antimicrobial activity assay; (ii) in distilled water at a concentration of 8 mg/mL for the evaluation of cytotoxic, hepatotoxic, and anti-inflammatory activity; and (iii) in a hydroethanolic solution (ethanol/water; 80:20, v/v), in a concentration of 5 mg/mL, for antioxidant activity evaluation. These solutions were diluted successively in order to obtain the working concentrations.

2.6.1. Antimicrobial Activity

The antibacterial activity was evaluated using Gram-negative (*Enterobacter cloacae* (ATCC—American type culture collection 35030), *Escherichia coli* (ATCC 35210), and *Salmonella* Typhimurium (ATCC 13311)) and Gram-positive bacteria strains (*Listeria monocytogenes* (NCTC—National collection of type cultures 7973) and *Staphylococcus aureus* (ATCC 6538)). The minimum inhibitory (MIC) and minimum bactericidal (MBC) concentrations were used to estimate antimicrobial potential; the microdilution method was applied, and the results were expressed in mg/mL [21].

For the antifungal activity, *Aspergillus fumigatus* (ATCC 1022), *Aspergillus niger* (ATCC 6275), *Aspergillus versicolor* (ATCC 11730), *Penicillium funiculosum* (ATCC 36839), *Penicillium ochrochloron* (ATCC 9112), and *Trichoderma viride* (IAM 5061) strains were used. Minimum inhibitory concentration (MIC) and minimum fungicidal concentration (MFC) were determined using a modified microdilution method, and the results were also expressed in mg/mL [21].

2.6.2. Cytotoxic, Hepatotoxic, and Anti-Inflammatory Activity

In order to evaluate the cytotoxicity of the extracts, the sulforhodamine B assay was applied using methodology previously described by Guimarães et al. [22]. For that, four human tumor cell lines, acquired from Leibniz-Institut DSMZ, HeLa (cervical carcinoma),

HepG2 (hepatocellular carcinoma), MCF-7 (breast adenocarcinoma), and NCI-H460 (non-small cell lung cancer) were used, as well as a primary cell culture, PLP2, obtained from a freshly harvested porcine liver (purchased from a local slaughter house) to test the hepato-toxic activity. Ellipticine was used as a positive control and the results were expressed as GI_{50} values (the extract concentration that inhibits 50% of the net cell growth).

For the anti-inflammatory activity evaluation, a process previously mentioned in Jabeur et al. [18] was performed. For that purpose, a mouse macrophage-like cell line RAW 264.7 stimulated with LPS (Lipopolysaccharides) was used, and the extracts concentration ranged between 400–125 µg/mL. Nitric oxide (NO) production was studied with a Griess reagent system kit. Dexamethasone (50 µM) was used as a positive control and the results were expressed as EC_{50} values (µg/mL) equal to the sample concentration providing a 50% inhibition of NO production.

2.6.3. Antioxidant Activity

For the antioxidant activity evaluation, two colorimetric assays were used, using methodologies previously described [23–25]: the cell-based assays of thiobarbituric acid reactive substances (TBARS) formation inhibition and oxidative hemolysis inhibition (OxHLIA). The extracts capacity to inhibit the formation of TBARS was assessed using porcine brain cells as oxidizable biological substrates and the results were expressed as EC_{50} values (µg/mL), which translate the extract concentration responsible for 50% of antioxidant activity. In turn, the extracts capacity to inhibit the oxidative hemolysis was tested using sheep blood erythrocytes as ex vivo models and the extract concentration able to promote a Δt hemolysis delay of 60 min was calculated based on the Ht_{50} values of the hemolytic curves of each extract concentration. The results were expressed as IC_{50} values (µg/mL), which translate the extract concentration required to keep 50% of the erythrocyte population intact for 60 min. Trolox was used as a positive control in both assays.

2.7. Optimization of the Extraction Process to Obtain a Natural Colorant Rich in Anthocyanins, from the Epicarp of Solanum melongena L. Fruits

In order to optimize the extraction of anthocyanin compounds from the eggplant epicarp, ultrasound-assisted extraction was used and an experimental design called the central composite design (CCD) was developed, applying response surface methodology (RSM), aiming to simplify and reduce the operational costs of the process, shorten the extraction process times, and reduce the energy spent as well as the solvent consumption [14,26]. In order to obtain the extract rich in anthocyanins, it was essential to consider several factors that affect its stability to achieve the maximum extraction efficiency [27]. Hence, RSM methodology was used to optimize relevant factors (time, percentage of solvent, and solid/liquid ratio) simultaneously, obtaining graphic models and polynomial equations that allow one to describe the ideal conditions that maximize the response criteria [14].

2.7.1. Experimental Design and Extraction Procedure Assisted by Ultrasound

Screening tests were performed based on individual analysis of the variables and those that caused significant effects were selected along with the relevant intervals (Table 1). The effects of the three defined variables were studied using a CCD, associated with five levels [28], which generated 20 run combinations, performed randomly in order to obtain better predictive capacity of the model. The extraction procedure was performed according to Lopez et al. [29], using ultrasound equipment (sonicators QSonica, model CL-334, Newtown, Connecticut, USA). The samples were extracted with a fix volume of 25 mL of solvent (acidified with 0.05% citric acid). Variables and intervals employed were: time (t or X_1 0.5 to 5.5 min), the ethanol/water extraction solvent (X_2 8 to 92%), and the solid/liquid ratio (X_3 5 to 65 g/L), particular levels analyzed within the ranges mentioned before are described in Table 1. Temperature was monitored and controlled below 30 °C.

Table 1. Coded experimental design of the independent variables used in the 5-level CCD design.

Extraction	Time (min)	Solvent (%)	Ratio (g/L)
1	−1	−1	−1
2	−1	−1	1
3	−1	1	−1
4	−1	1	1
5	1	−1	−1
6	1	−1	1
7	1	1	−1
8	1	1	1
9	−1.68	0	0
10	1.68	0	0
11	0	−1.68	0
12	0	1.68	0
13	0	0	−1.68
14	0	0	1.68
15	0	0	0
16	0	0	0
17	0	0	0
18	0	0	0
19	0	0	0
20	0	0	0

2.7.2. Preparation of Extracts Obtained by Ultrasound-Assisted Extraction

After the extraction procedure, the samples were centrifuged (5000 rpm for 20 min at 10 °C) and filtered through filter paper (Whatman n° 4). The supernatant was collected and divided into two fractions: one for HPLC-DAD analysis and the second for determining the extraction yield. The fraction separated for HPLC analysis (2 mL) was filtered through an LC syringe filter (0.22 μm) and then injected; the second fraction, used to determine the extraction yield (5 mL), was subjected to drying at a temperature of 105 °C for 48 h, for subsequent weighing of the solid extract.

2.7.3. Identification and Quantification of Anthocyanin Compounds by HPLC-DAD

The extracts obtained in the previous process were analyzed using a high-performance liquid chromatography system (HPLC, Dionex UltiMate 3000 UPLC, Thermo Scientific). The anthocyanins present in the samples were characterized according to their UV-Vis spectra and mass and retention times compared to the identification obtained in Section 2.5.2. The quantitative analysis of anthocyanin compounds was obtained through a calibration curve with the injection of a standard of known concentrations (200–0.25 μg/mL) of the compound cyanidin-3-O-glucoside (y = 97787x − 743469; R^2 = 0.993).

2.7.4. Response Format Used for Analytical Processes

The yield (%) and the content of the detected anthocyanin compound were used as responses. The results were expressed in three response formats (Y): Y_1, in mg of anthocyanins per gram of dry weight (mg/g dw), Y_2 in mg of anthocyanins per g of extracted residue (mg/g E), which was used to evaluate the purity of anthocyanins in the extracts, and Y_3 the amount of residue extracted expressed by yield (%).

2.7.5. Analysis of the Mathematical Model, Procedure for Optimization of Variables and Numerical Methods, Statistical Analysis, and Graphic Illustrations

The RSM data were adjusted by calculating least squares, using the following second order polynomial equation with complex interaction terms, as described by Pinela et al. [30]. The responses (Y) applied were the extraction yield (yield, %), the content of anthocyanins per gram of dry sample (mg/g dw), and the content of anthocyanins per g of extracted residue (mg/g E).

In order to optimize the forecasting model and, consequently, maximize the extraction yield, a simplex method was used, which allows solving non-linear problems [31]. In order to avoid variables with unnatural and unrealistic physical conditions, some limitations were imposed on the coded variables (namely t $\geq$ 0, 0 > S > 100).

The adjustment procedures, coefficient estimates, and statistical calculations of the experimental results were performed according to a procedure previously described by Melgar et al. [32]. The experimental data were fitted to the second-order polynomial model (Equation (1)) to obtain the regression coefficients (*b*) using Statgraphics Centurion XVI (StatPoint Technologies, Inc. Warrenton, VA, USA) and Design expert 12.0.1. (Stat-Ease, Inc., Minneapolis, MN, USA) softwares. The generalized second-order polynomial model used in the response surface analysis was the following:

$$Y = b_0 + \sum_{i=1}^{k} b_i X_i + \sum_{i=1}^{k} b_{ii} X_i^2 + \sum_{j=i+1}^{k} b_{ij} X_i X_j \tag{1}$$

where Y is the dependent variable (response variable) to be modelled, b_0 is a constant coefficient (intercept); b_i, b_{ii} and b_{ij} are the coefficients of the linear, quadratic, and interactive terms, respectively; k is the number of tested variables ($k = 3$); and X_i and X_j are the independent variables.

2.8. Preparation of the Optimal Extract Rich in Anthocyanins from the Epicarp of Solanum melongena L. Fruit

After the optimization studies, an ultrasound-assisted extraction was performed, from the epicarp of the *S. melongena* fruits, applying the optimum variable values, obtained in Section 2.7.1.

The sample was extracted with ethanol/water solvent (68:32; v/v) acidified with 0.05% citric acid (until pH = 3), in the proportion of 5 g/L, during 5.5 min. Then, the samples were centrifuged (Centurion K24OR, West Sussex, United Kingdom) at 5000 rpm for 5 min at 10 °C and filtered through paper filter (Whatman n° 4). The supernatant was evaporated on a rotary evaporator at 35 °C (Büchi R-210, Flawil, Switzerland) in order to remove the ethanolic fraction. Finally, the aqueous fraction obtained was lyophilized (FreeZone 4.5, Labconco), obtaining a pigmented extract.

2.9. Statistical Analysis

The described experiments, except for RSM, were performed in triplicate and the results were expressed as mean ± standard deviation (SD). The data for the evaluation of *S. melongena* fruit with regard to color parameters, nutritional parameters, chemical composition, bioactive potential, and optimization of anthocyanin extraction were analyzed through the application of different statistical tests. The Student *t*-test was applied in order to determine the significant differences between two different types of eggplant samples (whole fruit and pulp); in addition to that, the statistical analysis for the evaluation of the results of the three different types of eggplant samples (epicarp, whole fruit, and pulp), was carried out through the application of one-way analysis of variance (ANOVA) with *p*-value = 0.05 (SPSS v. 23.0; IBM Corp., Armonk, New York, NY, USA).

The tests of the optimization of the extraction process to obtain an extract rich in anthocyanins were analyzed using ANOVA and cluster analysis through the software Design-Expert 12.0.1 and Statgraphics 18.

3. Results and Discussions

3.1. Color Evaluation of the Epicarp, Pulp, and Whole Fruit of S. melongena

The color evaluation in the CIE L^* a^* b^* color space was performed on fresh and lyophilized eggplant epicarp and the results are shown in Table 2. The luminosity scale, represented by the parameter L^*, varies between 0 and 100, and the parameters a^* (green to red) and b^* (blue to yellow) vary between −120 and 120 [33]. In the color evaluation of the fresh epicarp, for the parameters L^*, a^*, and b^*, were obtained the mean values of 26.2 ± 0.5; 5.1 ± 0.2 and 1.5 ± 0.1, respectively. On the other hand, in the lyophilized eggplant epicarp the parameter L^* showed a value of 34 ± 2, and the values in parameters a^* and b^* were 2.8 ± 0.1 and -0.17 ± 0.01, respectively.

Table 2. Physical parameters (color-CIE $L^*\, a^*\, b^*$) of the fresh and lyophilized fruit epicarp of *Solanum melongena* L. (mean $\pm$ SD).

Color Evaluation Parameters	Fresh Epicarp	Lyophilized Epicarp	*p*-Value
L^*	26.2 ± 0.5	34 ± 2	<0.01
a^*	5.1 ± 0.2	2.8 ± 0.1	<0.01
b^*	1.5 ± 0.1	-0.17 ± 0.01	<0.01
Epicarp color			

L^* luminosity; a^* chromatic axis from green ($-$) to red ($+$); b^*, chromatic axis from blue ($-$) to yellow ($+$). The statistical differences were obtained through the Student *t*-test; *p* value < 0.05 means significant statistical variance.

Through the statistical analysis of the results, it was verified that the process of dehydration (by lyophilization) of the eggplant epicarp had significant influence ($p < 0.05$) in all the evaluated color parameters. The L^* parameter had lower values in the fresh epicarp, indicating a lighter hue when compared to the lyophilized sample. In contrast, the parameter a^* showed higher values in the fresh epicarp, showing a color nearest to the red hue. The same decrease, after the lyophilization process, was verified in parameter b^*, which revealed that the lyophilized epicarp presents a close hue to the blue color.

According to several results available in the literature, it is possible to verify that different dehydration processes can modify the physicochemical properties of the fruits [34]. In the present study, it was evident that the lyophilization process led to a decrease in the purple color of the epicarp. This may be associated with the degradation of anthocyanins present in the sample, mainly verified by the decrease in the value presented in parameter a^* [35]. Other authors [36–38] also studied this process and demonstrated that the color change after dehydration processes is associated with the degradation of anthocyanin pigments.

However, it is important to mention that the applied dehydration process must be chosen considering the degree of degradation of anthocyanins, in order to preserve the original physical and chemical characteristics of the samples as much as possible. According to literature, one of the methodologies that shows less associated degradation is the lyophilization process [39]. In fact, Badulescu et al. [40] studied the effect of dehydration by lyophilization and hot-air drying on the content of anthocyanins in raspberries and the results showed that the lyophilization process was more efficient in preserving the anthocyanin compounds in the samples.

3.2. Nutritional Characterization of Eggplant (Whole Fruit and Pulp)

The results obtained in the nutritional characterization of the whole fruit and the pulp of *Solanum melongena* L. are shown in Table 3.

Table 3. Nutritional parameters of the whole fruit and pulp of *Solanum melongena* L. (mean $\pm$ SD).

Nutritional Value	Whole Fruit	Pulp	*p*-Value
Humidity (g/100 g fw)	91.3 ± 0.3	92.9 ± 0.4	0.944
Ashes (g/100 g fw)	0.56 ± 0.01	0.56 ± 0.02	0.862
Proteins (g/100 g fw)	0.86 ± 0.02	0.78 ± 0.03	0.729
Fat (g/100 g fw)	0.050 ± 0.002	0.040 ± 0.001	<0.01
Carbohydrates (g/100 g fw)	3.0 ± 0.1	2.89 ± 0.01	0.893
Energy (kcal/100 g fw)	32.6 ± 1.1	27.5 ± 0.6	0.869
Energy (KJ/100 g fw)	137 ± 4	115 ± 2	0.869

fw: fresh weight. The statistical differences were obtained through the Student *t*-test; *p* value < 0.05 means a significant statistical difference.

The moisture content showed values of 91.3 ± 0.3 g/100 g fw (fresh weight) for the whole fruit samples and 92.9 ± 0.4 g/100 g fw for the pulp samples, indicating a high percentage of water in its composition. On the other hand, considerably lower levels were evident in the values of the remaining nutrients and non-nutrients evaluated in these samples. In the evaluation of the carbohydrate content, values of 3.0 ± 0.1 g/100 g fw were obtained for the whole fruit and 2.89 ± 0.01 g/100 g fw for the pulp; in fat content, lower concentrations were obtained, ranging from 0.050 ± 0.002 g/100 g fw (whole fruit) to 0.040 ± 0.001 g/100 g fw (pulp); the protein concentration was 0.86 ± 0.02 g/100 g fw (whole fruit) and 0.78 ± 0.03 g/100 g fw (pulp) and, finally, the ash content was 0.56 ± 0.01 g/100 g fw and 0.56 ± 0.02 g/100 g fw for the whole fruit and the pulp, respectively.

The energy value showed satisfactory values of 32.6 ± 1.1 kcal/100 g fw in the whole fruit and 27.5 ± 0.6 kcal/100 g fw in the pulp. The values obtained in the present study explain the high interest in the consumption of this food, namely, in diets with caloric restriction.

Through the statistical analysis performed, there was an absence of significant difference ($p > 0.05$) in most of the tested parameters (moisture, ash, proteins, carbohydrates, and energy), with the exception of the fat content, which showed a significant difference ($p < 0.05$), being higher in whole eggplant samples.

Regarding the data available in the literature, the United States Department of Agriculture (USDA) [41] presented in a report 11,209 similar concentrations to those achieved in the present study, namely, in the moisture (92.3 g/100 g fw), ash (0.66 g/100 g fw), and protein content (0.98 g/100 g fw). However, the remaining macronutrients showed diverse values, such as fat and carbohydrate content with values of 0.18 g/100 g fw and 5.88 g/100 g fw, respectively; while the amount of energy presented by the USDA was 25 kcal/100 g fw.

Kandoliya et al. [42] studied the composition of six eggplant varieties and found that the moisture content obtained in the pulp was 91.1 and 93.0 ± 0.6%, while the protein content was 0.66 to 1.28 ± 0.05%, these concentrations being in agreement with the values obtained in the present study.

In addition, Agoreyo et al. [43] evaluated the nutritional content of two unripened varieties of *Solanum melongena* (round and oval). The results obtained revealed moisture contents of 78.4 ± 0.49% and 72.93 ± 0.76%, protein contents of 5.79 ± 0.22 g/100 g fw and 4.58 ± 0.40 g/100 g fw, ash contents of 1.96 ± 0.12 g/100 g fw and 3.15 ± 1.54 g/100 g fw, and, finally, a carbohydrate concentration of 11.77 ± 1.55 g/100 g fw and 15.42 ± 0.69 g/100 g fw. This disparity in results, compared to the present study, may be related to several factors that directly interfere with the nutritional composition of foods, particularly: geographical location of production, cultivation conditions, climatic conditions, harvest conditions, or state of maturity of the raw material, among others [44].

Ossamulu et al. [45] also compared the nutritional values of four species of eggplants fruits, namely *Solanum marcrocarpon* (round), *Solanum aetheopicum*, *Solanum marcrocapon* (oval), and *Solanum gilo*. The results presented similarity of values in some parameters, among them the moisture (88.31 ± 0.23% to 91.94 ± 0.11%), ash content (0.41 ± 0.01 g/100 g fw to 0.56 ± 0.02 g/100 g fw), and energy value (34.02 ± 0.95 kcal/100 g fw to 22.90 ± 0.46 kcal/100 g fw). Otherwise, there were some discrepancies in protein (1.21 ± 0.02 g/100 g fw to 2.36 ± 0.03 g/100 g fw), fat (0.24 ± 0.01 g/100 g fw to 0.42 ± 0.02 g/100 g fw), and carbohydrates concentration (4.06 ± 0.19 g/100 g fw to 6.03 ± 0.19 g/100 g fw).

3.3. Chemical Composition of Whole Fruit and Pulp of S. melongena

3.3.1. Content of Sugars, Organic Acids, and Fatty Acids

The chemical composition of the whole fruit and pulp of *Solanum melongena* L. was evaluated and the results of hydrophilic compounds (free sugars and organic acids) and lipophilic compounds (fatty acids) are represented in Table 4.

Table 4. Free sugars, organic acids, and fatty acids profile of in *Solanum melongena* L. whole fruit and pulp (mean ± SD).

Free Sugars (g/100 g fw)			
	Whole Fruit	**Pulp**	***p*-Value**
Fructose	1.26 ± 0.01	1.26 ± 0.1	0.093
Glucose	1.29 ± 0.03	1.25 ± 0.01	<0.01
Sucrose	0.43 ± 0.02	0.260 ± 0.004	<0.01
Trehalose	0.100 ± 0.001	0.080 ± 0.003	0.201
Total sugars	3.0 ± 0.1	2.89 ± 0.01	<0.01
Organic acids (g/100 g fw)			
Oxalic	0.381 ± 0.005	0.38 ± 0.01	0.966
Quinic	0.13 ± 0.02	0.108 ± 0.002	<0.01
Malic	0.349 ± 0.004	0.33 ± 0.02	<0.01
Fumaric	tr	nd	-
Total organic acids	0.86 ± 0.03	0.82 ± 0.2	<0.01
Fatty acids (%)			
C6:0	0.56 ± 0.01	0.59 ± 0.02	<0.01
C8:0	0.67 ± 0.03	0.84 ± 0.01	<0.01
C10:0	1.61 ± 0.04	2.165 ± 0.002	<0.01
C11:0	1.1 ± 0.1	0.58 ± 0.03	<0.01
C12:0	1.25 ± 0.03	1.12 ± 0.04	0.516
C14:0	2.63 ± 0.03	2.7 ± 0.1	0.017
C15:0	0.56 ± 0.03	0.54 ± 0.01	0.077
C16:0	40.5 ± 0.9	44.8 ± 0.2	<0.01
C17:0	1.103 ± 0.002	0.99 ± 0.02	<0.01
C18:0	24.4 ± 0.2	24.4 ± 0.4	0.443
C18:1n9c	4.5 ± 0.2	5.5 ± 0.3	0.462
C18:2n6c	8.0 ± 0.4	4.10 ± 0.05	<0.01
C18:3n3	3.8 ± 0.1	0.79 ± 0.01	<0.01
C20:0	4.6 ± 0.2	5.2 ± 0.1	0.153
C22:0	1.60 ± 0.04	1.69 ± 0.03	0.489
C23:0	0.91 ± 0.03	1.12 ± 0.02	0.213
C24:0	2.3 ± 0.1	2.86 ± 0.04	<0.01
SFA	**83.8 ± 0.6**	**89.6 ± 0.3**	0.082
MUFA	**4.5 ± 0.2**	**5.5 ± 0.3**	0.462
PUFA	**11.8 ± 0.4**	**4.89 ± 0.04**	<0.01

fw: fresh weight; tr: traces; nd: not detected. Caproic acid (C6:0); caprylic acid (C8:0); capric acid (C10:0); hendecanoic acid (C11:0); lauric acid (C12:0); miristic acid (C14:0); pentadecyl acid (C15:0); palmitic acid (C16:0); margheric acid (C17:0); stearic acid (C18:0); oleic acid (C18:ln9c); linoleic acid (C18:2n6c); α-linolenic acid (C18:3n3); arachidic acid (C20:0); behenic acid (C22:0); tricosanoic acid (C23:0); lignoceric acid (C24:0); SFA (Saturated Fatty Acids); MUFA (Monounsaturated Fatty Acids); PUFA (Polyunsaturated Fatty Acids); nd: not detected. The statistical differences were obtained through the Student *t*-test; *p* value < 0.05 means significant statistical difference.

Concerning the composition in free sugars, the studied samples revealed the presence of two monosaccharides (fructose and glucose) and two disaccharides (sucrose and trehalose). In both samples, the monosaccharides were found in higher quantities, with fructose concentrations of 1.26 g/100 g fw in the whole fruit and pulp; and glucose concentrations of 1.25 ± 0.03 and 1.29 ± 0.01 g/100 g fw in the whole fruit and pulp, respectively.

Regarding the statistical analysis of the data, it was clear that the whole fruit and the pulp showed the absence of significant differences in the concentration of fructose (*p* = 0.093) and trehalose (*p* = 0.201), contrary to the molecules of glucose and sucrose, and total sugars content, where significant differences were evident (*p* < 0.05), exhibiting the whole fruit with higher values.

Other authors [46] also studied the individual sugar profile in the pulp of seven eggplant cultivars (Aydin Siyahi, Pala 49, Süper pala, Kemer 27, Kadife Kemer, Topan, and Kadife) and the results showed the presence of fructose (1.242 ± 0.028 g/100 g fw to 1.379 ± 0.051 g/100 g fw), glucose (1.275 ± 0.015 g/100 g fw to 1.327 ± 0.047 g/100 g fw), and sucrose (0.109 ± 0.036 g/100 g fw to 0.494 ± 0.048 g/100 g fw) in similar concentrations compared to the present study.

The composition in organic acids was also evaluated, and revealed, in both samples, the presence of several molecules of interest, particularly oxalic, quinic, and malic acids. In general, it was evident that the whole fruit and the pulp showed the presence of the same molecules, with the exception of fumaric acid, which was only detected in the whole fruit, but in trace amounts. Whole fruit samples showed a higher amount of oxalic, quinic, and malic acids, and total concentration of organic acids, compared to the pulp samples, where oxalic, quinic, and malic acid values, and total organic acid concentration obtained were lower.

In both samples, oxalic acid stood out as the major organic acid. According to Kayashima and Katayama [47], this molecule is present in several plants and is characterized by its antioxidant action and its ability to preserve oxidized materials. In addition, its potential in inducing resistance against pathogens [48] and the ability to exercise different functions in plants and fungi, including protection against insects, has been proven. Its use covers several areas, being widely used in the food sector as a preservative of fruits and vegetables, in order to extend the shelf life [49].

Furthermore, the malic acid appears in similar concentrations. This molecule stands out for its therapeutic capacity, namely in helping muscle recovery (being used in the treatment of fibromyalgia), fighting fatigue, and increasing energy (when acting on the Krebs cycle). Moreover, it is also commonly used by the food industry as an acidulant, flavoring, and stabilizer, and by the pharmaceutical industry in cleaning and regenerating wounds and burns [50].

Quinic acid, although present in lower concentration, is also characterized by its therapeutic properties, namely by its radioprotector, antioxidant, and anti-inflammatory activity [51].

Regarding the statistical evaluation of the results, significant variations were visible in all the detected molecules ($p < 0.05$), except for oxalic acid ($p = 0.966$).

Other studies have been carried out for the valorization of this food matrix, evaluating the composition in functional compounds. Ayas et al. [46] studied seven varieties of eggplant fruit (Aydin Siyahi, Pala 49, Süper pala, Kemer 27, Kadife Kemer, Topan, and Kadife) and, in addition to malic acid (129.87 ± 6.78 mg/100 g fw to 181.06 ± 19.46 mg/100 g fw), they also identified the presence of other organic acids that were not detected in the present study, namely ascorbic (7.69 ± 1.21 mg/100 g fw to 11.74 ± 1.33 mg/100 g fw) and citric acids (18.98 ± 0.80 mg/100 g fw at 21.63 ± 0.35 mg/100 g fw).

As with other molecules, the discrepancy in results obtained in the detection of organic acids can be explained due to differences in fruit varieties, cultivation conditions, and the geographical area where the *S. melongena* varieties were produced.

In the evaluation of the fatty acid composition, the presence of 17 molecules was evident, with C16:0 standing out as the major fatty acid, with values of $40.5 \pm 0.9\%$ and $44.8 \pm 0.2\%$ in the whole fruit and pulp samples, respectively. C18:0 also presented relevant concentrations.

Saturated fatty acids were the main group present, with values varying between $83.8 \pm 0.6\%$ in the whole fruit and $89.6 \pm 0.3\%$ in the pulp; followed by polyunsaturated fatty acids ($11.8 \pm 0.4\%$ whole fruit and $4.89 \pm 0.04\%$ in the pulp), and finally, monounsaturated fatty acids with concentrations of $4.5 \pm 0.2\%$ for the whole fruit and $5.5 \pm 0.3\%$ for the pulp. However, it is important to note that, although this group stands out with more pronounced concentrations, they are still low values that are unable to cause adverse effects on consumers' health [52].

Regarding the statistical analysis, a significant variation ($p < 0.05$) was noticed in C6:0, C8:0, C10:0, C11:0, C:14, C16:0, C17:0, C18: 2n6c, C18: 3n3, and C21:0. Contrarily, in the molecules of C12:0, C15:0, C18:0, C18:1n9c, C20:0, C22:0, and C23: 0, small oscillations were observed between both studied samples, verifying the absence of statistical significance ($p > 0.05$).

Other authors have also studied the fatty acids profile of eggplant fruit samples. In a study conducted by Hanifah et al. [53], eggplant fruits of 21 different morphological varieties (color and shape) were evaluated and the results revealed the presence of some fatty acid molecules similar to the present work but the composition was mostly different. This can be justified by the different varieties of the fruit assessed and the application of another extraction method.

The fatty acids content was also studied by Ayas et al. [46], when evaluating seven eggplant cultivars. The presence of six fatty acid molecules was detected (C16:0; C18:0; C18:1; C18:2; C18:3, and C20:0), as well as the composition in SFA (Saturated Fatty Acids) of 31.44% to 36.92%, in MUFA (monounsaturated fatty acids) of 3.04% to 11.71%, and in PUFA (Polyunsaturated Fatty Acids) of 53, 66% to 62.71%, presenting, therefore, different percentages comparing with the present study.

3.3.2. Profile in Non-Anthocyanin and Anthocyanin Phenolic Compounds

The detailed profile of the phenolic compounds present in the eggplant samples (whole fruit, pulp, and epicarp) is shown in Table 5. Four phenolic compounds were identified, three being non-anthocyanin and one anthocyanin. Considering the non-anthocyanin compounds (Figure 1A), three phenolic acids were identified, namely caffeic acid hexoxide, protocatechuic acid, and 5-*O*-caffeoylquinic acid. Concerning the anthocyanin compound, a glycosylated derivative of delphinidin was identified only in the epicarp of the fruit, as expected considering its purple color.

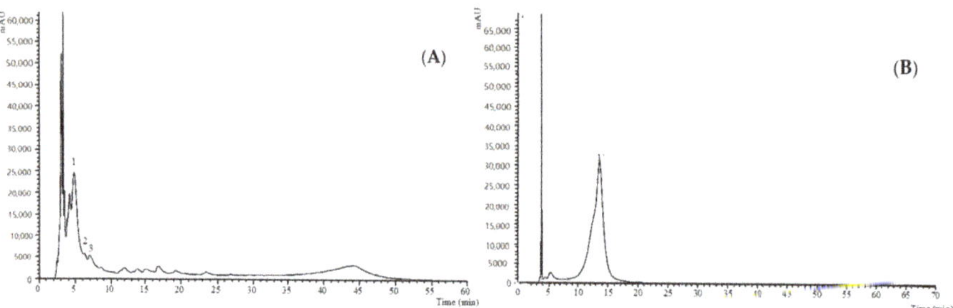

Figure 1. Profile of non-anthocyanin (**A**) and anthocyanin (**B**) phenolic compounds from the hydroethanolic extracts of the *S. melongena* L. (whole fruit and epicarp, respectively) recorded at 280 nm (i) and 520 nm (ii). The peak number correspondence is shown in Table 6.

Regarding the identified phenolic acids, compound 1 ($(M-H)^-$ at m/z 341) was identified as caffeic acid hexoside based on its UV spectrum (λ_{max} of 322) and the production of the MS^2 fragment at m/z 179 and 135, being detected only in the whole fruit and pulp. Compound 2 ($(M-H)^-$ at m/z 153) was positively identified as protocatechuic acid based on its UV spectrum (λ_{max} of 294) and the production of the MS^2 fragment at m/z 108, which presented similar characteristics to the commercial standard. This compound was detected only in the eggplant epicarp. Finally, compound 3 ($(M-H)^-$ at m/z 353) was also positively identified as 5-*O*-caffeoylquinic acid based on its UV spectrum (λ_{max} of 325) and the production of the MS^2 fragment at m/z 191, 179, and 135, in comparison with the commercial standard. The latter stands out as the major compound, as well as the only one present in all studied samples, with values of 0.0159 ± 0.0002, 0.0337 ± 0.0004, and 0.96 ± 0.01 mg/g of extract for the whole fruit, pulp, and epicarp, respectively. The remaining compounds showed lower concentrations, notably the caffeic acid hexoside, with levels of 0.0025 ± 0.0001 mg/g of extract (whole fruit), 0.0062 ± 0.0004 mg/g of extract (pulp), and protocatechuic acid with a concentration of 0.37 ± 0.02 mg/g of extract in the epicarp.

Table 5. Retention time (Tr), maximum absorption wavelengths in the UV-Vis region (λ_{max}), attempt to identify and quantify phenolic compounds in the hydroethanolic extract of the pulp, epicarp, and the whole fruit of *Solanum melongena* L. (mean ± SD).

Peak	Tr (min)	λ_{max} (nm)	$(M-H)^-$	ESI-MSn (Intensity (%))	Attempted Identification	Quantification (mg/g Extract)		
						Whole Fruit	Pulp	Epicarp
1	4.85	322	341	179(100),135(14)	Caffeic acid hexoside [1]	0.0025 ± 0.0001 [b]	0.0062 ± 0.0004 [a]	nd
2	5.33	294	153	108(100)	Protocatechuic acid [2]	nd	nd	0.37 ± 0.02
3	6.92	325	353	191(100),179(12),135(1)	5-O-caffeoylquinic acid [3]	0.0159 ± 0.0002 [c]	0.0337 ± 0.0004 [b]	0.96 ± 0.01 [a]
				Total phenolic acids		0.0184 ± 0.0003 [c]	0.0399 ± 0.0008 [b]	1.34 ± 0.03 [a]

Peak	Tr (min)	λ_{max} (nm)	$(M+H)^+$	ESI-MSn (Intensity (%))	Attempted Identification	Quantification (mg/g Extract)		
						Whole Fruit	Pulp	Epicarp
1′	13.56	523	611	303(100)	Delphinidin-O-rutinoside [4]	nd	nd	9.2 ± 0.2

nd: not detected; Tr-retention time. Calibration curves used: 1-caffeic acid (y = 388345x + 406369); 2-protocatequic acid (y = 214168x + 27102); 3-chlorogenic acid (y = 168823x − 161172); 4-cyanidin-3-O-glucoside (y = 97787x − 743469; R^2 = 0.9993). The statistical differences of the means were obtained through one-way analysis of variance (ANOVA). In each row, for each type of extract (whole fruit, pulp, and epicarp) different letters mean significant differences between the total amounts of compounds ($p < 0.05$).

Table 6. Antioxidant activity of the hydroethanolic extracts of the whole fruit, pulp, and epicarp of *S. melongena.* (mean ± SD).

Antioxidant Activity	Concentration		
	Whole Fruit	Pulp	Epicarp
TBARS (Values of EC_{50}, µg/mL)	8941 ± 284 [a]	4291 ± 178 [b]	135 ± 6 [c]
OxHLIA (Values of EC_{50}, µg/mL)	119 ± 7 [a]	82 ± 3 [b]	34 ± 1 [c]

EC_{50} values: Extract concentration corresponding to 50% of antioxidant activity. Trolox (positive control) $EC_{50} = 23$ µg/mL (TBARS inhibition) and 85 µg/mL (OxHLIA). The statistical differences in the means were obtained through one-way analysis of variance (ANOVA). In each line, for each type of extract (whole fruit, pulp, and epicarp) different letters means significant differences ($p < 0.05$).

The phenolic acids detected are described in the literature as having properties of high therapeutic interest. Caffeic acid is a hydroxycinnamic acid, found in several fruits and vegetables, characterized mainly by its anti-inflammatory and antioxidant potential, as well as by its therapeutic capacity in the prevention of cardiovascular diseases, hypoglycemic, and anti-cancerous activity [54,55]. On the other hand, protocatechuic acid is one of the main bioactive compounds present in some medicinal plants, and is characterized by its cardioprotective, neuroprotective, antibacterial, antidiabetic, antiviral, analgesic, chemotherapeutic, anti-inflammatory, and antioxidant potential [56]. The 5-*O*-caffeoylquinic acid is used mainly in the pharmaceutical, cosmetic, and food industry, and is distinguished by its antimicrobial, antioxidant, anti-tumor, and antihistamine activity [57].

In addition to non-anthocyanin phenolic compounds, in the epicarp of the fruit an anthocyanin phenolic compound, identified as delphinidin-*O*-rutinoside (1′) ($(M+H)^+$ at *m/z* 611), was also detected. This compound was identified based on its UV spectrum (λ_{max} of 523) and the production of the MS^2 fragment at *m/z* 303, in comparison to the commercial standard (Figure 1B) and presented a concentration of 18.4 ± 0.7 mg/g extract.

The delphinidins evidenced, in previous studies, several bioactive properties, being the anthocyanin compound with the greatest beneficial effect in preventing cancer, being related to the death of apoptotic or autophagic cells in various types of cancer. In addition, it has antioxidant and anti-inflammatory action [58,59].

Significant differences ($p < 0.05$) were found in all compounds, with the epicarp having a higher concentration in all the identified compounds.

Other authors have also studied the individual profile of phenolic compounds in eggplant samples. In a review study by Niño-Medina et al. [60], the authors reported that the phenolic acids, commonly present in eggplant fruits, are caffeic and ferulic acids, and their derivatives. Wu and Prior [61] studied the anthocyanin composition of the lyophilized eggplant fruit, and the results obtained allowed the identification of four derivatives of delphinidin in eggplant fruits: delphinidin-3-*O*-rutinoside, delphinidin-3-*O*-rutinoside-5-*O*-glucoside, delphinidin-3-*O*-glucoside, and delphinidin-3-*O*-rutinoside-5-*O*-galactoside.

García-Salas et al. [62] evaluated three eggplant cultivars (PSE: striped purple, LE: long, and RE: rounded), grown in Spain in two different seasons (spring and summer). The results showed the identification of 25 compounds, most of which are derived from caffeoylquinic acid. The detected anthocyanin compound was identified as delphinidin rutinoside. These authors also evaluated the effect of the growing season on the composition of the fruits, and it was concluded that the high summer temperatures negatively affected the composition of phenolic compounds in the eggplant fruits. The results obtained presented a higher concentration of phenolic compounds compared to the samples studied in the present study, which can be justified by the difference in the cultivation region, fruit varieties, and/or the extraction solvents used.

Salermo et al. [63] evaluated the content of phenolic compounds in samples of whole fruit, pulp, and epicarp of *Solanum melongena* L. grown in Italy. The results showed a higher concentration of phenolic compounds in the epicarp of the lyophilized fruit, using water: ethanol as extraction solvent (about 14.25 ± 0.06 mg GAE (Gallic Acid Equivalent) 1.5 g of extract).

Another work, performed by Ferarsa et al. [7], evaluated the anthocyanin composition in the epicarp of *S. melongena* L. fruits, grown in France. The results indicated that the use of acidified water (pH 2.0) showed greater extraction power of the phenolic compounds, and that the yield is better when using hydroethanolic extracts (50 and 75%). Concerning the temperature during extraction, it was observed that the increase in this parameter led to an increase in the content of phenolic compounds detected, with the optimum temperature being 75 °C. In this study, five anthocyanin molecules were identified in the epicarp: delphinidin-3-*O*-rutinoside, delphinidin-3-*O*-rutinoside-5-*O*-glucoside, petunidine-3-*O*-rutinoside, malvidine-3-*O*-rutinoside-5-*O*-glucoside, and cyanidin-3-*O*-rutinoside.

Finally, a different study carried out by Horincar et al. [64] evaluated the composition in phenolic compounds of the extract of the epicarp of the fruits of *Solanum melongena* L., targeting its application in the nutritional enrichment of beer. The results showed the presence of five anthocyanin compounds: delphinidin-3-*O*-rutinoside (82.39%), delphinidin-3-*O*-glucoside (11.36%), delphinidin-3-*O*-rutinoside-5-*O*-glucoside (2.37%), cyanidin-3-*O*-rutinoside, and petunidine-3-*O*-rutinoside.

As previously mentioned, there are several external factors that directly interfere with the nutritional and chemical composition of foods, changing not only the concentration of compounds, but also their variety and presence. The differences observed in the phenolic profile identified in the present study compared to previous studies can be explained by several factors, namely different cultivation conditions, storage conditions after harvest, climate, soil type, season, and geographic origin [59]. Differences in the extraction methods used and the type of solvent can also be an important factor. In addition, the content of phenolic compounds can also vary under the effect of stresses such as air pollution, exposure to extreme temperatures, periods of drought, and high light intensity, among others [65].

3.4. Antioxidant, Antimicrobial, Cytotoxic, and Hepatotoxic Activity of the Hydroethanolic Extract Obtained from the S. melongena Fruit by Conventional Method

In order to evaluate the antioxidant activity of the hydroethanolic extract obtained from *S. melongena* (whole fruit, pulp, and epicarp), two in vitro colorimetric methods were applied (inhibition of lipid peroxidation—TBARS and inhibition of oxidative hemolysis—OxHLIA) and the results are presented in Table 6. The extract showed antioxidant potential in all tests performed, however the lowest values (best antioxidant activity) were obtained in the OxHLIA assay.

In the TBARS evaluation, the EC_{50} values ranged between 135 ± 6 and 8941 ± 284 μg/mL, and according to the statistical analysis, showed the best activity in the epicarp, followed by the pulp, and, finally, the whole fruit. In the OxHLIA assay, the hydroethanolic extracts also showed better antioxidant capacity in the epicarp sample, followed by the pulp and, finally, the whole fruit.

Several studies [42,66,67] have been performed with the aim of studying the antioxidant potential (through in vitro tests) of different parts of eggplant and positive and promising results were observed.

Table 7 shows the results obtained for the antimicrobial activity of the hydroethanolic extract obtained from *S. melongena* (whole fruit, pulp, and epicarp). The results showed bacteriostatic (MIC) and bactericidal (MBC) activity in all bacterial cultures tested, highlighting the extracts obtained from the epicarp and the whole fruit with the best results. In all cases, *Escherichia coli* (*E.c.*) stood out as the most susceptible strain to the bactericidal and bacteriostatic potential of all extracts (whole fruit, pulp, and epicarp).

Table 7. Antibacterial (MIC and MBC mg/mL) and antifungal (MIC and MFC mg/mL) activity of *S. melongena* extracts (whole fruit, epicarp, and pulp).

Antibacterial Activity						
Extracts		*S.a.*	*L.m.*	*E.c.*	*En.cl.*	*S.t.*
Epicarp	MIC	4.00	4.00	2.00	4.00	4.00
	MBC	8.00	8.00	4.00	8.00	8.00
Whole fruit	MIC	4.00	4.00	2.00	4.00	4.00
	MBC	8.00	8.00	4.00	8.00	8.00
Pulp	MIC	4.00	4.00	2.00	8.00	8.00
	MBC	8.00	8.00	4.00	8.00	8.00
Ampicillin	MIC	0.012	0.40	0.40	0.006	0.75
	MBC	0.025	0.50	0.50	0.012	1.20
Antifungal Activity						
Extracts		*A.fum.*	*A.v.*	*A.n.*	*P.f.*	*P.o.*	*T.v.*
Epicarp	MIC	>8.00	>8.00	>8.00	>8.00	>8.00	>8.00
	MFC	>8.00	>8.00	>8.00	>8.00	>8.00	>8.00
Whole fruit	MIC	>8.00	>8.00	>8.00	>8.00	>8.00	>8.00
	MFC	>8.00	>8.00	>8.00	>8.00	>8.00	>8.00
Pulp	MIC	>8.00	>8.00	>8.00	8.00	1.00	>8.00
	MFC	>8.00	>8.00	>8.00	8.00	1.00	>8.00
Ketoconazole	MIC	0.20	0.20	0.20	0.20	0.20	0.20
	MFC	0.50	0.50	0.50	0.50	0.50	0.30

S.a.: *Staphylococcus aureus*; L.m.: *Listeria monocytogenes*; E.c.: *Escherichia coli*; En.cl.: *Enterobacter cloacae*; S.t.: *Salmonella* Typhimurium; Afun.: *Aspergillus fumigatus*; Av.: *Aspergillus versicolor*; A.n.: *Aspergillus niger*; P.f.: *Penicillium funiculosum*; P.o.: *Penicillium ochrochloron*; T.v.: *Trichoderma viride*. MIC: minimum inhibitory concentration, MBC: minimum bactericidal concentration; MFC: minimum fungicidal concentration.

Otherwise, in the evaluation of the antifungal potential, only the pulp showed fungicidal (MIC) and fungiostatic (MFC) activity against some strains, namely *Penicillium fumiculosum* (*P.f.*) and *Penicillium ochrochloron* (*P.o.*). This sample showed MIC and MFC values of 8.00 and 1.00 mg/mL for *P. fumiculosum* and *P. ochrochloron*, respectively.

Other authors [10,67] studied the antimicrobial action of *S. melongena* (fruits and/or leaves) and the results evidenced antibacterial and antifungal potential.

Regarding the results obtained for the cytotoxicity and hepatotoxicity assays (Table 8), it was clear that the epicarp of *S. melongena* fruit was the only hydroethanolic extract that revealed an inhibitory potential (GI_{50} < 400 µg/mL), showing its anti-proliferative capacity in the HeLa, NCI H460, and HepG2 tumor lines. This potential was demonstrated by GI_{50} values of 337 ± 7 µg/mL for the HeLa cell line, 338 ± 16 µg/mL for the NCI H460 cell line and, finally, 284 ± 10 µg/mL for the HepG2 cell line. The absence of toxicity of all the studied samples (GI_{50} > 400 µg/mL) was also evident, through the assessment of hepatotoxicity where a culture of primary non-tumor cells (PLP2) was used.

Akanitapichat et al. [66] also studied this matrix and evaluated the hepatoprotective potential of five varieties of *S. melongena* obtained in Thailand (SM1: uniform, purple, moderate size; SM2: white and green, moderate size; SM3: elongated and green; SM4: striped, green, moderate size; and SM5: uniform, light green, small size). For this purpose, the HepG2 cell line was used and the results demonstrated a hepatoprotective effect in all samples evaluated.

Table 8. Cytotoxic activity of *S. melongena* extracts (whole fruit, epicarp, and pulp) (mean $\pm$ SD).

	Cytotoxic Activity		
GI$_{50}$ (µg/mL)	Whole Fruit	Pulp	Epicarp
HeLa	>400	>400	337 ± 7
NCI-H460	>400	>400	338 ± 16
MCF 7	>400	>400	>400
HepG2	>400	>400	284 ± 10
PLP2	>400	>400	>400

HeLa: cervical carcinoma; NCI-H460: lung carcinoma; MCF7: breast carcinoma; HepG2: hepatocellular carcinoma; PLP2: primary culture of pig liver cells.

3.5. Optimization of the Process of Obtaining a Natural Colorant Extract Based on Anthocyanins from the S. melongena Epicarp

3.5.1. Effect of Solvent Concentration

In the optimization studies, considering the evaluation of the solvent concentration, it was possible to prove that the concentration of ethanol directly influences the polarity and viscosity of the extractor in physical-chemical terms. However, the percentage used will affect the mass transfer of the target compounds in the extraction process. In this context, the effects of ethanol concentration on the extraction yields of biomolecules were analyzed in a range of 8 to 92% concentration (v/v). From the visual representation of Figure 2A, the extraction yields of the desired constituents increased, with an increase in the concentration of ethanol in the range of 54–68% (v/v), but no more than that, since the decrease in recovery is evident in all cases. The reason for this behavior in the concentration of ethanol can be attributed to the polarity of the solvent, which eases the dissolution of the specific anthocyanins analyzed. As a result, 54–68% of the ethanol concentration achieved higher extraction yields, similar to the results shown in a study carried out by Muangrat et al. [68] in a different food matrix.

3.5.2. Effect of the Solid-Liquid Ratio

Considering the evaluation of solid-liquid ratio, the effects of the ratio of powdered solid dissolved in a specific percentage of a fixed volume of liquid on the extraction yields of anthocyanins was investigated in the range of 5 to 65 g/L. The obtained results are illustrated in Figure 2B. The results indicated that lower amounts of solid employed had greater impact on the extraction yields of the biomolecules. The anthocyanins yield increased remarkably with the increase of the extractant volume. Lower amounts with higher volumes would accelerate the diffusion of compounds from the sample due to bigger contact surface area. In fact, the amounts of anthocyanins extracted by ultrasound assisted extraction decreased when the amount of matrix analyzed increased, almost in a linear manner, until reaching the saturation point. Large amounts of solid in respect to liquid would easily form aggregates and lead to contact surface decrease, resulting in an obstacle to mass transfer [69]. Thus, 5 g of matrix per liter of solvent was chosen as the optimal solid-liquid ratio in this optimization process.

3.5.3. Effect of Extraction Time

Finally, in the evaluation of the extraction time, it was possible to observe that this is a key parameter in the extraction yield of flavonoids, being thus also investigated in this work, in an interval of 0.5 to 5.5 min. In Figure 2C, it is important to note that, in addition to the interaction effect in the RSM graph, we can also observe that interactions were found between time and solvent and, although the time factor is not statistically significant, it is important to include the terms of the quadratic equation, at least for the Y_3 response (yield), where this phenomenon is expressed widely. Initially, different extraction times were tested before the execution of this CCD, and longer times were discarded, showing non-significant differences. Although short times have been used, we can observe, as well as Melgar et al. [32], that the main effect of ultrasound-assisted extraction was produced within a few seconds. Apparently, the combination of the cavitation effect with the size of the fine particles and the percentage of solvent, has a very pronounced reaction on the extraction of the bioactive molecules. Thus, recognizing the relevance of ultrasound-assisted extraction, it is possible to save a lot of time, energy, and cost, increasing the importance of optimizing the extraction process.

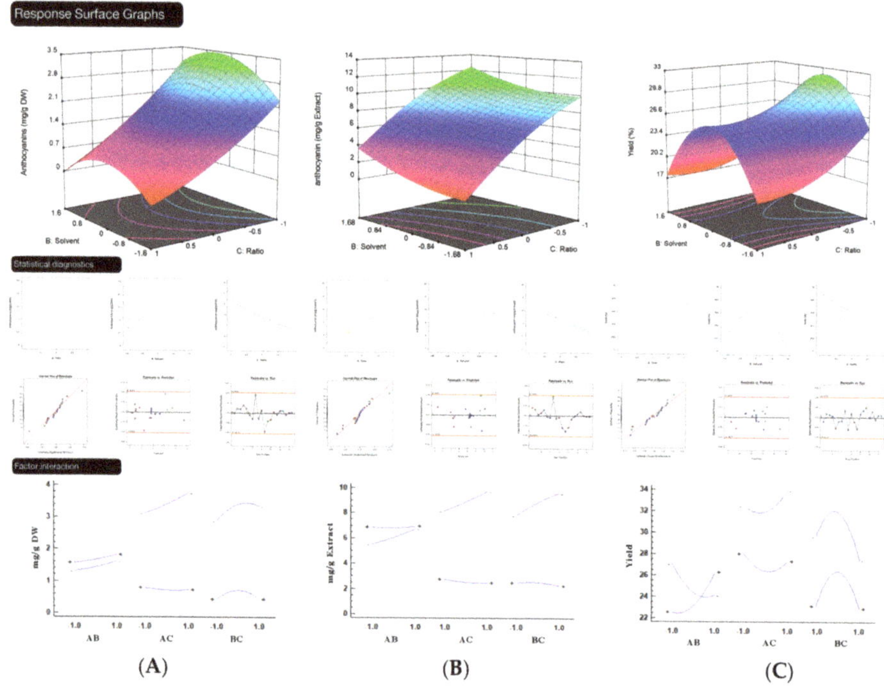

Figure 2. Graphical representation of response surface methodology (RSM) for: (**A**) response Y_1 (anthocyanins mg/g per dry mass). (From left to right and top to bottom: Modeling of the representation of the data obtained with time zero factor; single factorial behavior of time; single factorial behavior of the solvent; single factorial behavior of the ratio; normal residual plot; residuals versus predicted values; predicted values versus execution; and factorial interaction); for (**B**) the Y_2 response (mg/g extract). (From left to right and top to bottom: Modeling data representation obtained with time zero factor; single factorial behavior of time; single factorial behavior of the solvent; single factorial behavior of the ratio; normal residual graph; residuals versus predicted values; predicted values versus execution; and factorial interaction); for (**C**) the Y_3 response (yield). (From left to right and top to bottom: Modeling data representation obtained with time zero factor; single factorial behavior of time; single factorial behavior of the solvent; single factorial behavior of the ratio; normal residual graph; residuals versus predicted values; predicted values versus execution; and factorial interaction.).

3.5.4. Statistical Analysis and Model Adjustment

In Table 9, ANOVA analysis of the experimental data from the 20 extractions was executed, a second order polynomial equation was obtained from each response regression coefficient using the Design-Expert and Statgraphics software. The second order polynomial models for the extraction of epicarp (Y_1), extract (Y_2), and recovered yield (Y_3) are presented by the following equations:

$$Y1 = 1.83 + 0.14 * time + 0.12 * solv - 1.34 * ratio - 0.29 * sov\char94 2 + 0.23 * ratio\char94 2$$
$$Y2 = 0.61 + 0.44 * time + 0.72 * solv - 1.86 * ratio - 0.32 * solv\char94 2 + 1.05 * ratio\char94 2$$
$$Y3 = 27.34 + 0.20 * time - 0.62 * solv - 5.10 * ratio + 1.66 * time * solv - 4.62 * solv\char94 2 + 6.61 * ratio\char94 2$$

Table 9. Statistical analysis (ANOVA) of the central composite design (CCD), including response terms for building the predictive models and optimal response values for the parametric response criteria.

		Extraction		
		Y_1	Y_2	Y_3
Model		<0.0001	0.00	<0.0001
Intercept	b_0	1.83 (***)	6.61 (***)	27.34 (***)
Linear	b_1	0.1398 (NS)	0.4445 (NS)	0.1962 (NS)
	b_2	0.1163 (NS)	−0.7166 (*)	−0.6195 (NS)
	b_3	−1.34 (***)	−1.86 (***)	−5.1 (***)
Interaction	b_{12}	−0.0113 (NS)	−0.2858 (NS)	1.66 (NS)
	b_{13}	−0.1738 (NS)	−0.4649 (*)	−0.5325 (NS)
	b_{23}	−0.1163 (NS)	−0.4955 (*)	0.4913 (NS)
Quadratic	b_{11}	0.0387 (NS)	−0.0827 (NS)	0.028 (NS)
	b_{22}	−0.290 (**)	−0.3197 (NS)	−4.62 (***)
	b_{33}	0.2261 (*)	−1.05 (***)	6.61 (***)
Adaptation analysis statistical information				
Observations		20	20	20
R^2		0.96	0.94	0.94
R^2adj		0.93	0.98	0.88
Factorial optimization response				
		max	max	max
Optimal value		5.64	12.15	43.04
X_1		0.54	0.52	0.55
X_2		64.2398	58.4479	54.1773
X_3		4.65966	4.65966	4.66865
General optimization				
		max	max	max
Optimal response (Y)		5.53	11.96	40.96
Desirability		1		
Optimal factorial value		*$X_1 = 0.5$*	*$X_2 = 68.2$*	*$X_3 = 5$*

Significance values within parenthesis: * $p < 0.05$; ** $p < 0.01$; *** $p < 0.001$, NS: Not significant. R^2: coefficients of determination; R^2adj. adjusted coefficients of determination, coefficients b1: time, b2: temp, and b3: solv. The responses are summarized as Y_1: total anthocyanins (mg/g dry matter), Y_2: total anthocyanins (mg/g extract), Y_3: yield (%). Fixed variables are summarized as X_1 = time (min); X_2 = concentration of ethanol (%); and X_3 = solid/liquid ratio (g/L).

The significance of each coefficient was determined in the regression models and in the analysis of variance (ANOVA), from which only the significant numerical terms were displayed in previous equations. The ANOVA results can be used to assess whether the regression models were successfully established by p-values (<0.05) at a confidence level along with the lack of fit analysis ($p > 0.05$), which implied that the lack of adjustment was not significant in relation to the pure error, revealing that the elaborated models are useful to travel the design within the analyzed values. In addition, the coefficients of determination of the second order model (R^2) higher than 0.94 demonstrated that the models with high statistical significance were well adjusted to the experiments. Thus, based on the statistical analysis above, the models were adequate to predict the yields of anthocyanin extraction.

3.5.5. Optimization of the Analysis of the Response Surface Methodology

The optimal values obtained by optimizing the extraction using the ultrasound technique are divided into specific response values: Y_1 (solvent = 64%; ratio = 5 g/L; and time = 0.5 min), Y_2 = (solvent = 58%; ratio = 5 g/L; and time = 0.5 min), and Y_3

(solvent = 54%; ratio = 5 g/L; and time = 0.5 min); and in general optimization values for the three responses involved: with an optimal disability value of 0.99, 68.2% ethanol concentration (v/v), 5 g of *S. melongena* per liter of solvent, and 0.5 min extraction time.

4. Conclusions

Considering the nutritional and bioactive potentialities of the pulp of the fruits of *Solanum melongena* L. and the coloring compounds present in the epicarp, the present study aimed to characterize, chemically and nutritionally, the pulp, epicarp, and whole fruit of *S. melongena*. An evaluation was also made of its bioactive properties and optimization of methodologies for obtaining an extract rich in anthocyanins from the epicarp of *S. melongena*, in order to be used as a natural coloring ingredient.

This study showed that the lyophilization process is an adequate dehydration process, due to the small changes it causes in the sample. Otherwise, the nutritional and chemical evaluation showed a low energy and fat content, the high moisture content, and the presence of bioactive molecules confirming the interest of this fruit in calorie restricted diets. In addition, the antioxidant, bacteriostatic, and bactericidal activities were also observed in all tested samples.

The optimization studies allow to understand the behavior of the main variables and also the strength of the interaction, which provided a visual representation that helped to select the ideal points of extraction. In general terms, factors ratio and solvent were the ones that showed a bigger magnitude in their responses, therefore, particular handling of these two is strongly advisable; additionally, not strong interaction within the stronger factor tested was detected in the responses analyzed, which suggests the independence of them and how we can be sure of modifying any of them within the range tested without affecting the predicted responses. Finally, the statistical diagnostic showed good fitting of data, hence, moving within the levels of the experiment would still draw great outputs if we wanted to modify parameters, because the equations provided will continue displaying strong predictions.

Therefore, these results confirm the excellent chemical and nutritional composition of the eggplant, which, together with the bioactive properties identified, make it an excellent option to incorporate into the daily diet. Moreover, considering the presence of anthocyanins in the epicarp, it also has potential application in the industrial sector, namely through the use of its natural coloring pigment. Thus, chemical and nutritional studies have shown the dietary benefits of this food, considering the different forms of consumption; and the optimization study carried out allows the industry to carry out the pigment extraction process, without preliminary studies and experiments, avoiding economic and operational expenses.

In this sense, this study presented a topic with high relevance for several sectors, such as industrial, environmental, scientific, economic, and social. This wide relevance is justified by the proposal to develop strategies that reduce food waste, improving the quality of some commercial products. Furthermore, it leads to the application of innovative extraction technologies, an increase in the quality of products available on the market, an increase in consumer preference for these products, and, consequently, to an improvement in the environmental impact, economic growth, and the development of a sustainable economy.

Author Contributions: Conceptualization, I.C.F.R.F. and L.B.; formal analysis, G.F.P.S., E.P., R.C.C., C.P., R.M.V.A., and L.B.; funding acquisition, I.C.F.R.F. and L.B.; investigation, G.F.P.S., E.P., B.M., D.S., R.C.C., C.P., and R.M.V.A.; methodology, G.F.P.S., E.P., B.M., D.S., M.S., R.C.C., and C.P.; supervision, E.P., I.C.F.R.F., and L.B.; writing—original draft, G.F.P.S., E.P., B.M., C.P., and R.M.V.A.; writing—review and editing, D.S., M.S., I.C.F.R.F., and L.B. All authors have read and agreed to the published version of the manuscript.

Funding: The authors are grateful to the Foundation for Science and Technology (FCT, Portugal) and European Regional Development Fund (ERDF) under Program PT2020 for financial support to CIMO (UID/AGR/00690/2020); this work was also funded by ERDF through the Regional Operational Program North 2020, within the scope of project "Mobilizador" Norte-01-0247-FEDER-

024479: ValorNatural®. The authors are also grateful to Interreg España-Portugal for financial support through the project 0377_Iberphenol_6_E and TRANSCoLAB 0612_TRANS_CO_LAB_2_P; to Ministry of Education, Science and Technological Development, Republic of Serbia.

Acknowledgments: The authors are grateful to the national funding by FCT, P.I., through the institutional scientific employment program-contract for L. Barros and R. Calhelha's contract; C. Pereira's contract though the celebration of program-contract foreseen in No. 4, 5, and 6 of article 23° of Decree-Law No. 57/2016, of 29 August, amended by Law No. 57/2017, of 19 July.

Conflicts of Interest: The authors declare no conflict of interest.

References

1. Woiciechowski, A.L.; de Carvalho, J.C.; Spier, M.R.; Habu, S.; Yamaguishi, C.T.; Ghiggi, V.; Soccol, C.R. *Use of Agro-Industrial Residues in Food Bioprocesses*; Food Biotechnology, Science, Technology, Food Engineering and Nutrition Collection; Atheneu Publisher: Sao Paulo, Brazil; Rio de Janeiro, Brazil; Belo Horizonte, Brazil, 2013; Volume 12, pp. 143–172.
2. Toop, T.A.; Ward, S.; Oldfield, T.; Hull, M.; Kirby, M.E.; Theodorou, M.K. AgroCycle—Developing a circular economy in agriculture. *Energy Procedia* **2017**, *123*, 76–78. [CrossRef]
3. Gontard, N.; Sonesson, U.; Birkved, M.; Majone, M.; Bolzonella, D.; Celli, A.; Angellier-Coussy, H.; Jang, G.W.; Verniquet, A.; Broeze, J.; et al. A Research Challenge Vision Regarding Management of Agricultural Waste in a Circular Bio-Based Economy. *Crit. Rev. Environ. Sci. Technol.* **2018**, *48*, 614–654. [CrossRef]
4. Gürbüz, N.; Uluişik, S.; Frary, A.; Frary, A.; Doğanlar, S. Health Benefits and Bioactive Compounds of Eggplant. *Food Chem.* **2018**, *268*, 602–610. [CrossRef] [PubMed]
5. Radicetti, E.; Massantini, R.; Campiglia, E.; Mancinelli, R.; Ferri, S.; Moscetti, R. Yield and quality of eggplant (*Solanum melongena* L.) as affected bycover crop species and residue management. *Sci. Hortic.* **2016**, *204*, 161–171. [CrossRef]
6. Rodriguez-Jimenez, J.R.; Amaya-Guerra, C.A.; Baez-Gonzalez, J.G.; Aguilera-Gonzalez, C.; Urias-Orona, V.; Nino-Medina, G. Physicochemical, Functional, and Nutraceutical Properties of Eggplant Flours Obtained by Different Drying Methods. *Molecules* **2018**, *23*, 3210. [CrossRef]
7. Ferarsa, S.; Zhang, W.; Moulai-Mostefa, N.; Ding, L.; Jaffrin, M.Y.; Grimi, N. Recovery of Anthocyanins and Other Phenolic Compounds from Purple Eggplant Peels and Pulps Using Ultrasonic-Assisted Extraction. *Food Bioprod. Process.* **2018**, *109*, 19–28. [CrossRef]
8. Bobbio, P.A.; Bobbio, F.O. *Química do Processamento de Alimentos: Pigmentos.*, 2nd ed.; Livraria Varela: Sao Paulo, Brazil, 1995; pp. 105–120. ISBN 85-85519-12-6.
9. Kong, J.M.; Chia, L.S.; Goh, N.K.; Chia, T.F.; Brouillard, R. Analysis and Biological Activities of Anthocyanins. *Phytochemistry* **2003**, *64*, 923–933. [CrossRef]
10. Das, S.; Raychaudhuri, U.; Falchi, M.; Bertelli, A.; Braga, P.C.; Das, D.K. Cardioprotective properties of raw and cooked eggplant (*Solanum melongena* L). *Food Funct.* **2011**, *2*, 395–399. [CrossRef]
11. Jung, E.J.; Bae, M.S.; Jo, E.K.; Jo, Y.H.; Lee, S.C. Antioxidant Activity of Different Parts of Eggplant. *J. Med. Plants Res.* **2011**, *5*, 4610–4615.
12. Kaume, L.; Howard, L.R.; Devareddy, L. The Blackberry Fruit: A Review on Its Composition and Chemistry, Metabolism and Bioavailability, and Health Benefits. *J. Agric. Food Chem.* **2012**, *60*, 5716–5727. [CrossRef]
13. Dranca, F.; Oroian, M. Optimization of Ultrasound-Assisted Extraction of Total Monomeric Anthocyanin (TMA) and Total Phenolic Content (TPC) from Eggplant (*Solanum melongena* L.) Peel. *Ultrason. Sonochem.* **2016**, *31*, 637–646. [CrossRef] [PubMed]
14. Roriz, C.L.; Barros, L.; Prieto, M.A.; Morales, P.; Ferreira, I.C.F.R. Floral Parts of *Gomphrena globosa* L. as a Novel Alternative Source of Betacyanins: Optimization of the Extraction Using Response Surface Methodology. *Food Chem.* **2017**, *229*, 223–234. [CrossRef] [PubMed]
15. Association of Official Analytical Chemist (AOAC). *Official Methods of Analysis of AOAC International*; AOAC: Washinton, DC, USA, 2016; ISBN 0935584870.
16. Barros, L.; Pereira, E.; Calhelha, R.C.; Dueñas, M.; Carvalho, A.M.; Santos-Buelga, C.; Ferreira, I.C.F.R. Bioactivity and Chemical Characterization in Hydrophilic and Lipophilic Compounds of *Chenopodium ambrosioides* L. *J. Funct. Foods* **2013**, *5*, 1732–1740. [CrossRef]
17. Bessada, S.M.F.; Barreira, J.C.M.; Barros, L.; Ferreira, I.C.F.R.; Oliveira, M.B.P.P. Phenolic Profile and Antioxidant Activity of *Coleostephus myconis* (L.) Rchb.f.: An Underexploited and Highly Disseminated Species. *Ind. Crop. Prod.* **2016**, *89*, 45–51. [CrossRef]
18. Jabeur, I.; Tobaldini, F.; Martins, N.; Barros, L.; Martins, I.; Calhelha, R.C.; Henriques, M.; Silva, S.; Achour, L.; Santos-Buelga, C.; et al. Bioactive Properties and Functional Constituents of *Hypericum androsaemum* L.: A Focus on the Phenolic Profile. *Food Res. Int.* **2016**, *89*, 422–431. [CrossRef]
19. Gonçalves, L.C.P.; Marcato, A.C.; Rodrigues, A.C.B.; Pagano, A.P.E.; Freitas, B.C.; de Machado, C.O.; Nakashima, K.K.; Esteves, L.C.; Lopes, N.B.; Bastos, E.L. Betalaínas: Das cores das beterrabas à fluorescência das flores. *Rev. Virtual Química* **2015**, *7*, 292–309.
20. Albuquerque, B.R.; Prieto, M.A.; Barreiro, M.F.; Rodrigues, A.E.; Curran, T.P.; Barros, L.; Ferreira, I.C.F.R. Catechin-Based Extract Optimization Obtained from *Arbutus unedo* L. Fruits Using Maceration/Microwave/Ultrasound Extraction Techniques. *Ind. Crop. Prod.* **2017**, *95*, 404–415. [CrossRef]

21. Carocho, M.; Barros, L.; Calhelha, R.C.; Ćirić, A.; Soković, M.; Santos-Buelga, C.; Morales, P.; Ferreira, I.C.F.R. *Melissa officinalis* L. decoctions as functional beverages: A bioactive approach and chemical characterization. *Food Funct.* **2015**, *6*, 2240–2248.
22. Guimarães, R.; Barros, L.; Dueñas, M.; Calhelha, R.C.; Carvalho, A.M.; Santos-Buelga, C.; Queiroz, M.J.R.P.; Ferreira, I.C.F.R. Infusion and Decoction of Wild German Chamomile: Bioactivity and Characterization of Organic Acids and Phenolic Compounds. *Food Chem.* **2013**, *136*, 947–954. [CrossRef]
23. Gutteridge, J.M.C. Lipid Peroxidation and Antioxidants as Biomarkers of Tissue Damage. *Clin. Chem.* **1995**, *41*, 1819–1828. [CrossRef]
24. Ng, T.B.; Liu, F.; Wang, Z.T. Antioxidant activity of natural products from plants. *Life Sci.* **2000**, *66*, 709–723. [CrossRef]
25. Lockowandt, L.; Pinela, J.; Roriz, C.L.; Pereira, C.; Abreu, R.M.V.; Calhelha, R.C.; Alves, M.J.; Barros, L.; Bredol, M.; Ferreira, I.C.F.R. Chemical Features and Bioactivities of Cornflower (*Centaurea cyanus* L.) Capitula: The Blue Flowers and the Unexplored Non-Edible Part. *Ind. Crop. Prod.* **2019**, *128*, 496–503. [CrossRef]
26. Zhu, Z.; He, J.; Liu, G.; Barba, F.J.; Koubaa, M.; Ding, L.; Vorobiev, E. Recent insights for the green recovery of inulin from plant food materials using non-conventional extraction technologies: A review. *Innov. Food Sci. Emerg. Technol.* **2016**, *33*, 1–9. [CrossRef]
27. Jiménez, L.C.; Caleja, C.; Prieto, M.A.; Barreiro, M.F.; Barros, L.; Ferreira, I.C.F.R. Optimization and comparison of heat and ultrasound assisted extraction techniques to obtain anthocyanin compounds from *Arbutus unedo* L. fruits. *Food Chem.* **2018**, *264*, 81–91.
28. Heleno, S.A.; Prieto, M.A.; Barros, L.; Rodrigues, A.; Barreiro, M.F.; Ferreira, I.C.F.R. Optimization of microwave-assisted extraction of ergosterol from *Agaricus bisporus* L. byproducts using response surface methodology. *Food Bioprod. Process.* **2016**, *100*, 25–35. [CrossRef]
29. López, C.J. Desarrollo de un Aditivo Colorante Natural a Base de Cianidina Obtenido a Partir de Frutos de *Arbutus unedo* L.: Optimización de la Extracción y Estudio de su Aplicación en Gofres. Master's Thesis, Polythecnic Institute of Bragança, Bragança, Portugal, 2017.
30. Pinela, J.; Prieto, M.A.; Carvalho, A.M.; Barreiro, M.F.; Oliveira, M.B.P.P.; Barros, L.; Ferreira, I.C.F.R. Microwave-Assisted Extraction of Phenolic Acids and Flavonoids and Production of Antioxidant Ingredients from Tomato: A Nutraceutical-Oriented Optimization Study. *Sep. Purif. Technol.* **2016**, *164*, 114–124. [CrossRef]
31. Vieira, V.; Prieto, M.A.; Barros, L.; Coutinho, J.A.P.; Ferreira, O.; Ferreira, I.C.F.R. Optimization and Comparison of Maceration and Microwave Extraction Systems for the Production of Phenolic Compounds from *Juglans regia* L. for the Valorization of Walnut Leaves. *Ind. Crop. Prod.* **2017**, *107*, 341–352. [CrossRef]
32. Melgar, B.; Dias, M.I.; Barros, L.; Ferreira, I.C.F.R.; Rodriguez-Lopez, A.D.; Garcia-Castello, E.M. Ultrasound and Microwave Assisted Extraction of Opuntia Fruit Peels Biocompounds: Optimization and Comparison Using RSM-CCD. *Molecules* **2019**, *24*, 3618. [CrossRef]
33. Xu, M.; Du, C.; Zhang, N.; Shi, X.; Wu, Z.; Qiao, Y. Color Spaces of Safflower (*Carthamus tinctorius* L.) for Quality Assessment. *J. Tradit. Chin. Med. Sci.* **2016**, *3*, 168–175. [CrossRef]
34. Russo, P.; Adiletta, G.; Di Matteo, M. The Influence of Drying Air Temperature on the Physical Properties of Dried and Rehydrated Eggplant. *Food Bioprod. Process.* **2013**, *91*, 249–256. [CrossRef]
35. Jiang, T.; Mao, Y.; Sui, L.; Yang, N.; Li, S.; Zhu, Z.; Wang, C.; Yin, S.; He, J.; He, Y. Degradation of Anthocyanins and Polymeric Color Formation during Heat Treatment of Purple Sweet Potato Extract at Different PH. *Food Chem.* **2019**, *274*, 460–470. [CrossRef] [PubMed]
36. Lao, F.; Giusti, M.M. The Effect of Pigment Matrix, Temperature and Amount of Carrier on the Yield and Final Color Properties of Spray Dried Purple Corn (*Zea mays* L.) Cob Anthocyanin Powders. *Food Chem.* **2017**, *227*, 376–382. [CrossRef] [PubMed]
37. Weber, F.; Boch, K.; Schieber, A. Influence of Copigmentation on the Stability of Spray Dried Anthocyanins from Blackberry. *LWT Food Sci. Technol.* **2017**, *75*, 72–77. [CrossRef]
38. Nemzer, B.; Vargas, L.; Xia, X.; Sintara, M.; Feng, H. Phytochemical and Physical Properties of Blueberries, Tart Cherries, Strawberries, and Cranberries as Affected by Different Drying Methods. *Food Chem.* **2018**, *262*, 242–250. [CrossRef]
39. Lima-Corrêa, R.A.B.; Andrade, M.S.; da Silva, M.F.d.G.F.; Freire, J.T.; Ferreira, M.d.C. Thin-Layer and Vibrofluidized Drying of Basil Leaves (*Ocimum basilicum* L.): Analysis of Drying Homogeneity and Influence of Drying Conditions on the Composition of Essential Oil and Leaf Colour. *J. Appl. Res. Med. Aromat. Plants* **2017**, *7*, 54–63.
40. Badulescu, L.; Dobrin, A.; Stan, A.; Mot, A.; Bujor, O.C. Drying treatment effects on anthocyanins of organic raspberry (cv. Heritage) fruit. In Proceedings of the Third Nordic Baltic Drying Conference, Saint Petersburg, Russia, 12–14 June 2019; pp. 48–53.
41. USDA (United States Department of Agriculture Agricultural Research Service), 11209 Report, Raw Eggplant. Available online: https://ndb.nal.usda.gov/ndb/foods/show/2962 (accessed on 17 April 2019).
42. Kandoliya, U.K.; Bajaniya, V.K.; Bhadja, N.K.; Bodar, N.P.; Golakiya, B.A. Antioxidant and Nutritional of Egg Plant (*Solanun melongena* L.) Fruit Grown in Saurastra Region. *Intern. J. Curr. Microbiol. Appl. Sci.* **2015**, *4*, 806–813.
43. Agoreyo, B.O.; Obansa, E.S.; Obanor, E.O. Comparative nutritional and phytochemical analyses of two varieties of *Solanum melongena*. *Sci. World J.* **2012**, *7*, 5–8.
44. Raigón, M.D.; Prohens, J.; Muñoz-Falcón, J.E.; Nuez, F. Comparison of Eggplant Landraces and Commercial Varieties for Fruit Content of Phenolics, Minerals, Dry Matter and Protein. *J. Food Compos. Anal.* **2008**, *21*, 370–376. [CrossRef]

45. Ossamulu, I.; Akanya, H.; Jigam, A.; Egwim, E. Evaluation of Nutrient and Phytochemical Constituents of Four Eggplant Cultivars. *Elixir Food Sci.* **2014**, *73*, 26424–26428.
46. Ayas, F.A.; Colak, N.; Topuz, M.; Tarkowski, P.; Jaworek, P.; Seiler, G.; Inceer, H. Comparison of Nutrient Content in Fruit of Commercial Cultivars of Eggplant (*Solanum melongena* L.). *Pol. J. Food Nutr. Sci.* **2015**, *65*, 251–259.
47. Kayashima, T.; Katayama, T. Oxalic acid is available as a natural antioxidant in some systems. *Biochim. Biophys. Acta* **2002**, *1573*, 1–3. [CrossRef]
48. Wang, Q.; Lai, T.; Qin, G.; Tian, S. Response of Jujube Fruits to Exogenous Oxalic Acid Treatment Based on Proteomic Analysis. *Plant. Cell Physiol.* **2009**, *50*, 230–242. [CrossRef] [PubMed]
49. Barberis, A.; Cefola, M.; Pace, B.; Azara, E.; Spissu, Y.; Serra, P.A.; Logrieco, A.F.; D'hallewin, G.; Fadda, A. Postharvest Application of Oxalic Acid to Preserve Overall Appearance and Nutritional Quality of Fresh-Cut Green and Purple Asparagus during Cold Storage: A Combined Electrochemical and Mass-Spectrometry Analysis Approach. *Postharvest Biol. Technol.* **2019**, *148*, 158–167. [CrossRef]
50. Cheng, C.; Zhou, Y.; Lin, M.; Wei, P.; Yang, S.T. Polymalic Acid Fermentation by Aureobasidium Pullulans for Malic Acid Production from Soybean Hull and Soy Molasses: Fermentation Kinetics and Economic Analysis. *Bioresour. Technol.* **2017**, *223*, 166–174. [CrossRef] [PubMed]
51. Jang, S.A.; Park, D.W.; Kwon, J.E.; Song, H.S.; Park, B.; Jeon, H.; Sohn, E.H.; Koo, H.J.; Kang, S.C. Quinic Acid Inhibits Vascular Inflammation in TNF-α-Stimulated Vascular Smooth Muscle Cells. *Biomed. Pharm.* **2017**, *96*, 563–571. [CrossRef] [PubMed]
52. Karam, J.; del Mar Bibiloni, M.; Pons, A.; Tur, J.A. Total Fat and Fatty Acid Intakes and Food Sources in Mediterranean Older Adults Requires Education to Improve Health. *Nutr. Res.* **2020**, *73*, 67–74. [CrossRef]
53. Hanifah, A.; Maharijaya, A.; Putri, S.P.; Laviña, W.A.; Sobir. Untargeted Metabolomics Analysis of Eggplant (*Solanum melongena* L.) Fruit and Its Correlation to Fruit Morphologies. *Metabolites* **2018**, *8*, 49. [CrossRef] [PubMed]
54. Park, S.H.; Min, T.S. Caffeic Acid Phenethyl Ester Ameliorates Changes in IGFs Secretion and Gene Expression in Streptozotocin-Induced Diabetic Rats. *Life Sci.* **2006**, *78*, 1741–1747. [CrossRef]
55. Araújo, V.M. Estudo do Potencial Terapêutico do Ácido Cafeico em Protocolos de Diabetes e Dislipidemia em Camundongos. Master's Thesis, Universidade Federal do Ceará, Fortaleza, Brazil, 2014; 99p.
56. Kakkar, S.; Bais, S. A Review on Protocatechuic Acid and Its Pharmacological Potential. *ISRN Pharm.* **2014**, *2014*, 1–9. [CrossRef]
57. Forino, M.; Tenore, G.C.; Tartaglione, L.; Carmela, D.; Novellino, E.; Ciminiello, P. (1S,3R,4S,5R)5-O-Caffeoylquinic Acid: Isolation, Stereo-Structure Characterization and Biological Activity. *Food Chem.* **2015**, *178*, 306–310. [CrossRef]
58. Song, S.E.; Jo, H.J.; Kim, Y.W.; Cho, Y.J.; Kim, J.R.; Park, S.Y. Delphinidin Prevents High Glucose-Induced Cell Proliferation and Collagen Synthesis by Inhibition of NOX-1 and Mitochondrial Superoxide in Mesangial Cells. *J. Pharm. Sci.* **2016**, *130*, 235–243. [CrossRef] [PubMed]
59. Lee, D.Y.; Park, Y.J.; Hwang, S.C.; Kim, K.D.; Moon, D.K.; Kim, D.H. Cytotoxic Effects of Delphinidin in Human Osteosarcoma Cells. Acta Orthop. *Traumatol. Turc.* **2018**, *52*, 58–64.
60. Niño-Medina, G.; Urías-Orona, V.; Muy-Rangel, M.D.; Heredia, J.B. Structure and Content of Phenolics in Eggplant (*Solanum melongena*)-A Review. *South. Afr. J. Bot.* **2017**, *111*, 161–169. [CrossRef]
61. Wu, X.; Prior, R.L. Identification and Characterization of Anthocyanins by High-Performance Liquid Chromatography-Electrospray Ionization-Tandem Mass Spectrometry in Common Foods in the United States: Vegetables, Nuts, and Grains. *J. Agric. Food Chem.* **2005**, *53*, 3101–3113. [CrossRef] [PubMed]
62. García-Salas, P.; Gómez-Caravaca, A.M.; Morales-Soto, A.; Segura-Carretero, A.; Fernández Gutiérrez, A. Identification and quantification of phenolic compounds in diverse cultivars of eggplant grown in different seasons by high performance liquid chromatography coupled to diode array detector and electrospray-quadrupole-time of flight-mass spectrometry. *Food Res. Int.* **2014**, *57*, 114–122. [CrossRef]
63. Salerno, L.; Modica, M.N.; Pittalà, V.; Romeo, G.; Siracusa, M.A.; Di Giacomo, C.; Sorrenti, V.; Acquaviva, R. Antioxidant Activity and Phenolic Content of Microwave-Assisted *Solanum melongena* Extracts. *Sci. World J.* **2014**, *2014*. [CrossRef] [PubMed]
64. Horincar, G.; Enachi, E.; Bolea, C.; Râpeanu, G.; Aprodu, I. Value-Added Lager Beer Enriched with Eggplant (*Solanum melongena* L.) Peel Extract. *Molecules* **2020**, *25*, 731. [CrossRef]
65. Bartwal, A.; Mall, R.; Lohani, P.; Guru, S.K.; Arora, S. Role of secondary metabolites and brassinosteroids in plant defense against environmental stresses. *J. Plant Growth Regul.* **2013**, *32*, 216–232. [CrossRef]
66. Akanitapichat, P.; Phraibung, K.; Nuchklang, K.; Prompitakkul, S. Antioxidant and Hepatoprotective Activities of Five Eggplant Varieties. *Food Chem. Toxicol.* **2010**, *48*, 3017–3021. [CrossRef]
67. Hong, H.; Lee, J.H.; Kim, S.K. Phytochemicals and Antioxidant Capacity of Some Tropical Edible Plants. *Asian Australas. J. Anim. Sci.* **2018**, *31*, 1677–1684. [CrossRef]
68. Muangrat, R.; Pongsirikul, I.; Blanco, P.H. Ultrasound assisted extraction of anthocyanins and total phenolic compounds from dried cob of purple waxy corn using response surface methodology. *J. Food Process. Pres.* **2018**, *42*, 1–11. [CrossRef]
69. Taofiq, O.; Corrêa, R.C.G.; Barros, L.; Prieto, M.A.; Bracht, A.; Peralta, R.M.; González-Paramás, A.M.; Barreiro, M.F.; Ferreira, I.C.F.R. A Comparative Study between Conventional and Non-Conventional Extraction Techniques for the Recovery of Ergosterol from *Agaricus blazei* Murrill. *Food Res. Int.* **2019**, *125*, 108541. [CrossRef] [PubMed]

Table 1. *Cont.*

Parameters	Values	Reference
Organic matter (%)	57–62	
Total carbon (%)	2.0–3.3	[26,30,31]
Total nitrogen (g/L)	2.0–2.4	[21]
Total sugar (g/L)	5.0–12.0	[16,21,26,32–34]
Total fat (%)	1.0–23	[35]
Total suspended solids (g/L)	25–30	[36,37]
Polyalcohol (%)	9.0–15	[21,35,38]
Total phenols (g/L)	0.5–6.1	[6,26,29,33,39–41]

2.2. Microbiological Traits of OMWW

The microbial community present in OMWW is strongly influenced by several parameters, among which the ripeness state and the olive variety are the most influential [42,43]. The microbial density in OMWW varies between 10^5 and 10^6 CFU/mL (CFU: colony-forming unit) and is mainly composed of yeasts, bacteria, and molds [42–45]. The yeast population includes species belonging to *Pichia, Candida,* and *Saccharomyces* genera [44,45]. A survey carried out on OMWW revealed the presence of over 100 identified fungi, mainly belonging to the genera *Acremonium, Alternaria, Aspergillus, Bionectria, Byssochlamys, Chalara, Cerrena, Fusarium, Lasiodiplodia, Lecythophora, Paecilomyces, Penicillium, Phycomyces, Phoma, Rhinocladiella,* and *Scopulariopsis* [46].

Although many studies report that the culturable microbial population is represented by only a few bacterial communities, such as: *Firmicutes, Actinobacteria, Alphaproteobacteria, Betaproteobacteria,* and *Gammaproteobacteria,* recently, microarray analyses have revealed a high-density of a larger microbial population, including *Proteobacteria, Bacteroidetes, Chloroflexi, Cyanobacteria,* and *Actinobacteria.* However, the most commonly reported microbial communities, representing 50% of the 16S rRNA gene sequences deposited in GenBank, include *Gammaproteobacteria (Enterobacteriaceae, Moraxellaceae, Xanthomonadaceae,* and *Pseudomonadaceae)* with a percentage of almost 30%, and *Betaproteobacteria (Oxalobacteraceae* and *Comamonadaceae)* with a percentage of 21.5% [46]. *Alphaproteobacteria* and *Actinobacteria (Micrococcaceae, Microbacteriaceae,* and *Propionibacteriaceae)* together comprised 20%, whereas *Firmicutes (Bacillaceae, Clostridiaceae, Lactobacillaceae,* and *Paenibacillaceae)* and *Bacterioides (Prevotellaceae, Porphyromonadaceae,* and *Sphingobatteriaceae)* phyla accounted for approximately 6.8%, respectively. Furthermore, differences in microbial population have been detected, highlighting that only 15% of operational taxonomic units (OTUs) are commonly detected [47]. In addition, high densities of enteric bacteria belonging to *Porphyromonadaceae, Prevotellaceae, Lachnospiraceae, Eubacteriaceae, Peptococcaceae, Peptostreptococcaceae,* and *Ruminococcaceae* spp. or to genera *Acinetobacter, Enterobacter* spp., *Pseudomonas, Citrobacter, Escherichia, Klebsiella,* and *Serratia* spp. have been reported [48].

3. Reuse of OMWW

3.1. OMWW Management and Bioremediation

The implementation of any treatment based on the circular economy approach and "waste" reuse concept represents a competitive and innovative choice for agro-food companies for achieving a reduction in cost management and environmental impact.

According to Tsagaraki and co-workers [39], 1 m^3 of OMWW corresponds to 100–200 m^3 of domestic wastewater. The COD and BOD$_5$ values of OMWW are very consistent and even higher when obtained by conventional systems (150 g O$_2$/L COD and 90 g O$_2$/L BOD$_5$ vs. 90 g O$_2$/L COD and 30 g O$_2$/L BOD$_5$ for conventional and two-phase extraction systems, respectively).

According to the European Directive 2000/60/CE [49], the OMWW requires specific treatment prior to direct discharge to ensure environmental protection and for regenerated wastewater. Indeed, the disposal of untreated OMWW on agricultural soil causes severe environmental damages, such as altering the color of natural water sources and exercising toxic effects on aquatic life and soil quality. OMWW is characterized by a high content of components with low biodegradability (e.g., long-chain fatty acids, lipids, and simple and complex sugars). Therefore, the most common applied systems for OMWW reuse are concerned with lowering the pollutant load and/or extracting bioactive compounds for different applications [25].

A plethora of physicochemical treatments has been developed in order to remove the phenolic compounds. However, in the majority of the studies no ecotoxicological evaluation has been reported and the success of treatment is mainly based on the reduction of color, COD, phenol content, etc. The most relevant parameter used to evaluate compost phytotoxicity is the germination index (GI). Low GI values could be attributed to the fact that at the starting stage, substrates have high concentrations of water-soluble organic substances, toxic constituents such as alcohols, organic fatty acids and phenolic compounds, high C/N ratios due to the presence of ammonia and other toxic nitrogen-based products, as well as high heavy metal and mineral salt contents [50,51].

In addition to traditional settling (conducted in tanks called "hell"), various treatments have been proposed: physicochemical, biological, or a combination.

Physicochemical systems include different methods based on the use of flocculants, coagulants, membrane filtration and reverse osmosis [52], or applying oxidation cryogenesis, electrocoagulation [53,54], or a photochemical system [55]. Generally, after these treatments, the resulting products can be spread on agricultural soil as an organic fertilizer or simply subjected to evaporation in open tanks [56]. However, these practices are expensive as they produce matrices, such as sludge, that must either undergo further treatments or be disposed of.

Several reports confirm that microorganisms can be proposed as a promising alternative for bioremediation of OMWW [51]. Biological methods, involving anaerobic or aerobic digestion and composting, have been applied to break complex organic compounds into simpler molecules and may lead to the production of proteins, exopolysaccharides, or energy [57,58]. The main interest in anaerobic digestion is the production of energy and reuse of the effluent for irrigation purposes [59]. However, the leading limitation is the inhibition of methanogenic bacteria by both phenolic and organic acids compounds [60]. According to Azbar and co-workers [36], anaerobic filters or up flow anaerobic sludge bed reactors are suitable systems to remove unwanted compounds from OMWW. Filidei et al. [61] proposed sedimentation–filtration treatment of OMWW prior to anaerobic digestion as a useful method for its disposal. On the other hand, aerobic treatment is used to reduce the polluting load, responsible for certain biostatic and phytotoxic effects. Aerobic treatment has also been applied to reduce the polluting effect of municipal wastewater, focusing on the degradation of phenolic compounds. Several microorganisms, such as *Pleurotus ostreatus, Bacillus pumilus, Yarrowia lipolytica*, etc., have been tested [62–64]. Furthermore, a pool of *Candida boidinii* and *Pichia holstii* strains has been selected for its ability to reduce (up to 40%) the phenolic content of OMWW combined with 6.0 g/L of $(NH_4)_2SO_4$ at 10 °C [65].

OMWW has been proposed [66,67] as a growth substrate for *Azotobacter vinelandii* and the resultant effluents applied to cropland as fertilizer. Therefore, recent studies have shown that the biotechnological potential of indigenous microbiota should be further exploited with respect to bioremediation of OMWW and inactivation of plant and human pathogens.

3.2. OMWW Phenolic Compounds for Agricultural Use

Phenolic compounds from OMWW might be used for integrated pest management programs. Several studies have reported the use of microorganisms (as single or consortia) to degrade organic compounds in effluents [68,69]. Although OMWWs do not contain toxic substances, they are characterized by high COD values and a high concentration of compounds with biostatic activity. Recently, increasing attention has been focused on the degrading properties of microorganisms and biological aerobic treatments using yeasts and filamentous fungi, which have emerged as suitable biofertilization methods for conducting residues with lower toxicity, COD, and phenolic contents. Aissam et al. [70] treated OMWW with microorganisms isolated from the same source, such as *Candida boidinii*, *Geotrichum candidum*, *Penicillium* sp. and *Aspergillus niger*, obtaining a 40–73% reduction in phenols and a 45–78% reduction in COD value. Bleve et al. [45] identified several strains belonging to the genera *Geotrichum*, *Saccharomyces*, *Pichia*, *Rhodotorula*, and *Candida* that showed strain-dependent phenol removal efficiency, decreasing phenolic and COD values, regardless of initial phenolic concentrations. In particular, *G. candidum*, both as free and Ca-alginate immobilized cells, showed the best degradation performance, and when immobilized showed a double reduction rate ability. Indeed, Ca-alginate improved the proteolytic stability of the enzymes responsible for the degradation process. Maza-Márquez et al. [69] demonstrated that the use of a microalgal–bacterial consortium, in a photo-bioreactor, induces a decrease in pollutant load by affecting COD, BOD_5, phenolic compounds, color, and turbidity values of OMWW. The dominant green microalgae *Scenedesmus obliquus*, *Chlorella vulgaris* along with cyanobacteria *Anabaena* sp., showed a synergistic effect on resistance to toxic pollutants, leading to their decomposition. In addition, the effect of *Lactiplantibacillus plantarum* strains on decolorization and biodegradation of phenolic compounds was evaluated [69], highlighting strains able to decrease the OMWW pH within 6 days. Growth of *L. plantarum* induced the depolymerization of high molecular weight phenols, resulting in discoloration of fresh OMWW and a significant reduction in total phenols [71]. Approximately 58% of the color, 55% of the COD, and 46% of the phenolic compounds were removed when OMWW was diluted tenfold before *L. plantarum* addition.

Furthermore, OMWW has also been proposed for biopesticide and compost production. The OMWW application on soil and crops resulted in a growth suppression of most phytopathogenic bacteria and fungi and weed species without any effect on crop growth. However, certain measures should be adhered to when OMWW is used as a biopesticide, especially regarding dose and timing of use [47].

4. OMWW as a Source of Biopolymers and Bio-Energy Production

4.1. Enzyme and Exopolysaccharide Production

OMWW represents a suitable substrate for the production of enzymes by fungi. Fungi are microorganisms known for their ability to synthesize different biological catalysts that can be used in different areas. In particular, Ntougias and co-workers [46] demonstrated that ligninolytic fungi are a useful source of phenoloxidase, polyphenoloxidase, and peroxidase useful for removing recalcitrant compounds in OMWW.

Several yeast strains have been characterized as highly pectolytic, xylanolytic, provided with cellulase, β-glucanase, β-glucosidase, peroxidase, and polygalacturonase activities, which could effectively degrade the complex compounds responsible for OMWW toxicity [65,72].

Several yeasts have been described as able to reduce phenolics and sugars present in OMWW, although white-rot fungi appear to contribute more to discoloration [73]. Moreover, Giannoutsou and co-workers [74] isolated six phenotypically distinct groups of yeasts and three selected isolates were identified through biochemical tests and partial 18S rDNA gene sequence analysis as most closely related to *Saccharomyces* spp., *Candida boidinii*, and *G. candidum*. These fungal genera have been reported as able to degrade the phenolic content present in OMWW [75,76].

Several reports also propose strains belonging to different species, such as *Panus trigrinus*, *Hericium erinaceus*, and *Pleurotus citrinopileatus* for laccase (Lac) and manganese peroxidase (Mnp) production [77–79]. Filamentous fungi, such as *Aspergillus oryzae*, *A. niger*, *Aspergillus ibericus*, *Aspergillus uvarum*, *G. candidum*, *Rizhopus oryzae*, *Rhizopus arrhizus*, and *Penicillium citrinum*, have been described as lipolytic reservoirs due to their ability to produce lipase [80]. These enzymes have been used in different industries, such as dairy and pharmaceutical [81]. Moreover, OMWW has been confirmed as a suitable substrate for production of pectinase, with *Cryptococcus albidus* var. *albidus* IMAT 473 showing the best biotechnological aptitude. This enzyme, compared with other products on the market, showed a broad spectrum endopolygalacturonase activity [82–84].

Besides enzyme production, OMWWs have also been evaluated as a source of polysaccharides, especially exopolysaccharides (ESP) [85] with glucose as the main monosaccharide, followed by galactose, arabinose, rhamnose, and galacturonic acid. Xanthan, a glucose-mannose and glucuronic acid repeating unit compound, is the main ESP used in different products, such as in cosmetic formulations or as a supplement and thickening compound [86]. However, the EPS production through a fermentation process depends on the type of microorganism. The first production of EPS in OMWW (used at 30% *v/v*) was obtained through a strain of *Xanthomonas campestris* that showed a productive capacity of 4 g/L [87]. Similarly, *Paenibacillus jamilae* sp. highlighted, on OMWW, the production of an EPS consisting of fucose, xylose, rhamnose, arabinose, mannose, galactose, and glucose. Morillo et al. [88] reported that *P. jamilae* CECT 5266 strain (in a 80% *v/v* of OMWW) produced an EPS consisting of glucose, galactose, mannose, arabinose, rhamnose, hexosamine, and uronic acid, in agreement with results previously reported by Ruiz-Bravo et al. [89] using the strain *P. jamilae* CP-7.

4.2. Production of Bio-Energy and Biofuels

The need to reduce dependence on conventional fossil fuels in favor of new alternative energy resources is a top global priority. Green energies could contribute to the reduction of greenhouse gas emissions and their consequent unfavorable impacts on global warming and climate change [90]. The high content of organic matter and the low content of nitrogen, volatile acid sugars, polyalcohols, and fats, make OMWW an attractive resource for the production of bioenergy and alternative biofuels, such as methane or ethanol [6,51]. Several microorganisms are used for biohydrogen production, through single or combined catabolic pathways (e.g., *Rhodobacter sphaeroides*, *Rhodopseudomonas palustris*, and *Chlamydomonas reinhardtii*). The production of these substances takes place through a process of anaerobic digestion, which consists of two phases. During the first phase, macromolecules, such as carbohydrates, proteins, and lipids, are transformed by hydrolytic and acidogenic fermentative bacteria into simple or intermediate organic compounds, volatile organic acids (acetic, propionic, and butyric acids), alcohols (ethanol), ketones, CO_2, and hydrogen. In the second step, through interactions between methanogenic and acetogenic microorganisms, these metabolites are transformed into CH_4 and CO_2 [91]. However due to the presence of oily residues or phenols responsible for antimicrobial activity, OMWW must be first treated or diluted [92]. As already known, before implementing an anaerobic digestion process, the treatment of OMWW with some fungi, such as *A. niger*, *Aspergillus terreus*, and *Pleurotus sajor-caju* play a key role in order to increase the final production of the reference bioenergy compound. Hamdi et al. [57] and Borja and co-workers [93], through a comparative kinetic study, demonstrated that the pretreatment of OMWW with *A. niger* and *A. terreus* increased the methane yield. Massadeh and Modallal [94] evaluated the ability of a *P. sajor-caju* strain to degrade the phenols of OMWW producing ethanol. For this purpose, the authors examined the effects of dilution with water (in a 1:1 ratio), heat treatment (at 100 °C), and treatment with H_2O_2. The results showed that the degradation of phenols by *P. sajor-caju* reached a level of 50% in heat-treated OMWW, 53% in heat-treated OMWW pretreated with H_2O_2, and 58% in undiluted heat-treated OMWW. The highest ethanol yield was obtained in samples pretreated with *P. sajor-caju* and after 48 h of fermentation with 50% diluted and heat-treated OMWW. Further biological treatment was carried out with *Saccharomyces cerevisiae*. Sarris et al. [95] and Nikolaou et al. [96] confirmed the aptitude of *S. cerevisiae* to produce ethanol and optimal fermentation parameters were detected using the 1:1 OMWW/water mixture ratio. The fermentation kinetics of molasses mixed with OMWW where *S. cerevisiae* was immobilized affected the ethanol yield, reaching values up to 67.8 g/L per day. Moreover, Zanichelli et al. [97] proposed a multiphase treatment using *S. cerevisiae* added to OMWW with glucose, to a final sugar concentration of 200 g/L, with *A. niger* extract to hydrolyze the present polysaccharides. Although *S. cerevisiae* showed low fermentative performance, indigenous strains belonging to *Pichia fermentans* and *Candida* spp. reduced phenolic content up to 15% and 18%, respectively, without any addition or pretreatment [98]. Furthermore, Sarris et al. [99] demonstrated the ability of *Y. lipolytica* strain ACA-DC 5029 to grow on media containing a low concentration of crude glycerol and OMWW, producing a significant amount of citric acid and erythritol. In the presence of high glycerol concentration, a shift towards erythritol production was observed, simultaneously with high amounts of citric acid production. The strain showed promising characteristics to be used in the biotransformation of biodiesel derived from the combination of crude glycerol and OMWW and the subsequent production of added-value chemical compounds.

4.3. Use of OMWW in Feed Formulation

The use of agro-industrial by-products in animal feed can represent an economically and environmentally advantageous solution for the livestock sector, increasing its profitability and sustainability [100]. Olive oil by-products have been tested for the formulation of feed for lambs, pigs, and chickens by evaluating the antioxidant activity on animals and on final products. Makri et al. [101] evaluated the effect of OMWW addition in a silage formulation for lambs, containing 52.5% of solids, 7.5% of OMWW, and 40% of water. The administration of OMWW-containing silage was found effective in improving animal welfare and productivity. Furthermore, several authors tested the effectiveness of a reduction in oxidative stress and in the stimulation of the immune response of the same extract for pigs. Gerasopoulos et al. [102] studied the antioxidant effect of the addition of 4% of OMWW (representing the retentate obtained by microfiltration) in silage. Piglets fed with the fortified formulation showed an increase in tested biomarkers (as total antioxidant capacity: TAC; glutathione: GSH; catalase activity: CAT; protein carbonyls: CARB; and reactive thiobarbituric acids: TBARS) in blood and tissues and a decrease in oxidative stress, with an overall increase in productivity. In addition, Varricchio et al. [103] evaluated the antioxidant activity in piglets fed with phenol extracts, and results highlighted an increase in leukocytes and cyclooxygenase-2 (COX-2), known as markers of inflammation. Gerasopoulos et al. [104] repeated the test in chickens, highlighting markers of antioxidant activity with the same silage formulation proposed for piglet feeding. The results confirmed that such supplementation lowers the levels of lipid peroxidation and protein oxidation by increasing the total antioxidant capacity in plasma confirming that both OMWW and oil by-products (leaves and olive pomace) can be a viable alternatives to fortify animal feeds.

5. Bioactive Properties of OMWW

Olive oil by-products are rich in bioactive compounds with potential health benefits [41]. Ciriminna et al. [105] investigated the relationship between phenolics and health benefits on food, pharmaceutical, and cosmetic applications. Regarding the food sector, the addition of phenols from OMWW seems very interesting not only to strengthen the beneficial effects of foods themselves, but also to extend their shelf life. In the U.S., olive pulp extracts have been approved by the Food and Drug Administration (FDA) with GRAS (Generally Recognized as Safe) (GRN No. 459) status as antioxidants in baked goods, beverages, cereals, sauces and dressings, condiments, and snacks, at a final concentration of up to 3 g/kg [106,107]. Commercial OMWW implementation in food and recovery of phenols is of great interest [108,109] and at least five companies worldwide recover phenols from OMWWs [110] to sell them as natural preservatives or bioactive additives in food products [111].

5.1. Antioxidant Properties

Many reports, both in humans and animals, confirmed that most degenerative diseases, such as cancer and cardiovascular diseases [112] are related to oxidative stress, which has also been identified as a causative agent for declining immune function and atherosclerosis [113]. Several nutraceuticals aimed to reduce the oxidative stress are currently available on the market [114]. Phenols are recognized as the main compounds responsible for the health effects of the Mediterranean diet in prevention of chronic diseases and diet-associated diseases (DRDs), such as obesity, metabolic syndrome, type 2 diabetes (T2D), cardiovascular disease (CVD), hypertension, and some cancers. Their role has been clearly recognized by the European Food Safety Authority [115] with the health claim: *"Olive oil polyphenols contribute to the protection of blood lipids from oxidative stress."*

In recent years, an increased interest in the extraction of phenols from OMWW has been registered and different extraction techniques have been proposed [116]. Phenols are active ingredients of many medicinal plants and the mechanisms of their pharmacological activity are not yet fully understood. Beyond the mechanism of protection, based on antioxidant activities, phenols have highlighted: scavenger property against free radicals and reactive oxygen forms (ROS); ability to act as chelators of heavy metals (especially iron) and capability to inhibit lipoxygenase, involved in inflammatory processes. The main radical species, involved in diseases, responsible of cytotoxic effect and in damaging membranes' lipids, are superoxide anion (O^{2-}), hydrogen peroxide (H_2O_2), and hydroxyl radical (OH^-) [117].

5.2. Antimicrobial Properties

The main phenolic compounds present in OMWW are those derived by oleuropein hydrolysis, as hydroxytyrosol, tyrosol and elenolic acid, but also other phenols: caffeic acid, p-coumaric acid, vanillic acid, syringic acid, gallic acid, luteolin, quercetin, cyanidin, verbascoside, and other polymeric compounds [10,11]. Marković et al. [118] demonstrated that hydroxytyrosol, tyrosol, oleuropein, and oleocanthal present a wide spectrum of biological effects on physiological processes, being antiatherogenic, cardioprotective, anticancer, neuroprotective, antidiabetic, anti-obesity compounds. Furneri and co-workers [119] revealed that oleuropein was also effective against *Mycoplasma fermentans* and *Mycoplasma hominis*, which are naturally resistant to erythromycin and often also to tetracycline. Biocompounds of olive products, such as aliphatic aldehydes [120], have also been shown to inhibit or retard the growth of a range of bacteria and yeasts and could be considered as an alternative for the prevention or treatment of infections. Moreover, they have been evaluated for drug formulations to reduce the spread of antimicrobial resistance bacteria [121]. Bisignano et al. [122] demonstrated that hydroxytyrosol possesses an in vitro antimicrobial property against respiratory and gastrointestinal infectious agents, such as *Vibrio parahaemolyticus*, *Vibrio cholerae*, Salmonella Typhi, *Haemophilus influenzae*, *Staphylococcus aureus*, and *Moraxella catarrhalis*, at low concentrations.

6. OMWW as Replacer of Synthetic Additives

The strong demand for adequate nutrition is accompanied by the concern for environmental pollution with a considerable emphasis on the recovery and recycling of food by-products and wastes [9].

Several studies have focused on replacing synthetic additives with natural substances, mainly derived from plants and agro-industry by-products [123,124] with promising results. The addition of such substances not only inhibits the growth of pathogens but also prolongs the shelf life of food products. OMWWs are added as such or as extracts, concentrated and stabilized and, in some cases, microencapsulated. Specifically, encapsulation protects them from degradation due to different factors reducing the amount of compounds required to be efficient and controlling their release into the food matrix [125].

Besides therapeutic benefits, biophenols present in OMWW have been explored for their antimicrobial, antifungal, and antiviral properties. Obied et al. [14] reported that the phenolic fraction of OMWW shows antibacterial activity against several species, particularly *S. aureus*, *Bacillus subtilis*, *Escherichia coli*, and *Pseudomonas aeruginosa*. However, the antimicrobial activity was found to be higher when the whole phenolic content is used, compared with the activity of the single phenolic compound [126]. In particular, Serra et al. [127] showed that natural OMWW extracts exhibited a higher antimicrobial activity compared with the three individual biophenols (quercetin, hydroxytyrosol, and oleuropein), suggesting a synergic effect among molecules. In most cases, to inhibit the growth of target strains, the effective tested dose was 1000 μg/mL. In addition, it has been shown that individual phenolic compounds, used at low concentrations, were not able to inhibit the growth of *E. coli*, *Klebsiella pneumoniae*, *S. aureus*, and *Staphylococcus pyogenes*, while whole OMWW was effective in inhibition of both Gram-positive and Gram-negative bacteria [121]. Other authors, however, reported that the bactericidal and fungicidal activities of OMWW are mainly related to the content of phenolic monomers, such as hydroxytyrosol and tyrosol [128]. Hydroxytyrosol was found to also be active against foodborne pathogens such as *Listeria monocytogenes*, *S. aureus*, Salmonella Enterica, and *Yersinia* spp. [129] and against beneficial microorganisms, such as *L. acidophilus* and *Bifidobacterium bifidum*. In addition, Fasolato et al. [130] confirmed the bactericidal effect of phenol extract purified from OMWWs. In particular, *S. aureus* and *L. monocytogenes* showed the lowest level of resistance (minimum bactericidal concentration MBC = 1.5–3.0 mg/mL) while Gram-negative bacteria (e.g., Salmonella Typhimurium and *Pseudomonas* spp.) showed higher resistance, with MBC values ranging from 6 to 12 mg/mL. In the same study, among the tested starter species, the growth of *Staphylococcus xylosus* and *L. curvatus* was drastically reduced (at concentrations of 0.75 and 1.5 mg/mL MBC, respectively).

Application of OMWW as Food Supplement

In several studies, olive oil by-products have been added as concentrates or ingredients in the formulation of novel foods in different agro-food supply chains (Table 2). In a review, Galanakis [131] collected data related to the addition of OMWW extracts (but also of other oil industrial by-products) to fortify meat and meat products. The results showed that the obtained antioxidants induce an improvement of hygienic conditions and rheological characteristics of the final products. Olive phenols have shown better performance in raw meat treatment [132] as they were able to hinder lipid oxidation. To evaluate such an effect, an oxidation test with a thiobarbituric acid reaction (TBAR) was applied for a storage period of 72 h at 4 °C. Results in limiting lipid oxidation appear to be dependent on the concentration of phenols (500 mg ascorbic acid or catechin/L and 100 mg olive phenols/L). Lopez et al. [133] and Veneziani et al. [111] recently applied OMWW-extracted polyphenols in fermented sausages and white meat burgers, improving quality parameters and extending their shelf life. In particular, the addition of the extracts inhibited the fungal growth and spore germination in fermented sausages by performing a dose- and species -dependent activity both in vitro and in situ tests. In particular, the treatment with 2.5% of OMWW extract strongly inhibited in situ growth of *Cladosporium cladosporioides*, *Penicillium aurantiogriseum*, *Penicillium commune*, and *Eurotium amstelodami*. Veneziani et al. [111] evaluated the effect of OMWW-extracted polyphenols in white meat burgers, wrapped in PVC, on improving sensorial and hygienic characteristics. The addition of the phenolic extract at different concentrations (0.75 and 1.50 g/kg) delayed the growth of mesophilic aerobic bacteria, highlighting a dose-dependent behavior, with a 24 h extension of shelf life, compared with both the control and sample treated with the lowest concentration. In addition, Fasolato et al. [130], according to Servili et al. [134], found that a 38.6 g/L concentration of phenolic extract was effective in increasing fresh chicken breast shelf life. Samples were immersed in a solution containing the extract for a few seconds, before packing and storage at 4 °C. The results showed a delay of growth of both Enterobacteriaceae and *Pseudomonas* spp. with at least a 2 day increase in shelf life, compared with the control. In addition, the treatments were shown to positively affect the odor of meat, decreasing the TBARS value. De Leonardis et al. [135] proposed the addition of lard with olive phenols as a "novel food", showing that the natural antioxidants of OMWW were highly effective in oxidative stabilization of lard. The phenol extract significantly increased the oxidative stability of lard, and the applied doses (100–200 ppm) were not cytotoxic when tested on mouse cell lines (embryonic fibroblasts). In addition, several studies have tested phenol extracts in dairy products to enhance antioxidant activity and better stabilize the products. Troise et al. [136] tested the antioxidant activity of OMWW phenolic extract in UHT milk samples, on inhibition of the Maillard reaction (MR), by adding phenolic extract at 0.1 and 0.05% *w/v*, revealing the reduction of reactive carbonyl species formation in samples before heat treatment, inducing a greater stability without any detrimental sensorial effects. Phenol extracts (100 and 200 mg/L) from OMWW have also been added in a functional milk drink (similar to yogurt) and fermented with a GABA-producing strain (*L. plantarum* C48) and a LAB strain of human origin (*L. paracasei* 15N). The results showed that the addition of phenolic compounds did not interfere with either the fermentation process or the activities of functional LAB [134]. The addition of extracts of both OMWW and olive pomace, at different concentrations (2, 4, 6, and 8 mg/100 g of butter) was tested in a butter formulation [137], revealing that the highest concentration confers resistance to oxidative stress during storage at 25 °C for 3 months, inhibiting the growth of *S. aureus*, total coliforms, yeast, and molds. Roila et al. [138] added biophenol extract (at 250 µg/mL and 500 µg/mL) to mozzarella cheese retarding the growth of *Pseudomonas fluorescens* and Enterobacteriaceae. The shelf life was directly proportional to the concentration, increasing by 2 and 4 days, respectively. Galanakis et al. [139] tested the antioxidant effect of OMWW phenolic extracts in combination with other antioxidants, demonstrating a reduction in oxidative deterioration during baking of bread and rusks and showing an antimicrobial effect against *S. aureus*, *B. subtilis*, *E. coli*, and *P. aeruginosa* (at 200 mg/Kg of flour). Recently, Cedola et al. [140] enriched bakery products by adding OMWW and OP previously subjected to ultrafiltration and evaluated the quality traits of final products from both a chemical and sensory point of view. Ultrafiltered OMWW, was used both in bread dough (1500 g of wheat flour, 900 g of OMWW, 45 g of fresh compressed yeast) and for the formulation of spaghetti at a final concentration of 30% *w/w*. The results showed that the addition of OMWW into bread and pasta slightly increased the chemical quality of bread and pasta without compromising their sensory traits. Zbakh et al. [141] proposed the exploitation of OMWW for setting up a functional beverage. Commercial products can include different additives, such as ascorbic acid as antioxidants, chelators including ethylenediaminetetraacetic acid (EDTA) and acidifiers, such as citric acid or carbon dioxide. The use of additional antioxidants was not required in beverages when OMWW extract were applied. Recently, a certain interest has developed in new beverages using aqueous extracts obtained with olive leaves, characterized by a high concentration of biophenols. Some of these products are already in the market and sold as integrators for human consumption. Further studies are required to investigate the effects of different formulations on the bioavailability of OMWW phenols and on their beneficial effects. These biological properties can have a significant impact on human health through reducing the incidence of many diseases, especially cardiovascular and chronic degenerative diseases.

As previously reported by Zbakh et al. [141], which confirmed that OMWW phenolic compounds are highly bioavailable and safe, the potential application of OMWW for setting up functional beverages as a natural concentrate of substances with antioxidant action could be a promising opportunity. To date, on the market there are beverages containing water extracts with different pharmacological indications: antioxidant, blood pressure regulator, and incidence on the metabolism of lipids and carbohydrates, although no reference legislation for the use of olive water for human consumption is currently available.

Table 2. Application of OMWW in agro-food chains.

Agro-Food Chain	Food	Quantity	Activity	Results	Reference
	Fermented sausages	2.5%	Antifungal	Inhibition of *C. cladosporioides, P. aurantiogriseum, P. commune,* and *E. amstelodami* growth	[133]
	White meat burgers	0.75–1.50 g/kg	Antimicrobial	Retarding the growth of aerobic mesophilic bacteria	[111]
	Lard	100–200 ppm	Natural antioxidant	Stabilization in oxidative of lard	[135]
	Meat	75 to 100 mg/L	Natural antioxidant	Extension of shelf life: color retaining, inhibition of microbial growth and fat deterioration	[132]
	Fresh chicken	38.6 g/L	Antimicrobial	Delay in Enterobacteriaceae and *Pseudomonas* spp. growth	[130,134]
	Milk	0.1–0.05% *w/v*	Functional ingredient	Increasing product stability	[136]
	Functional milk, fortified beverage	100–200 mg/L	Beverage fortification	Formulation of functional milk	[134]
	Butter	80 mg/kg	Natural antioxidants	Confering resistance to oxidative stress	[137]
	'Fior di latte' cheese	250–500 µg/mL	Antimicrobial	Increasing shelf life	[138]
	Bread and rusks	200 mg/Kg of flour	Antimicrobial and natural antioxidants	Inhibition of *S. aureus, B. subtilis, E. coli,* and *P. aeruginosa* growth and reducing oxidative deterioration during cooking	[131]
	Bread and pasta	900 g of OMWW (for bread) and 30% *w/w* (for pasta)	Antioxidant and food fortification	Food fortification: enhancing chemical composition without compromising the sensory characteristics	[140]

7. Conclusions and Future Perspectives

According to *The future of Food and Agriculture: trends and challenges* [142], about one third of all produced food is lost or wasted along the food chain, from production to consumption, highlighting an inefficiency of current food systems. The valorization and reuse of food by-products can create a virtuous recycling system in accordance with the Global Food 2030 objectives.

Although the two-phase extraction system is slowly replacing the traditional olive oil extraction techniques, the disposal of OMWW remains a problem for many small olive oil mills in Italy and in other Mediterranean countries and the valorization of such a by-product appears more important than ever for the agro-food industry.

The chance for agro-food companies to implement a circular economy strategy has offered new choices in by-product valorization. Despite several chemical characterizations of olive by-products, further searches are needed to fully understand the resources of such an interesting valuable raw material. Future olive oil waste management strategies should include a combination of physical and biotechnological processes, followed by further treatments, for producing valuable by-products with high functional activities. In this way, costs of treatments could be compensated by the income from useful by-products.

Author Contributions: Conceptualization, C.C. and C.L.R.; methodology, F.V.R., N.R. and A.P.; software, P.F. and N.R.; validation, C.L.R., C.C. and A.V.; investigation, P.F. and A.V.; resources, C.C. and C.L.R.; data curation, P.F., A.V. and A.P.; writing—original draft preparation, P.F. and F.V.R.; writing—review and editing, C.C. and C.L.R.; supervision, C.C. and F.V.R.; funding acquisition, C.C. All authors have read and agreed to the published version of the manuscript.

Funding: PON "RICERCA E INNOVAZIONE" 2014–2020, Azione II—Obiettivo Specifico 1b—Progetto "Miglioramento delle produzioni agroalimentari mediterranee in condizioni di carenza di risorse idriche—WATER4AGRIFOOD. Project number: ARS01_00825. CUP B64I20000160005.

Institutional Review Board Statement: Not applicable.

Informed Consent Statement: Not applicable.

Acknowledgments: This study was supported by "Miglioramento delle produzioni agroalimentari mediterranee in condizioni di carenza di risorse idriche"—WATER4AGRIFOOD-PON Ricerca e Innovazione 2014–2020 and conducted within a Ph.D. research program in Biotecnologie (XXXV cycle) by Paola Foti who received a grant "Dottorato innovativo con caratterizzazione industriale, PON RI 2014–2010", titled "Olive oil by-products as a new functional food and source of nutritional food ingredients" from the Department of Agriculture, Food and Environment (Scientific Tutor: C.C. Cinzia Caggia).

Conflicts of Interest: The authors declare no conflict of interest.

References

1. FaoStat. Food and Agriculture Organization of the United Nations. *Crops* **2020**. Available online: http://www.fao.org/faostat/en/?#data/QC (accessed on 1 January 2021).
2. Roig, A.; Cayuela, M.L.; Sanchez-Monedero, M. An overview on olive mill wastes and their valorisation methods. *Waste Manag.* **2006**, *26*, 960–969. [CrossRef]
3. Casa, R.; D'Annibale, A.; Pieruccetti, F.; Stazi, S.; Sermanni, G.G.; Cascio, B.L. Reduction of the phenolic components in olive-mill wastewater by an enzymatic treatment and its impact on durum wheat (*Triticum durum Desf.*) germinability. *Chemosphere* **2003**, *50*, 959–966. [CrossRef]
4. Rinaldi, M.; Rana, G.; Introna, M. Olive-mill wastewater spreading in southern Italy: Effects on a durum wheat crop. *Field Crop. Res.* **2003**, *84*, 319–326. [CrossRef]
5. Fernandez-Bolaños, J.; Rodríguez, G.; Rodríguez, R.; Guillén, R.; Jimenez-Araujo, A. Extraction of interesting organic compounds from olive oil waste. *Grasas y Aceites* **2006**, *57*, 95–106. [CrossRef]
6. Dermeche, S.; Nadour, M.; Larroche, C.; Moulti-Mati, F.; Michaud, P. Olive mill wastes: Biochemical characterizations and valorization strategies. *Process. Biochem.* **2013**, *48*, 1532–1552. [CrossRef]
7. EU Directive 2018/851 of the European Parliament and of the Council of 30 May 2018 Amending Directive 2008/98/EC on Waste —European Environment Agency. Available online: https://eur-lex.europa.eu/legal-content/EN/TXT/?uri=uriserv:OJ.L_.2018.150.01.0109.01.ENG (accessed on 1 January 2021).
8. DECRETO LEGISLATIVO 3 aprile 2006, n. 152 Norme in Materia Ambientale. (GU Serie Generale n.88 del 14-04-2006-Suppl. Ordinario n. 96). Available online: https://www.camera.it/parlam/leggi/deleghe/06152dl.htm (accessed on 1 January 2021).
9. Di Nunzio, M.; Picone, G.; Pasini, F.; Chiarello, E.; Caboni, M.F.; Capozzi, F.; Gianotti, A.; Bordoni, A. Olive oil by-product as functional ingredient in bakery products. Influence of processing and evaluation of biological effects. *Food Res. Int.* **2019**, *131*, 108940. [CrossRef] [PubMed]
10. D'Antuono, I.; Kontogianni, V.G.; Kotsiou, K.; Linsalata, V.; Logrieco, A.F.; Tasioula-Margari, M.; Cardinali, A. Polyphenolic characterization of olive mill wastewaters, coming from Italian and Greek olive cultivars, after membrane technology. *Food Res. Int.* **2014**, *65*, 301–310. [CrossRef]
11. Obied, H.K.; Prenzler, P.; Konczak, I.; Rehman, A.-U.; Robards, K. Chemistry and bioactivity of olive biophenols in some antioxidant and antiproliferative in vitro bioassays. *Chem. Res. Toxicol.* **2008**, *22*, 227–234. [CrossRef] [PubMed]
12. De Marco, E.; Savarese, M.; Paduano, A.; Sacchi, R. Characterization and fractionation of phenolic compounds extracted from olive oil mill wastewaters. *Food Chem.* **2007**, *104*, 858–867. [CrossRef]
13. El Abbassi, A.; Kiai, H.; Hafidi, A. Phenolic profile and antioxidant activities of olive mill wastewater. *Food Chem.* **2012**, *132*, 406–412. [CrossRef]
14. Obied, H.K.; Prenzler, P.; Robards, K. Potent antioxidant biophenols from olive mill waste. *Food Chem.* **2008**, *111*, 171–178. [CrossRef]
15. WOS. 2021. Available online: https://wcs.webofknowledge.com (accessed on 1 January 2021).
16. Paredes, C.; Cegarra, J.; Roig, A.; Sanchez-Monedero, M.; Bernal, M.P. Characterization of olive mill wastewater (alpechin) and its sludge for agricultural purposes. *Bioresour. Technol.* **1999**, *67*, 111–115. [CrossRef]
17. Peri, C. *The Extra-Virgin Olive Oil Handbook*; Peri, C., Ed.; John Wiley & Sons, Ltd.: Oxford, UK, 2014.
18. Rodis, P.S.; Karathanos, V.T.; Mantzavinou, A. Partitioning of olive oil antioxidants between oil and water phases. *J. Agric. Food Chem.* **2002**, *50*, 596–601. [CrossRef]
19. Akar, T.; Tosun, I.; Kaynak, Z.; Ozkara, E.; Yeni, O.; Sahin, E.N.; Akar, S.T. An attractive agro-industrial byproduct in environmental cleanup: Dye biosorption potential of untreated olive pomance. *J. Hazard. Mater.* **2009**, *166*, 1217–1225. [CrossRef]
20. Baeta-Hall, L.; Sàágua, M.C.; Bartolomeu, M.L.; Anselmo, A.M.; Rosa, M.F. Biodegradation of olive oil husks in composting aerated piles. *Bioresour. Technol.* **2005**, *96*, 69–78. [CrossRef]
21. Pisante, M.; Inglese, P.; Lercker, G. *L'ulivo e l'olio*; Cultura&Cultura, Ed.; Bayer Crop Science: Bologna, Italy, 2009; pp. 691–692.
22. Galiatsatou, P.; Metaxas, M.; Arapoglou, D.; Kasselouri-Rigopoulou, V. Treatment of olive mill waste water with activated carbons from agriculturalby-products. *Waste Manag.* **2002**, *22*, 803–812. [CrossRef]
23. Al-Malah, K.; Azzam, M.O.; Abu-Lail, N.I. Olive mills effluent (OME) wastewater post-treatment using activated clay. *Sep. Purif. Technol.* **2000**, *20*, 225–234. [CrossRef]

24. Niaounakis, M.; Halvadakis, C.P. *Olive Processing Waste Management: Literature Review and Patent Survey*, 2nd ed.; Elsevier Ltd.: Oxford, UK, 2006.

25. Caporaso, N.; Formisano, D.; Genovese, A. Use of phenolic compounds from olive mill wastewater as valuable ingredients for functional foods. *Crit. Rev. Food Sci. Nutr.* **2017**, *58*, 2829–2841. [CrossRef] [PubMed]

26. Sierra, J.; Martí, E.; Montserrat, G.; Cruañas, R.; Garau, M. Characterisation and evolution of a soil affected by olive oil mill wastewater disposal. *Sci. Total. Environ.* **2001**, *279*, 207–214. [CrossRef]

27. Vlyssides, A.; Loizidou, M.; Zorpas, A.A. Characteristics of solid residues from olive oil processing as bulking material for co?composting with industrial wastewaters. *J. Environ. Sci. Health Part A* **1999**, *34*, 737–748. [CrossRef]

28. Martin-Garcia, A.I.; Moumen, A.; Ruiz, D.Y.; Alcaide, E.M. Chemical composition and nutrients availability for goats and sheep of two-stage olive cake and olive leaves. *Anim. Feed Sci. Technol.* **2003**, *107*, 61–74. [CrossRef]

29. Lafka, T.-I.; Lazou, A.E.; Sinanoglou, V.J.; Lazos, E.S. Phenolic and antioxidant potential of olive oil mill wastes. *Food Chem.* **2011**, *125*, 92–98. [CrossRef]

30. Di Giovacchino, L.; Costantini, N.; Serraiocco, A.; Surrichio, G.; Basti, C. Natural antioxidants and volatile compounds of virgin olive oils obtained by the two or three-phase centrifugal decanters. *Eur. J. Lipid Sci. Technol.* **2001**, *103*, 279–285. [CrossRef]

31. Garcia-Castello, E.M.; Cassano, A.; Criscuoli, A.; Conidi, C.; Drioli, E. Recovery and concentration of polyphenols from olive mill wastewaters by integrated membrane system. *Water Res.* **2010**, *44*, 3883–3892. [CrossRef]

32. García García, I.; Jiménez Peña, P.R.; Bonilla Venceslada, J.L.; Martín Martín, A.; Martín Santos, M.A.; Ramos Gómez, E. Removal of phenol compounds from olive mill wastewater using *Phanerochaete chrysosporium, Aspergillus niger, Aspergillus terreus* and *Geotrichum Candidum*. *Process Biochem.* **2000**, *35*, 751–758. [CrossRef]

33. Vlyssides, A.; Loizides, M.; Karlis, P. Integrated strategic approach for reusing olive oil extraction by-products. *J. Clean. Prod.* **2004**, *12*, 603–611. [CrossRef]

34. Caputo, A.C.; Scacchia, F.; Pelagagge, P.M. Disposal of by-products in olive oil industry: Waste-to-energy solutions. *Appl. Therm. Eng.* **2003**, *23*, 197–214. [CrossRef]

35. Alburquerque, J.A.; Gonzálvez, J.; García, D.; Cegarra, J. Agrochemical characterisation of "alperujo", a solid by-product of the two-phase centrifugation method for olive oil extraction. *Bioresour. Technol.* **2004**, *91*, 195–200. [CrossRef]

36. Azbar, N.; Bayram, A.; Filibeli, A.; Müezzinoğlu, A.; Sengul, F.; Ozer, A. A Review of waste management options in olive oil production. *Crit. Rev. Environ. Sci. Technol.* **2004**, *34*, 209–247. [CrossRef]

37. Fiestas, R.; De Ursinos, J.A.; Borja-Padilla, R. Biomethanization. *Int. Biodeter. Biodegr.* **1996**, *38*, 145–153. [CrossRef]

38. Khoufi, S.; Hamza, M.; Sayadi, S. Enzymatic hydrolysis of olive wastewater for hydroxytyrosol enrichment. *Bioresour. Technol.* **2011**, *102*, 9050–9058. [CrossRef]

39. Tsagaraki, E.; Lazarides, H.N.; Petrotos, K. Olive mill wastewater treatment. In *Utilization of By-Products and Treatment of Waste in the Food Industry*; Oreopoulou, V., Russ, V., Eds.; Springer: Boston, MA, USA, 2007; pp. 133–157. [CrossRef]

40. Yangui, T.; Dhouib, A.; Rhouma, A.; Sayadi, S. Potential of hydroxytyrosol-rich composition from olive mill wastewater as a natural disinfectant and its effect on seeds vigour response. *Food Chem.* **2009**, *117*, 1–8. [CrossRef]

41. Obied, H.; Allen, M.S.; Bedgood, D.; Prenzler, P.; Robards, K.; Stockmann, R. Bioactivity and analysis of biophenols recovered from olive mill waste. *J. Agric. Food Chem.* **2005**, *53*, 823–837. [CrossRef]

42. Kavroulakis, N.; Ntougias, S. Bacterial and β-proteobacterial diversity in *Olea europaea* var. *mastoidis-* and *O. europaea* var. *koroneiki*-generated olive mill wastewaters: Influence of cultivation and harvesting practice on bacterial community structure. *World J. Microbiol. Biotechnol.* **2010**, *27*, 57–66. [CrossRef]

43. Tsiamis, G.; Tzagkaraki, G.; Chamalaki, A.; Xypteras, N.; Andersen, G.; Vayenas, D.; Bourtzis, K. Olive-mill wastewater bacterial communities display a cultivar specific profile. *Curr. Microbiol.* **2011**, *64*, 197–203. [CrossRef]

44. Ben Sassi, A.; Ouazzani, N.; Walker, G.M.; Ibnsouda, S.; El Mzibri, M.; Boussaid, A. Detoxification of olive mill wastewater by Maroccan yeast isolates. *Biodegradation* **2008**, *19*, 337–3461. [CrossRef]

45. Bleve, G.; Lezzi, C.; Chiriatti, M.; D'Ostuni, I.; Tristezza, M.; Di Venere, D.; Sergio, L.; Mita, G.; Grieco, F. Selection of non-conventional yeasts and their use in immobilized form for the bioremediation of olive oil mill wastewaters. *Bioresour. Technol.* **2011**, *102*, 982–989. [CrossRef] [PubMed]

46. Ntougias, S.; Bourtzis, K.; Tsiamis, G. The microbiology of olive mill wastes. *BioMed Res. Int.* **2013**, *2013*, 1–16. [CrossRef] [PubMed]

47. El-Abbassi, A.; Saadaoui, N.; Kiai, H.; Raiti, J.; Hafidi, A. Potential applications of olive mill wastewater as biopesticide for crops protection. *Sci. Total. Environ.* **2017**, *576*, 10–21. [CrossRef] [PubMed]

48. Venieri, D.; Rouvalis, A.; Iliopoulou-Georgudaki, J. Microbial and toxic evaluation of raw and treated olive oil mill wastewaters. *J. Chem. Technol. Biotechnol.* **2010**, *85*, 1380–1388. [CrossRef]

49. EC Directive, (2000), Directive 2000/60/EC of the European Parliament and of the Council of 23 October 2000 Establishing a Framework for Community Action in the Field of Water Policy, Official Journal, L 327, 22.12.2000. Available online: https://eur-lex.europa.eu/legal-content/EN/TXT/HTML/?uri=LEGISSUM:l28002b (accessed on 1 January 2021).

50. Said-Pullicino, D.; Gigliotti, G. Oxidative biodegradation of dissolved organic matter during composting. *Chemosphere* **2007**, *68*, 1030–1040. [CrossRef]

51. Ahmed, P.; Fernández, P.M.; Figueroa, L.I.C.; Pajot, H. Exploitation alternatives of olive mill wastewater: Production of value-added compounds useful for industry and agriculture. *Biofuel Res. J.* **2019**, *6*, 980–994. [CrossRef]

52. Paraskeva, C.; Papadakis, V.; Tsarouchi, E.; Kanellopoulou, D.; Koutsoukos, P. Membrane processing for olive mill wastewater fractionation. *Desalination* **2007**, *213*, 218–229. [CrossRef]
53. Adhoum, N.; Monser, L. Decolourization and removal of phenolic compounds from olive mill wastewater by electrocoagulation. *Chem. Eng. Process. Process. Intensif.* **2004**, *43*, 1281–1287. [CrossRef]
54. Ochando-Pulido, J.; Pimentel-Moral, S.; Verardo, V.; Martinez-Ferez, A. A focus on advanced physico-chemical processes for olive mill wastewater treatment. *Sep. Purif. Technol.* **2017**, *179*, 161–174. [CrossRef]
55. Cermola, F.; DellaGreca, M.; Iesce, M.; Montella, S.; Pollio, A.; Temussi, F. A mild photochemical approach to the degradation of phenols from olive oil mill wastewater. *Chemosphere* **2004**, *55*, 1035–1041. [CrossRef]
56. Belaqziz, M.; El-Abbassi, A.; Lakhal, E.K.; Agrafioti, E.; Galanakis, C.M. Agronomic application of olive mill wastewater: Effects on maize production and soil properties. *J. Environ. Manag.* **2016**, *171*, 158–165. [CrossRef]
57. Hamdi, M. Effects of agitation and pretreatment on the batch anaerobic digestion of olive mill wastewater. *Bioresour. Technol.* **1991**, *36*, 173–178. [CrossRef]
58. Hachicha, S.; Cegarra, J.; Sellami, F.; Hachicha, R.; Drira, N.; Medhioub, K.; Ammar, E. Elimination of polyphenols toxicity from olive mill wastewater sludge by its co-composting with sesame bark. *J. Hazard. Mater.* **2009**, *161*, 1131–1139. [CrossRef]
59. Koutsos, T.; Chatzistathis, T.; Balampekou, E. A new framework proposal, towards a common EU agricultural policy, with the best sustainable practices for the re-use of olive mill wastewater. *Sci. Total. Environ.* **2018**, *622–623*, 942–953. [CrossRef]
60. Hamdi, M. Anaerobic digestion of olive mill wastewaters. *Process. Biochem.* **1996**, *31*, 105–110. [CrossRef]
61. Filidei, S.; Masciandaro, G.; Ceccanti, B. Anaerobic digestion of olive oil mill effluents: Evaluation of wastewater organic load and phytotoxicity reduction. *Water Air Soil Pollut.* **2003**, *145*, 79–94. [CrossRef]
62. Tomati, U.; Galli, E.; Di Lena, G.; Buffone, R. Induction of laccase in *Pleurotus ostreatus* mycelium grown in olive oil waste waters. *Agrochimica* **1991**, *35*, 273–279.
63. Ramos-Cormenzana, A.; Juarez-Jimenez, B.; García-Pareja, M.P. Antimicrobial activity of olive mill wastewaters (alpechin) and biotransformed olive oil mill wastewater. *Int. Biodeter. Biodegr.* **1996**, *38*, 283–290. [CrossRef]
64. Scioli, C.; Vollaro, L. The use of *Yarrowia lipolytica* to reduce pollution in olive mill wastewaters. *Water Res.* **1997**, *31*, 2520–2524. [CrossRef]
65. Sinigaglia, M.; Di Benedetto, N.; Bevilacqua, A.; Corbo, M.R.; Capece, A.; Romano, P. Yeasts isolated from olive mill wastewaters from southern Italy: Technological characterization and potential use for phenol removal. *Appl. Microbiol. Biotechnol.* **2010**, *87*, 2345–2354. [CrossRef] [PubMed]
66. Ehaliotis, C.; Papadopoulou, K.; Kotsou, M.; Mari, I.; Balis, C. Adaptation andpopulation dynamics of *Azotobacter vinelandii* during aerobic biological treatment of olive-mill wastewater. *FEMS Microbiol. Ecol.* **1999**, *30*, 301–311. [CrossRef] [PubMed]
67. Piperidou, C.I.; Chiadou, C.I.; Stalikas, C.D. Bioremediation of olive oil millwastewater: Chemical alterations induced by *Azotobacter vinelandii*. *J. Agric. Food Chem.* **2000**, *48*, 1942–1948. [CrossRef] [PubMed]
68. Cerrone, F.; Barghini, P.; Pesciaroli, C.; Fenice, M. Efficient removal of pollutants from olive washing wastewater in bubble-column bioreactor by *Trametes versicolor*. *Chemosphere* **2011**, *84*, 254–259. [CrossRef] [PubMed]
69. Maza-Márquez, P.; Gonzalez-Martinez, A.; Martínez-Toledo, M.V.; Fenice, M.; Lasserrot, A.; González-López, J. Biotreatment of industrial olive washing water by synergetic association of microalgal-bacterial consortia in a photobioreactor. *Environ. Sci. Pollut. Res.* **2016**, *24*, 527–538. [CrossRef]
70. Aissam, H.; Penninckx, M.J.; Benlemlih, M. Reduction of phenolics content and COD in olive oil mill wastewaters by indigenous yeasts and fungi. *World J. Microbiol. Biotechnol.* **2007**, *23*, 1203–1208. [CrossRef]
71. Lamia, A.; Moktar, H. Fermentative decolorization of olive mill wastewater by *Lactobacillus plantarum*. *Process. Biochem.* **2003**, *39*, 59–65. [CrossRef]
72. Romo-Sánchez, S.; Alves-Baffi, M.; Arévalo-Villena, M.; Úbeda-Iranzo, J.; Briones-Pérez, A. Yeast biodiversity from oleic ecosystems: Study of their biotechnological properties. *Food Microbiol.* **2010**, *27*, 487–492. [CrossRef] [PubMed]
73. Abdelhadi, B.S.; Benlemilih, M.; Koraichi, S.I.; Ahansal, L.; Hammoumi, A.; Boussaid, A. Suitability of yeasts for the treatment of olive mill wastewater. *Aquat. Environ. Toxicol.* **2010**, *4*, 113–117.
74. Giannoutsou, E.; Meintanis, C.; Karagouni, A. Identification of yeast strains isolated from a two-phase decanter system olive oil waste and investigation of their ability for its fermentation. *Bioresour. Technol.* **2004**, *93*, 301–306. [CrossRef] [PubMed]
75. Mann, J.; Markham, J.L.; Peiris, P.; Nair, N.; Spooner-Hart, R.N.; Holford, P. Screening and selection of fungi for bioremediation of olive mill wastewater. *World J. Microbiol. Biotechnol.* **2009**, *26*, 567–571. [CrossRef]
76. Millán, B.; Lucas, R.; Robles, A.; García, T.; De Cienfuegos, G.A.; Gálvez, A. A study on the microbiota from olive-mill wastewater (OMW) disposal lagoons, with emphasis on filamentous fungi and their biodegradative potential. *Microbiol. Res.* **2000**, *155*, 143–147. [CrossRef]
77. Fenice, M.; Giovannozzisermanni, G.; Federici, F.; D'Annibale, A. Submerged and solid-state production of laccase and Mn-peroxidase by *Panus tigrinus* on olive mill wastewater-based media. *J. Biotechnol.* **2003**, *100*, 77–85. [CrossRef]
78. Koutrotsios, G.; Larou, E.; Mountzouris, K.C.; Zervakis, G.I. Detoxification of olive mill wastewater and bioconversion of olive crop residues into high-value-added biomass by the choice edible mushroom *Hericium erinaceus*. *Appl. Biochem. Biotechnol.* **2016**, *180*, 195–209. [CrossRef]
79. Zerva, A.; Zervakis, G.; Christakopoulos, P.; Topakas, E. Degradation of olive mill wastewater by the induced extracellular ligninolytic enzymes of two wood-rot fungi. *J. Environ. Manag.* **2017**, *203*, 791–798. [CrossRef] [PubMed]

80. Crognale, S.; D'Annibale, A.; Federici, F.; Fenice, M.; Quaratino, D.; Petruccioli, M. Olive oil mill wastewater valorisation by fungi. *J. Chem. Technol. Biotechnol.* **2006**, *81*, 1547–1555. [CrossRef]

81. Cordova, J.; Nemmaoui, M.; Ismaili-Alaoui, M.; Morin, A.; Roussos, S.; Raimbault, M.; Benjilali, B. Lipase production by solid state fermentation of olive cake and sugar cane bagasse. *J. Mol. Catal. B Enzym.* **1998**, *5*, 75–78. [CrossRef]

82. Federici, F. Production, purification and partial characterization of an endopolygalacturonase from *Cryptococcus albidus* var. *albidus. Antonie van Leeuwenhoek* **1985**, *51*, 139–150. [CrossRef] [PubMed]

83. Federici, F.; Montedoro, G.; Servili, M.; Petruccioli, M. Pectic enzyme production by *Cryptococcus albidus* var. *albidus* on olive vegetation waters enriched with sunflower calathide meal. *Biol. Wastes* **1988**, *25*, 291–301. [CrossRef]

84. Petruccioli, M.; Servili, M.; Montedoro, G.F.; Federici, F. Development of a recycle procedure for the utilisation of vegetation waters in the olive-oil extraction process. *Biotechnol. Lett.* **1988**, *10*, 55–60. [CrossRef]

85. Nadour, M.; Laroche, C.; Pierre, G.; Delattre, C.; Moulti-Mati, F.; Michaud, P. Structural characterization and biological activities of polysaccharides from olive mill wastewater. *Appl. Biochem. Biotechnol.* **2015**, *177*, 431–445. [CrossRef] [PubMed]

86. Petri, D. Xanthan gum: A versatile biopolymer for biomedical and technological applications. *J. Appl. Polym. Sci.* **2015**, *132*. [CrossRef]

87. Lopez, M.; Ramos-Cormenzana, A. Xanthan production from olive-mill wastewaters. *Int. Biodeterior. Biodegrad.* **1996**, *38*, 263–270. [CrossRef]

88. Morillo, J.A.; Del Águila, V.G.; Aguilera, M.; Ramos-Cormenzana, A.; Monteoliva-Sánchez, M. Production and characterization of the exopolysaccharide produced by *Paenibacillus jamilae* grown on olive mill-waste waters. *World J. Microbiol. Biotechnol.* **2007**, *23*, 1705–1710. [CrossRef] [PubMed]

89. Ruiz-Bravo, A.; Jimenez-Valera, M.; Moreno, E.; Guerra, V.; Ramos-Cormenzana, A. Biological response modifier activity of an exopolysaccharide from *Paenibacillus jamilae* CP-7. *Clin. Diagn. Lab. Immunol.* **2001**, *8*, 706–710. [CrossRef]

90. Hill, J. Environmental costs and benefits of transportation biofuel production from food-and lignocellulose-based energy crops: A Review. In *Sustainable Agriculture*; Lichtfouse, E., Navarrete, M., Debaeke, P., Véronique, S., Alberola, C., Eds.; Springer: Dordrecht, The Netherlands, 2009; pp. 125–139.

91. Moraes, B.; Zaiat, M.; Bonomi, A. Anaerobic digestion of vinasse from sugarcane ethanol production in Brazil: Challenges and perspectives. *Renew. Sustain. Energy Rev.* **2015**, *44*, 888–903. [CrossRef]

92. Lercker, G. Reflui oleari, i metodi per renderli "buoni". *Olivo E Olio* **2014**, *9*, 37–41.

93. Borja, R.; Alba, J.; Garrido, S.E.; Martinez, L.; Garcia, M.P.; Incerti, C.; Ramos-Cormenzana, A. Comparative study of anaerobic digestion of olive mill wastewater (OMW) and OMW previously fermented with *Aspergillus terreus. Bioprocess. Eng.* **1995**, *13*, 317–322. [CrossRef]

94. Massadeh, M.I.; Modallal, N. Ethanol production from olive mill wastewater (OMW) pretreated with *Pleurotus sajor-caju. Energy Fuels* **2007**, *22*, 150–154. [CrossRef]

95. Sarris, D.; Matsakas, L.; Aggelis, G.; Koutinas, A.A.; Papanikolaou, S. Aerated *vs* non-aerated conversions of molasses and olive mill wastewaters blends into bioethanol by *Saccharomyces cerevisiae* under non-aseptic conditions. *Ind. Crop. Prod.* **2014**, *56*, 83–93. [CrossRef]

96. Nikolaou, A.; Kourkoutas, Y. Exploitation of olive oil mill wastewaters and molasses for ethanol production using immobilized cells of *Saccharomyces cerevisiae. Environ. Sci. Pollut. Res.* **2017**, *25*, 7401–7408. [CrossRef]

97. Heinrich, A.; Zanichelli, D.; Carloni, F.; Hasanaji, E.; D'Andrea, N.; Filippini, A.; Setti, L. Production of ethanol by an integrated valorization of olive oil byproducts. The role of phenolic inhibition (2 pp). *Environ. Sci. Pollut. Res.* **2006**, *14*, 5–6. [CrossRef]

98. Taccari, M.; Ciani, M. Use of *Pichia fermentans* and *Candida* sp. strains for the biological treatment of stored olive mill wastewater. *Biotechnol. Lett.* **2011**, *33*, 2385–2390. [CrossRef]

99. Sarris, D.; Stoforos, N.G.; Mallouchos, A.; Kookos, I.; Koutinas, A.A.; Aggelis, G.; Papanikolaou, S. Production of added-value metabolites by *Yarrowia lipolytica* growing in olive mill wastewater-based media under aseptic and non-aseptic conditions. *Eng. Life Sci.* **2017**, *17*, 695–709. [CrossRef]

100. Berbel, J.; Posadillo, A. Opportunities for the bioeconomy of olive oil byproducts. *Biomed. J. Sci. Tech. Res.* **2018**, *2*, 2094–2096.

101. Makri, S.; Kafantaris, I.; Savva, S.; Ntanou, P.; Stagos, D.; Argyroulis, I.; Kotsampasi, B.; Christodoulou, V.; Gerasopoulos, K.; Petrotos, K.; et al. Novel feed including olive oil mill wastewater bioactive compounds enhanced the redox status of lambs. *In Vivo* **2018**, *32*, 291–302. [CrossRef]

102. Gerasopoulos, K.; Stagos, D.; Petrotos, K.; Kokkas, S.; Kantas, D.; Goulas, P.; Kouretas, D. Feed supplemented with polyphenolic byproduct from olive mill wastewater processing improves the redox status in blood and tissues of piglets. *Food Chem. Toxicol.* **2015**, *86*, 319–327. [CrossRef]

103. Varricchio, E.; Coccia, E.; Orso, G.; Lombardi, V.; Imperatore, R.; Vito, P.; Paolucci, M. Influence of polyphenols from olive mill wastewater on the gastrointestinal tract, alveolar macrophages and blood leukocytes of pigs. *Ital. J. Anim. Sci.* **2019**, *18*, 574–586. [CrossRef]

104. Gerasopoulos, K.; Stagos, D.; Kokkas, S.; Petrotos, K.; Kantas, D.; Goulas, P.; Kouretas, D. Feed supplemented with byproducts from olive oil mill wastewater processing increases antioxidant capacity in broiler chickens. *Food Chem. Toxicol.* **2015**, *82*, 42–49. [CrossRef]

105. Ciriminna, R.; Meneguzzo, F.; Fidalgo, A.; Ilharco, L.M.; Pagliaro, M. Extraction, benefits and valorization of olive polyphenols. *Eur. J. Lipid Sci. Technol.* **2015**, *118*, 503–511. [CrossRef]

106. Food and Drug Administration (FDA). 2014. Available online: http://www.accessdata.fda.gov/scripts/fdcc/?set=GRASNotices&id=459 (accessed on 1 January 2021).
107. Galanakis, C.M. *A Study for the Implementation of Polyphenols from Olive Mill Wastewater in Foodstuff and Cosmetics*; General Secretariat for Research and Technology (GSRT): Athens, Greece, 2015.
108. Galanakis, C.M. Recovery of high added-value components from food wastes: Conventional, emerging technologies and commercialized applications. *Trends Food Sci. Technol.* **2012**, *26*, 68–87. [CrossRef]
109. Rahmanian, N.; Jafari, S.M.; Galanakis, C.M. Recovery and removal of phenolic compounds from olive mill wastewater. *J. Am. Oil Chem. Soc.* **2013**, *91*, 1–18. [CrossRef]
110. Galanakis, C.M.; Schieber, A. Editorial. Special issue on recovery and utilization of valuable compounds from food processing by-products. *Food Res. Int.* **2014**, *65*, 299–300. [CrossRef]
111. Veneziani, G.; Novelli, E.; Esposto, S.; Taticchi, A.; Servili, M. Applications of recovered bioactive compounds in food products. In *Olive Mill Waste*; Academic Press: Cambridge, MA, USA, 2017; pp. 231–253. [CrossRef]
112. Pellegrini, N.; Serafini, M.; Colombi, B.; Del Rio, D.; Salvatore, S.; Bianchi, M.; Brighenti, F. Total antioxidant capacity of plant foods, beverages and oils consumed in Italy assessed by three different in vitro assays. *J. Nutr.* **2003**, *133*, 2812–2819. [CrossRef] [PubMed]
113. Meydani, M. Nutrition, immune cells, and atherosclerosis. *Nutr. Rev.* **1998**, *56*, S177–S182. [CrossRef]
114. Visioli, F.; Davalos, A.; Hazas, M.D.C.L.D.L.; Crespo, M.D.C.; Tomé-Carneiro, J. An overview of the pharmacology of olive oil and its active ingredients. *Br. J. Pharmacol.* **2019**, *177*, 1316–1330. [CrossRef]
115. European Food Safety Authority (EFSA). Scientific Opinion on the substantiation of health claims related to polyphenols in olive and protection of LDL particles from oxidative damage (ID **1333**, *1638*, 1639, 1696, 2865), maintenance of normal blood HDL cholesterol concentrations (ID 1639), maintenance of normal blood pressure (ID 3781), "anti-inflammatory properties" (ID 1882), "contributes to the upper respiratory tract health" (ID 3468), "can help to maintain a normal function of gastrointestinal tract" (3779), and "contributes to body defences against external agents" (ID 3467) pursuant to Article 13(1) of Regulation (EC) No 1924/2006. *EFSA J.* **2011**, *9*, 2033.
116. Aissa, I.; Kharrat, N.; Aloui, F.; Sellami, M.; Bouaziz, M.; Gargouri, Y. Valorization of antioxidants extracted from olive mill wastewater. *Biotechnol. Appl. Biochem.* **2017**, *64*, 579–589. [CrossRef]
117. Girotti, A.W. Lipid hydroperoxide generation, turnover, and effector action in biological systems. *J. Lipid Res.* **1998**, *39*, 1529–1542. [CrossRef]
118. Marković, A.K.; Torić, J.; Barbarić, M.; Brala, C.J. Hydroxytyrosol, tyrosol and derivatives and their potential effects on human health. *Molecules* **2019**, *24*, 2001. [CrossRef] [PubMed]
119. Furneri, P.M.; Piperno, A.; Sajia, A.; Bisignano, G. Antimycoplasmal activity of hydroxytyrosol. *Antimicrob. Agents Chemother.* **2004**, *48*, 4892–4894. [CrossRef]
120. Boudet, A.-M. Evolution and current status of research in phenolic compounds. *Phytochemistry* **2007**, *68*, 2722–2735. [CrossRef] [PubMed]
121. Tafesh, A.; Najami, N.; Jadoun, J.; Halahlih, F.; Riepl, H.; Azaizeh, H. Synergistic antibacterial effects of polyphenolic compounds from olive mill wastewater. *Evid. Based Complement. Altern. Med.* **2011**, *2011*, 1–9. [CrossRef] [PubMed]
122. Bisignano, G.; Tomaino, A.; Cascio, R.L.; Crisafi, G.; Uccella, N.; Saija, A. On the *in-vitro* antimicrobial activity of oleuropein and hydroxytyrosol. *J. Pharm. Pharmacol.* **1999**, *51*, 971–974. [CrossRef] [PubMed]
123. Farag, R.S.; Mahmoud, E.A.; Basuny, A. Use crude olive leaf juice as a natural antioxidant for the stability of sunflower oil during heating. *Int. J. Food Sci. Technol.* **2007**, *42*, 107–115. [CrossRef]
124. Jaber, H.; Ayadi, M.; Makni, J.; Rigane, G.; Sayadi, S.; Bouaziz, M. Stabilization of refined olive oil by enrichment with chlorophyll pigments extracted from Chemlali olive leaves. *Eur. J. Lipid Sci. Technol.* **2012**, *114*, 1274–1283. [CrossRef]
125. Mohammadi, A.; Jafari, S.M.; Assadpour, E.; Esfanjani, A.F. Nano-encapsulation of olive leaf phenolic compounds through WPC–pectin complexes and evaluating their release rate. *Int. J. Biol. Macromol.* **2016**, *82*, 816–822. [CrossRef]
126. Obied, H.K.; Karuso, P.; Prenzler, P.D.; Robards, K. Novel secoiridoids with antioxidant activity from Australian olive mill waste. *J. Agric. Food Chem.* **2007**, *55*, 2848–2853. [CrossRef]
127. Serra, A.T.; Matias, A.; Nunes, A.V.M.; Leitão, M.; Brito, D.; Bronze, M.; Silva, S.; Pires, A.; Crespo, M.T.B.; Romão, M.S.; et al. In vitro evaluation of olive- and grape-based natural extracts as potential preservatives for food. *Innov. Food Sci. Emerg. Technol.* **2008**, *9*, 311–319. [CrossRef]
128. Hazas, M.D.C.L.D.L.; Piñol, C.; Macià, A.; Romero-Fabregat, M.-P.; Pedret, A.; Solà, R.; Rubió, L.; Motilva, M.J. Differential absorption and metabolism of hydroxytyrosol and its precursors oleuropein and secoiridoids. *J. Funct. Foods* **2016**, *22*, 52–63. [CrossRef]
129. Medina, E.; De Castro, A.; Romero, C.; Brenes, M. Comparison of the concentrations of phenolic compounds in olive oils and other plant oils: Correlation with antimicrobial activity. *J. Agric. Food Chem.* **2006**, *54*, 4954–4961. [CrossRef]
130. Fasolato, L.; Cardazzo, B.; Balzan, S.; Carraro, L.; Taticchi, A.; Montemurro, F.; Novelli, E. Minimum bactericidal concentration of phenols extracted from oil vegetation water on spoilers, starters and food-borne bacteria. *Ital. J. Food Saf.* **2015**, *4*. [CrossRef]
131. Galanakis, C.M. Phenols recovered from olive mill wastewater as additives in meat products. *Trends Food Sci. Technol.* **2018**, *79*, 98–105. [CrossRef]

132. Barbier, C. The Behaviour of Some Antioxidants in an In Vitro Meat Model. Master's Thesis, Department of Food Technology, Engineering and Nutrition, Lund University, Lund, Sweden, 2009.
133. Lopez, C.C.; Serio, A.; Mazzarrino, G.; Martuscelli, M.; Scarpone, E.; Paparella, A. Control of household mycoflora in fermented sausages using phenolic fractions from olive mill wastewaters. *Int. J. Food Microbiol.* **2015**, *207*, 49–56. [CrossRef] [PubMed]
134. Servili, M.; Rizzello, C.G.; Taticchi, A.; Esposto, S.; Urbani, S.; Mazzacane, F.; Di Maio, I.; Selvaggini, R.; Gobbetti, M.; Di Cagno, R. Functional milk beverage fortified with phenolic compounds extracted from olive vegetation water, and fermented with c-aminobutyric acid (GABA)-producing and potential probiotic lactic acid bacteria. *Int. J. Food Microbiol.* **2011**, *147*, 45–52. [CrossRef]
135. De Leonardis, A.; Macciola, V.; Lembo, G.; Aretini, A.; Nag, A. Studies on oxidative stabilisation of lard by natural antioxidants recovered from olive-oil mill wastewater. *Food Chem.* **2007**, *100*, 998–1004. [CrossRef]
136. Troise, A.D.; Fiore, A.; Colantuono, A.; Kokkinidou, S.; Peterson, D.G.; Fogliano, V. Effect of olive mill wastewater phenol compounds on reactive carbonyl species and Maillard reaction end-products in ultrahigh-temperature-treated milk. *J. Agric. Food Chem.* **2014**, *62*, 10092–10100. [CrossRef] [PubMed]
137. Mikdame, H.; Kharmach, E.; Mtarfi, N.E.; Alaoui, K.; Ben Abbou, M.; Rokni, Y.; Majbar, Z.; Taleb, M.; Rais, Z. By-products of olive oil in the service of the deficiency of food antioxidants: The case of butter. *J. Food Qual.* **2020**, *2020*, 1–10. [CrossRef]
138. Roila, R.; Valiani, A.; Ranucci, D.; Ortenzi, R.; Servili, M.; Veneziani, G.; Branciari, R. Antimicrobial efficacy of a polyphenolic extract from olive oil by-product against "Fior di latte" cheese spoilage bacteria. *Int. J. Food Microbiol.* **2019**, *295*, 49–53. [CrossRef] [PubMed]
139. Galanakis, C.M.; Tsatalas, P.; Charalambous, Z.; Galanakis, I.M. Control of microbial growth in bakery products fortified with polyphenols recovered from olive mill wastewater. *Environ. Technol. Innov.* **2018**, *10*, 1–15. [CrossRef]
140. Cedola, A.; Cardinali, A.; D'Antuono, I.; Conte, A.; Del Nobile, M.A. Cereal foods fortified with by-products from the olive oil industry. *Food Biosci.* **2019**, *33*, 100490. [CrossRef]
141. Zbakh, H.; El Abbassi, A. Potential use of olive mill wastewater in the preparation of functional beverages: A review. *J. Funct. Foods* **2012**, *4*, 53–65. [CrossRef]
142. Food and Agriculture Organization. *The Future of Food and Agriculture, Trends and Challenges*; Food and Agriculture Organization: Rome, Italy, 2017; Volume 4, ISBN 9789251095515.

Review

Applications of Plant Polymer-Based Solid Foams: Current Trends in the Food Industry

Marcela Jarpa-Parra [1,*] and Lingyun Chen [2]

1   Research Direction, Universidad Adventista de Chile, Casilla 7-D, Chillán 3780000, Chile
2   Department of Agricultural, Food and Nutritional Science, Faculty of Agriculture, Life and Environmental Sciences, University of Alberta, Edmonton, AB T6G 2P5, Canada; lingyun1@ualberta.ca
*   Correspondence: marcelajarpa@unach.cl

Abstract: Foams are a type of material of great importance, having an extensive range of applications due to a combination of several characteristics, such as ultra-low density, tunable porous architecture, and outstanding mechanical properties. The production of polymer foams worldwide is dominated by those based on synthetic polymers, which might be biodegradable or non-biodegradable. The latter is a great environmental concern and has become a major waste management problem. Foams derived from renewable resources have aroused the interest of researchers, solid foams made from plant polymers in particular. This review focuses on the development of plant polymer-based solid foams and their applications in the food industry over the last fifteen years, highlighting the relationship between their material and structural properties. The applications of these foams fall mainly into two categories: edible foams and packaging materials. Most plant polymers utilized for edible applications are protein-based, while starch and cellulose are commonly used to produce food packaging materials because of their ready availability and low cost. However, plant polymer-based solid foams exhibit some drawbacks related to their high water absorbency and poor mechanical properties. Most research has concentrated on improving these two physical properties, though few studies give a solid understanding and comprehension of the micro- to macrostructural modifications that would allow for the proper handling and design of foaming processes. There are, therefore, several challenges to be faced, the control of solid foam structural properties being the main one.

Keywords: foams; cellulose; natural fibers; mechanical properties; microstructure

Citation: Jarpa-Parra, M.; Chen, L. Applications of Plant Polymer-Based Solid Foams: Current Trends in the Food Industry. *Appl. Sci.* **2021**, *11*, 9605. https://doi.org/10.3390/app11209605

Academic Editor: Isidoro Garcia-Garcia

Received: 8 September 2021
Accepted: 6 October 2021
Published: 15 October 2021

Publisher's Note: MDPI stays neutral with regard to jurisdictional claims in published maps and institutional affiliations.

1. Introduction

Foams are a type of material of great importance with an extensive variety of applications due to their combining several characteristics, such as ultra-low density and tunable porous architecture and mechanical properties. Industrial applications include their being used in kinetic energy absorbers, biomedical scaffolds, thermal insulation, construction, separations, and devices for the storage and generation of energy [1].

The production of polymer foams worldwide is currently dominated by those based on synthetic polymers. They are mostly inexpensive and tailor-made to serve several purposes, having a wide range of applications where mechanical function is important, both in the food and non-food industries [2,3]. Synthetic polymers might be biodegradable (e.g., poly(glycolic acid) (PGA), poly(lactic acid) (PLA) and their copolymers) or non-biodegradable [4]. The latter is a great environmental concern, since non-biodegradable polymer based-foams have become a significant part of countries' solid waste, posing a major waste management challenge [5,6]. Commonly, non-biodegradable synthetic polymer-based foams are petroleum-derived, such as polyethylene (PE), polystyrene (PS), polyurethanes (PU), and polypropylene (PP) [6,7]. Biodegradable polymers decompose into CO_2, water, inorganic compounds, and biomasses that finally form humus [8], which recommends them as alternatives to non-biodegradable polymer foams [8,9].

In this context of increased environmental concern, many researchers have developed foams from renewable resources, such as cellulosic materials, starch, proteins, and other biopolymers, for food, environmental, medical, and other applications [5,6]. Some of these have shown great short-term promise, particularly in medical and environmental contexts, and research into this area has been making great advances in the last decade. Due to the high number of them, not all the applications can be considered in this systematic review, which will restrict its focus to the development of solid foams produced from plant polymers with potential or direct applications in the food industry over the period from 2015 to 2021. A systematic literature search was performed across the Web of Science, Science Direct, and MDPI databases. The information is organized according to the main sources of the foaming agents, with a special emphasis on the foam structure–property relationships.

2. Food Industry Applications

Utilization of plant polymer-based solid foams in the food industry can be categorized into two main areas according to their final uses: (i) as an edible material, e.g., food matrix, or (ii) as packaging material. Although their final state is solid, solid foams start as wet foams before being solidified. They can be produced by different mechanical, physical, and chemical foaming processes [10]. In the wet stage, they are air-filled systems, in which foam structure stabilization is the result of molecules acting as surfactants or Pickering stabilizers. The transition from a wet to a solid state is usually accomplished by either cooling, heating, or curing the wet foam [11]. The processing techniques to solidify plant polymer-based foams may include baking, freeze-drying, extrusion, injection molding, or compression molding processes [12]. For packaging materials, supercritical fluids, such as carbon dioxide (CO_2) or nitrogen (N_2), can be used with molding techniques, producing stable foams that are lighter and of higher dimensions than their solid counterparts [10].

Edible solid foams constitute the fundamental basis of several food products, such as bread, meringue, and ice cream [13]. For a solid foam to be considered an edible material, the first consideration, obviously, is to be food grade. Additionally, some important properties to be evaluated are moisture adsorption capacity, mechanical and physical properties, sensory and organoleptic properties, cell morphology, digestibility, loading capacity, etc. The evaluation will depend on the targeted usage of the foam.

Some of the properties mentioned above will be relevant for foams intended to serve as a structural basis for packaging materials. However, certain mechanical and physical properties will be of special relevance to their suitability for this purpose. Thus, these foams must: (i) provide protection for fragile food products, (ii) show appropriate mechanical and barrier properties, (iii) be lightweight and (iv) non-toxic, and (v) have appropriate moisture adsorption capacity, amongst other things. In addition, the evaluated properties will depend on the food to be packaged, as well as other factors, such as shelf-life, storage conditions, etc. [14,15].

The present review article deals with the research and development of solid foams derived from plant polymers with potential or direct applications in the food industry over the last fifteen years. Additionally, this review will highlight details of the micro- and nanostructure of foam, the structure–property relationships between polymers, and the physicochemical characteristics elucidated in the studies consulted. It should be borne in mind that although the rheological properties before solidification are important for the physicochemical characteristics of solid foams, this matter will not be addressed in this review due to the depth of the matter and because it has already been touched upon in several other reviews. Readers are encouraged to seek out more detailed information in the articles by Dollet and Raufaste [16], Nastaj and Sołowiej [17] and Alavi et al. [18].

2.1. Plant Polymer-Based Foams as Edible Materials

Edible solid foams are of interest for a variety of applications in the food industries. Those produced from plant-derived compounds have been gaining in importance not only amongst vegan, vegetarian, and flexitarian consumers, but also amongst those who are concerned about carbon footprints. Despite the many deficiencies of early plant polymers, in terms of function, drawbacks or higher prices which limited their acceptance, the abundance of agricultural commodities and new regulations for material recycling and disposal have made them more desirable, as they are relatively inexpensive and ubiquitous [19]. Depending on the foaming agent, foam pore configuration, mechanical properties, and possible tunable structure, various edible plant polymer-based foams can be developed to serve different purposes. To describe those applications, the following information is organized according to the foaming agent, highlighting its role in the final product structure, and is summarized in Table 1, where further details about foam structure and polymer structure are given (for a list of some polymers, see Figure 1).

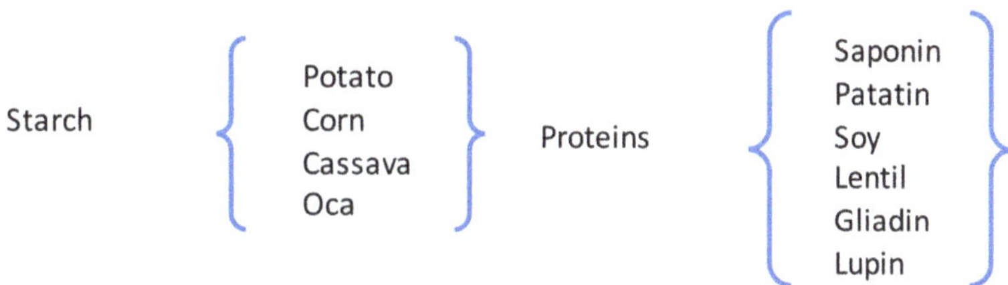

Figure 1. Examples of sources of plant polymers utilized to produce solid foams.

2.1.1. Saponins

Saponins (Figure 2) are amphiphilic glycosidic secondary metabolites produced by a wide variety of plants. Soapwort (*Saponaria officinalis*) is a natural source of saponins, which are known for their surface properties and capacity to form foams [20]. Jurado-Gonzalez and Sörensen [21] studied the chemical and physical properties of soapwort extract as well as its foaming properties under common food processing conditions, such as in the presence of sodium chloride and sucrose. The saponin extract exhibited high foaming capacity and stability. In addition, low pH did not significantly affect foam properties, while heating the extract improved the foaming capacity and stability. Testing the saponin extract at concentrations below 30% ethanol slowly lowered its foaming capacity. Meanwhile, heating increased foam capacity and stability. All these results confirm that the saponin extract from soapwort is a potential alternative foaming agent for use in several food systems, especially in hot food applications.

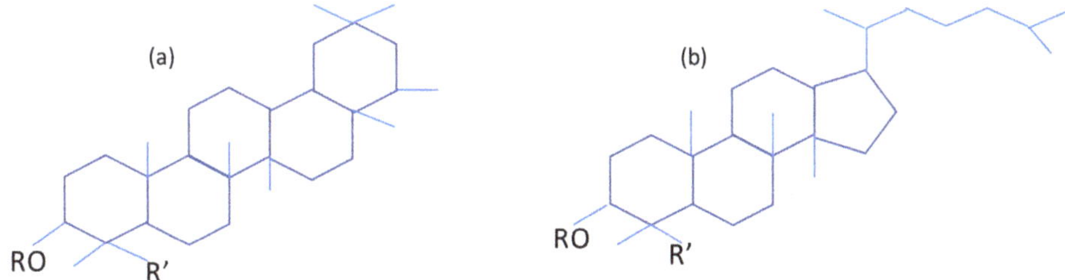

Figure 2. Saponin structure: (**a**) Triterpenoid type; and (**b**) Steroid type.

For example, the soapwort concentrate and extract powder enriched with saponin developed by Çam and Topuz [22] through optimized extraction, enrichment, and drying processes was used as a foaming agent in the production of traditional Turkish delight. The foams produced from saponin concentrate and extract powder showed good stability and resistance to heating process. In terms of chewiness, taste, and texture, they had similar results to commercial soapwort extract at the same soluble solid content, though these foams scored slightly lower than the latter. However, a positive aspect is that taste and odor problems caused by microbial proliferation, along with quality losses due to the lack of standardization in the extracting process, are less frequent than when commercial soapwort extract is used, which creates an opportunity for industrial production. Characterizations of the foam structure, which might help to improve the final product were not systematically conducted, thus further research is required for understanding the textural analysis results.

2.1.2. Potato Protein

Potato protein obtained as a by-product of starch production has several potential applications in the food industry. The most abundant protein fraction is patatin, which has shown good foaming properties, as demonstrated by Schmidt et al. [23]. The foaming capacity of patatin samples tested by Schmidt ranged from ~0.8–1.8 L/L, with the highest overrun value at pH 3 and reduced values at pH 5 and 7, respectively. Higher foam overrun at pH 3 is likely related to the unfolding of patatin at pH levels lower than 4.5. Foam stability of different potato protein fractions displayed a wide range of values. The relative foam stability was lowest at pH 3, ranging from 18–78% of the initial foam, while 67–80% was seen at pH 5 and 7. Descended stability value was especially pronounced for the patatin fraction, though the reason behind this is not known. It is suggested that phenolic compounds present in these fractions can alter the hydrophobic character of proteins. Thus, patatin fractions may show different stability values due to modifications in surface activity, foam height stability, and liquid drainage [23].

Ozcelik, Ambros, Morais, and Kulozik [24] examined the use of patatin as a foaming agent and pectin as a foam stabilizer to produce a snack from dried raspberry puree foam by freeze drying. They also compared the effects of using microwaves during freeze drying on the structure and on the storage behavior of the raspberry puree foam. The results showed that raspberry puree has a better storage stability under the foam structure during the long-term storage period at 37 °C. It was expected that the open porous structure could have increased the deterioration of the bioactives due to the higher surface area. However, by the end of the study, Ozcelik et al. hypothesized that hydrocolloids and potato protein produced a protective barrier as a dried lamella in the foam structure around the pores which resembles a glassy membrane structure and slows down the deterioration. The microwave-assisted freeze-drying process did not affect raspberry puree foam negatively during storage, as compared to the traditional freeze-drying technology used as a control. For example, there was no significant difference ($p > 0.05$) in the color among all samples by the end of storage. Furthermore, water activity was below 0.6 for all samples during the entire storage period; thus, microbiological stability was ensured.

2.1.3. Soy Protein

The quaternary and tertiary structures of native soy protein limit and hinder foaming properties for food applications because of the large size of the molecules and their compact tertiary structure. Thus, some treatments that modify structure, such as heating and hydrolysis, must be applied to allow soy protein to be used as a foaming agent [25].

Soy protein isolate (SPI) was utilized by Zhang et al. [26] to prepare a solid foam from freeze-dried O/W emulsions containing bacterial cellulose (BC) as Pickering particles. Using different oil fractions, the researchers modified pore size and density. Increasing the amount of oil, SPI–BC solid foams were produced, which exhibited uniform and smaller pores that displayed an open-cell structure with pore sizes of several dozen micrometers (<50 μm). This is likely because emulsion droplets gradually became smaller and more uniform, contributing to the construction of a denser network and increased viscosity to prevent droplet accumulation. Thus, the physical stability of the prepared emulsions was high before freeze-drying. Along with this tunable structure, SPI–BC solid foams showed improved mechanical properties, no cytotoxicity, and great biocompatibility, with potential for food industry applications [27].

Another way of using SPI as a foaming agent was tested by Thuwapanichayanan et al. [28] to produce a banana snack. SPI banana foam had a dense porous structure that was crispier than foams produced by fresh egg albumin (EA) or whey protein concentrate (WPC). It is probable that SPI could not be well dispersed in the banana puree during whipping and that the final interfacial tension at the air/liquid interface might not be low enough to produce a significant foaming of the banana puree. WPC and EA banana foams underwent less shrinkage because SPI-banana foam was less stable during drying, so its structure collapsed. Also, WPC and EA banana foams had fewer volatile substances due to shorter drying times.

A similar approach was attempted by Rajkumar et al. [29] using a combination of soy protein as a foaming agent and methyl cellulose as a stabilizer to produce a foamed mango pulp by the foam mat drying method. To obtain the same level of foam expansion, the optimum concentration of soy protein as foaming agent was 1% compared to 10% of egg albumin. Although biochemical and nutritional qualities in the final product were better when using egg albumin, the much lower concentration required for soy protein would be beneficial in terms of cost. It would be interesting to understand how the soy protein and methyl cellulose combination contributed to the positive results in foam expansion; however, this effect was not studied.

Similarly, blackcurrant berry pulp was foamed using SPI and carboxyl methyl cellulose (CMC) as foaming and stabilizer agents, respectively. In this study, Zheng, Liu, and Zhou [30] tested the effect of microwave-assisted foam mat drying on the vitamin C content, anthocyanin content, and moisture content of SPI blackcurrant foam. Several parameters of the microwave drying process, such as pulp load and drying time, had positive effects up to a certain level and then showed a negative effect on the content of both vitamin C and anthocyanin in blackcurrant pulp foam. At the lower pulp load condition, microwave power causes the temperature of blackcurrant pulp foam to rise, which resulted in vitamin C degradation. As for drying time, the longer the drying time, the greater the degradation of vitamin C in the blackcurrant pulp. The quality attributes in terms of color and appearance were better than in samples treated under traditional drying conditions.

2.1.4. Gliadin and Lupin

Ceresino et al. [31] studied the impact of glycerol (Gly), linoleic acid (LA), and transglutaminase (SB6) in two concentrations (1-TG-Glia and 2-TG-Glia) as plasticizers on the development of gliadin (Glia) solid edible foams. Nanomorphology, studied by SAXS, indicated that Gly impaired the unfolding of gliadin in the foam; however, no statistically significant impact of the glycerol on gliadin polymerization within the foam was observed. In addition, there was a notable variation regarding bubble size distribution and a weakening effect on foam stabilization. However, if small pore sizes are desired in aerated food, there is a potential use for glycerol as an alternative to sugar, as well as the use of glycerol as a co-surfactant.

The addition of linoleic acid (LA) in the foams caused the formation of an interrupted Glia network with large, sparsely located bubbles as revealed by X-ray tomograms. Overall, LA impacted gliadin polymerization and foam morphology by preventing the formation of S–S bonds and isopeptide bonds in the gliadin protein. The use of linolenic acid led to the formation of specific nanomorphologies in the foams, referred to as lamellar phases—a process that has been observed for the first time ever in this field of study. These processes suggest that the fortification of gliadin foam with linoleic acid, which is an essential fatty acid for humans, is possible in gliadin breakfast snack prototypes as a main ingredient to improve the nutritional profile of starch-rich foods, since those two components are "compatible" in structured foods.

However, comparison of the non-treated gliadin foam (0-Glia) with the 2-TG-Glia (1.17 U/g) showed that an increase in the size of the bubbles as well as improved bubble spatial homogeneity occurred after a great cross-linking and polymerization of gliadins in the foam. This resulted in a well-developed protein matrix and foam morphology, in comparison to the other sets of foams composed of food dispersants. The study also suggests that SB6 may have played a role in gliadin folding and unfolding, probably because of the deamidation reaction.

The results from this study showed that gliadin is a promising resource to create edible solid foams and has great foaming functionality.

In a similar study, Ceresino et al. [32] studied the impact of transglutaminases (TGs; SB6 and commercial), glycerol (Gly), soy lecithin (Lec) and linoleic acid (LA) on the micro- and nanostructure of solid foods for the creation of foams from LPI and fat blends that could be used to make aerated foods with an appealing texture. As in the previous study, 3-D tomographic images of LPI with TG revealed that SB6 contributed to an exceptional bubble spatial organization. Thus, SB6 significantly contributes to the foam's homogeneous periodic morphology. However, due to spray-drying preparation of Lupin, transglutaminases have limited influence on further cross-linking of the proteins. Nevertheless, the addition of lecithin promoted the formation of new hexagonal structures at the nanometric scale of the foam matrix, whereas linoleic acid contributed and led to the creation of lamellar structures and complex protein–lipid bonding at the nanometric and molecular range of the foam walls. Thus, both lecithin and linoleic acid generate less homogeneous foams.

Results suggest that hydrophobic interactions between lupin proteins, which favor the development of a stable cross-linked matrix and thick matrix walls due to protein bond rupture and rearrangements, promote the formation of a relatively continuous surface. It has also been noted that both LA and LPI are compatible in forming stable layers. This is a result of great significance which may lead to many applications in the confectionery industry.

As in the first study, glycerol produced a weakening effect on lupin foam stability, which might be associated with its effect on decreasing polymerization. Glycerol–Lupin foams were too brittle and unstable to withstand tomography and SEM analysis.

2.1.5. Starch

Starch is a low-cost, biodegradable, non-toxic, and readily available organic polysaccharide. As such, it is a valuable and convenient resource for the food industry. However, its use in edible solid foams is still not well explored.

One of the techniques that has been studied is that used in the production of extruded starch-based snacks. The extrusion process produces foam by the expansion of the ingredients' melted mixture. The expansion phenomenon occurs in different stages. The first stage is the expansion of the melt, which is often followed by a shrinkage phase before the melt solidifies. Then, the melt exits the die, followed by its expansion due to the sudden pressure drop. Internal moisture migrates quickly towards the surface, which drastically decreases the moisture content and temperature of the melt. The melt continues to expand until the melt temperature decreases below its glass transition temperature (Tg), where expansion ceases and the structure of the extrudate sets [33].

In Martínez-Sanz [34], foams were produced by extrusion cooking using Spirulina at 0, 1, 5, and 10 wt.% mixed with corn starch. During extrusion, starch granules are ruptured as they are subjected to high shear, pressure, and temperature. As a result, the crystalline regions of the granules are melted. A polymerization and rearrangement of amylose and amylopectin chains occurs before leaving the extruder die nozzle. Shorter amylose and amylopectin chains are associated by hydrogen bonds after exiting the nozzle if water or some other plasticizer is present [35]. The gelatinization of starch was produced during extrusion by raising the moisture content up to 30 wt.% and letting the samples equilibrate for 24 h before processing. As a result, starch foams were very amorphous, while Spirulina–starch foams showed a slightly more crystalline structure than the pure extruded starch, due to Spirulina complexation with amylose. Thus, Spirulina–starch foams showed more densely packed and well-connected porous structures, making the foam texture harder as Spirulina content increased. During storage, the free fatty acids from Spirulina re-crystallized and the resistant starch content in the 10% Spirulina foam rose. Nutritionally speaking, this result might be interesting for the food industry, which is always looking for a way to produce new healthy snacks.

Several extruded starch-based snacks have been studied in this category. For example, corn starch-based snacks with amaranth, quinoa and kañiwa [66]; rice starch and pea protein snacks [33]; and corn starch based-snacks with common bean [67]; among others. According to Zhang et al. [68], starch is suitable for this kind of application, since granule degradation by gelatinization allows the formation of a stable, expanded structure. Regarding proteins, during the extrusion process they are unfolded, realigned, hydrolysed, and can cross-link with other ingredients like starch. In addition, proteins affect extrudate expansion through their ability to influence the water distribution within the extruded melt. The presence of protein can disrupt the pore walls of the expanding melt, which results in reduced expansion and smaller sized, denser cell structures. On the other hand, protein content can increase the number of pores of starch-based extrudates at certain levels, as a result of a greater number of nucleation sites [34].

Table 1. Studies for food industry applications of plant polymesr-based solid foams.

Plant Polymer	Study	Type of Foam	Foam Characteristics	Polymer Characteristics
Proteins				
Saponin/Soapwort extract	• Foaming agent in the production of traditional Turkish delight [22]	Edible	Good stability and resistance to heating process	They are composed by an aglycone unit called sapogenin linked to one or more carbohydrate chains. The sapogenin unit consists of either a sterol or a triterpene unit, which is the more common. The carbohydrate side-chain is habitually attached to the 3 carbon of the sapogenin. The carbohydrate portion is water-soluble, whereas the sapogenin is fat-soluble; thus, saponins have surface-active properties [36]
	• Egg white replacement in sponge cake up to 75% [37]	Edible	Good stability and resistance to heating process. Similar behavior to egg white in sponge cake elaboration	
Patatin	• Foaming agent to produce a snack from dried raspberry puree foam by freeze drying [24]	Edible	Foam structure is open and resembles a glassy membrane structure around the pores	Patatin belongs to a family of 40–42 kDa glycoproteins with isoelectric point values between 4.5 and 5.2. It shows a secondary structure composed of 35% alpha-helixes, 45% beta-strands and 15% aperiodic. It has a denaturation temperature of 60 °C at pH 7.0 and a relatively low stability as a function of pH showing loss of structure at pH $\leq$ 4.5 [38]

Table 1. *Cont.*

Plant Polymer	Study	Type of Foam	Foam Characteristics	Polymer Characteristics
Proteins				
	• Preparation of a solid foam from O/W freeze-dried emulsions containing bacterial cellulose as Pickering particles. (Zhang et al. [26])	Edible and packaging material	Foams have a tunable structure, e.g., size and density tailoring. They also showed uniform and smaller pores with an open-cell structure and pore sizes of about <50 μm, as well as improved mechanical properties	
Soy protein isolate	• SPI-banana snack [28]	Edible	Foams with dense porous structure that renders a crispier texture	Soy protein is a globulin protein. Its polypeptide chains have a three-dimensional structure linked by disulfide and hydrogen bonds with a molecular weight ranging from 300,000 to 600,000 KDa. Proportion of two major protein polymers in soy protein are 35% conglycinin (7S) and 52% glycinin (11S), giving about 80% of the total soy protein [39,40]
	• Foamed mango–soy protein pulp [29]	Edible	Good foam expansion. There is no link given by the authors for foaming properties and structure	
	• Foamed blackcurrant–soy protein pulp [30]	Edible	There is no link given by the authors for foaming properties and structure and there is no description of the latter	
	• Replacement of eggs by soy protein isolate and mono- or diglycerides in yellow cake [41]	Edible	SPI by itself cannot guarantee a suitable foam structure to form the cake. SPI–MDG foams produce batters with correct specific density and appropriate nanostructure, though fewer and larger porosities are observed	
Lentil protein concentrate	• Replacement of eggs by lentil protein in angel food cake [42]	Edible	Lentil protein produced a foam that retains air bubbles due to strong networks around the air cells. The mean area of air cells is low, while the number of air cells per unit area is high	Lentil proteins are mainly comprised of albumins, (16%) and globulins (70%). Albumins have a molecular weight of about 20. Globulins contain both legumin- and vicilin-like proteins. The first group consists of six polypeptide pairs that interact noncovalently and have a molecular weight (Mw) of 320–380 kDa. Vicilins are trimers of glycosylated subunits with a Mw of 50–60 kDa [43]

Table 1. *Cont.*

Plant Polymer	Study	Type of Foam	Foam Characteristics	Polymer Characteristics
Polysaccharides				
Potato Starch	• Foam made from thermoplastic starch (TPS). TPS was made by acetylation and esterification with maleic anhydride of potato starch [44]	Packaging	TPS foam showed lower absorption with improved water resistance. The foam microstructure showed a sandwich-type structure, more or less dense outer layers, and a more compact cellular structure than pure TPS foam. Foams with more modified starch expanded more and became more porous	Potato starch granules are on average shorter than sweet potato starch granules, while bigger than rice starch granules. Amylose content is lower than wheat and corn starch and higher than tapioca and sweet potato starch. It also has the highest molecular weight and the lowest degree of branching. Amylopectin of potato is much less densely branched than other starches, it has much longer chains, and it carries mono-phosphate ester groups [45]
	• Foam made from silylated starch [46]	Packaging	Foams show the typical sandwich structure, with denser outer layers with small cells and an inner layer with larger and more expanded cells. Silylated starch foams have a more compact structure with thicker outer layers than traditional starch foams. They become mechanically more resistant and have less water absorption capacity	
	• Bioactive foams derived by thermopressing from sweet potato starch and essential oils [47]	Packaging	Foams with essential oils had small cracks and holes. They displayed a more irregular but denser surface due to starch-lipid complexes forming during the thermal process. Starch and essential oils also formed strong interactions, resulting in starch–essential oil complexes in the foam layers. Thus, essential oil drops were trapped within the starch granules. Foams presented a sandwich structure with two well-defined layers and the presence of air cells. Essential oil addition and type also affected the layer thickness and the air cell size between the foams.	
	• Foam plates prepared by baking potato starch, corn fibers, and poly(vinyl alcohol) (PVA) [48]	Packaging	The foam has a sandwich type structure with dense outer skins containing small cells comprising the surface of the foam. The interior of the foam has large cells with thin walls. Adding over 50% corn fiber, foamed trays contain few small cells in their outer skin. In the interior the cells are smaller, and the foam becomes denser. Trays containing only potato and PVA had thinner skins and larger cells with thicker walls. The outer skin of trays containing corn fiber show compressed and bounded fibers	

Table 1. *Cont.*

Plant Polymer	Study	Type of Foam	Foam Characteristics	Polymer Characteristics
Corn Starch	• Foams produced by extrusion cooking using corn starch mixed with Spirulina [34]	Edible	Starch foams were very amorphous. Spirulina–starch or hybrid foams showed a slightly more crystalline structure than the pure starch foam. Thus, hybrid foams showed more densely packed and well-connected porous structures, and foam texture is harder	Corn starch is, in general terms, similar to other cereal starches, and in specific properties has greatest similarity to its genetically closely related cousins, sorghum and the millets. Normal corn is composed of amylose and amylopectin. It is usually composed of 27% amylose and 73% amylopectin [49]. However, this amylose/amylopectin ratio varies slightly with different corn varieties, environmental and soil conditions. Waxy maize consists of amylopectin only, and high amylose corn contains amylose as high as 70% [50]
	• Glyoxal cross-linked starch-based foam without and with corn husk fiber, kaolin, and beeswax [51,52]	Packaging	Cross-linked starch foams had more expanded structures, and their cell walls were thinner than those of native foams. They showed areas of weak formation on the surface. The additives eliminated these zones. Addition of fiber, kaolin or beeswax increased the cell size in the center of the foams	
Cassava Starch	• Cotton-fiber-reinforced cassava starch foams prepared by compression molding [53]	Packaging	Foams showed a sandwich-type structure. The addition of cotton fibers, produced more dense structures, thicker cell walls, and lower area porosity	Cassava starch granules are round with a granule size between 5 and 35 µm. The starch has an A-type X-ray diffraction pattern, usually characteristic of cereals, and not the B type found in other root and tuber starches. The C-type spectrum, intermediate between A and B types, has also been reported. The nonglucosidic fraction of cassava starch is very low; the protein and lipid content are below 0.2%. There is thus no formation of an amylose complex with lipids in native starch. Amylose contents of 8–28% have been reported, but most values lie within the range of 16–18%. The starch gelatinizes at relatively low temperatures. Initial and final gelatinization occurs at 60 °C and 80 °C, respectively. The swelling power of the starch is also very high: 100 g of dry starch will absorb 120 g of water at 100 °C. At this temperature, over 50% of the starch is soluble [54]
	• Cassava starch foams added with sunflower proteins and cellulose fibers [55]	Packaging	Foams exhibited a more compact, homogeneous, and dense microstructure. The cells were of moderate size, with fibers homogeneously spread throughout the whole material. Baked foams that included proteins were practically devoid of inner open cells	
	• Cassava starch foams added with malt bagasse [56]	Packaging	Foams showed a sandwich-type structure with dense outer skins that enclose small cells. The interior of the foams had large air cells with thin walls. They have a good distribution of the malt bagasse throughout the polymeric matrix and showed good expansion with large air cells	
	• Cassava starch foams added with sesame cake [57]	Packaging	Foams exhibited sandwich-type structure with denser outer skins that enclose small cells whereas inner structure is less dense with large cells. They also showed good expansion	

Table 1. *Cont.*

Plant Polymer	Study	Type of Foam	Foam Characteristics	Polymer Characteristics
	• Cassava starch foams added with grape stalks [58]	Packaging	Foams present dense and homogeneous external walls, with small, closed cell structure. The interior shows a structure with large open cells and a sandwich-type structure typical of thermoplastic starch-based materials obtained by thermal expansion	
	• Cassava starch foams added with pineapple shell [59]	Packaging	Foams showed a good distribution of the pineapple shell fiber throughout the polymeric matrix and a semi-crystalline structure. They have a sandwich-type structure with dense outer skins and small cells comprising the surface of the foam and larger sized cells in the interior of the foam	
	• Cassava starch foams added with sugarcane bagasse [60]	Packaging	Foams have filler fibers well incorporated into the starch matrix and well distributed, making the material homogeneous	
	• Cassava starch foams added with organically modified montmorillonite and sugarcane bagasse [61,62]	Packaging	Foams exhibited sandwich-type structure with denser outer skins that enclose small cells whereas the inner structure is less dense, with large cells. They also showed good expansion	
	• Baked foams from citric acid modified cassava starch (CNS) and native cassava starch (NS) blends [63]	Packaging	SEM micrographs of foams showed that the cells formed were open with connectivity between cells. They had a sandwich-type structure composed of two layers. The outer layers had a smaller cell size but a denser structure, whereas the interior had a larger cell size and a more expanded structure. NS foam showed a thinner cell wall with a broad distribution of cell sizes. CNS foam, revealed a smaller cell size and a denser structure.	

Table 1. *Cont.*

Plant Polymer	Study	Type of Foam	Foam Characteristics	Polymer Characteristics
Oca Starch	• Oca starch foams added with sugarcane bagasse (SB) and asparagus peel fiber (AP) [64]	Packaging	Foams with addition of fibers showed a less compact structure and with distribution no homogenous of poreswhen compared to the control. The fiber distribution through the cellulose matrix was dissimilar for both SB and AP fiber. Trays with SB fiber had larger cells arranged in a thinner layer than those with AP fiber. Both exhibited the typical sandwich structure	Oca starch has a phosphorus content ~60% lower than potato starch. Its amylose content is approximately 21% (lower than that of maize and potato starches). Amylopectin is similar to that of potato amylopectin, with some differences in the length of its internal chain and amount of fingerprint B-chains. Oca starch granules had a volume moment mean size of 34.5 μm and B-type polymorph [65]

2.1.6. Plant Polymer-Based Egg Protein Replacers

In aerated foods such as meringues, marshmallow, bread, cakes, and soufflés, foams are responsible for the appropriate textures and a particular mouthfeel because of the little air bubbles trapped in the food system [1]. This is a complex process, where different physicochemical interactions and processing conditions play crucial roles to achieve the appropriate foamy structure. Polymer functional properties (foaming capacity and stability, as well as emulsification) are critical to achieve a solid foam-like structure. That is why egg proteins are important ingredients in the bakery industry and the main goal when using other polymers as substitutes is to replicate their functionalities because replacers influence the textural and physical properties of the final product. Several studies have been carried out using whey protein to prepare meringue and there has been an extensive gain of knowledge in the understanding of its rheology and surface properties, as shown by the works of Nastaj and Sołowiej [17] and Nastaj, Sołowiej, Terpiłowski, Mleko [69,70].

However, in the field of plant polymers, finding substitutes for egg proteins as foaming agents is not an easy task, though some advances have been made. For example, soapwort extract can be used to partially replace egg white proteins (EWP) as a foaming agent in sponge cake formulations. Rheological and physical properties of cake batters and physical and sensory properties of sponge cakes were analyzed to determine the effects of soapwort extract addition. Different formulations were produced replacing egg white proteins with soapwort extract by 25%, 50%, and 75% on weight basis. Replacement of the protein source up to 75% did not have a significant effect on the specific gravity of batters ($p > 0.05$). Likewise, the flow behavior indices (n) and the consistency indices (k) of cake batters were not affected by the addition of soapwort extract. Saponin content in the soapwort extract and saponin chemical structure containing polar (water-soluble side chains) and nonpolar (sapogenin) molecules facilitate incorporation of air into the batter. Additionally, physical properties of sponge cakes were not altered either. Regarding the sensory properties of sponge cakes, the results were also favorable. Sponge cakes formulated with 50% and 75% soapwort extract on weight basis received significantly higher chewiness scores than control cakes ($p < 0.05$) [37].

Likewise, lentil protein can totally or partially substitute egg and milk protein as a foaming and stabilizer agent in angel food cake, resulting in products with satisfactory quality [42]. In this study, increasing the amount of lentil protein will increase the viscosity of the batter due to the more entangled lentil protein structure at higher protein concentrations, which might help air retention in the aerated system. Less air escaped from the lentil protein-based foamy batter than the control during baking, probably due to the formation of strong networks around the air cells that prevented microstructure collapse during gas expansion when the wet foam was transformed into solid foam. Additionally, the mean area of air cells was reduced by the presence of lentil protein but the number of air cells per unit area increased; thus, the height of angel food cakes with lentil protein remained very close to that of the control formulations [42].

Likewise, yellow cake prepared from a mixture of soy protein isolate and 1% mono- or diglycerides that replaced eggs as a foaming agent yielded a similar specific volume, specific gravity, firmness, and moisture content [41]. The authors of this study presented a detailed explanation of cake structure based on an analysis of the relationship between SPI secondary structure and different additives. β-sheet structures were dominant among the secondary structures in all control and eggless groups. The content of these structures increased after the addition of SPI, suggesting an increase in the intermolecular interactions in gliadins.

On the other hand, adding Xanthan gum into the cakes with SPI and soy lecithin reduced the β-sheets significantly and converted them to random coils and helical structures. This might be attributed to the slightly charged property of soy lecithin, as it binds to proteins via hydrophobic interactions and the remaining phosphate groups repel each other to loosen the molecules, leading to weakened intermolecular interactions arising from β-sheets and looser molecules being converted into coils. Gliadin is rich in proline; therefore, the β-turn conformation could be more highly favored. They also proposed a schematic model for the gluten formed in the presence of SPI. The model was based on the structural analyses of gliadin and glutenin after the addition of the protein and additives. Analysis of starch suggested the existence of new cross-linking with ester bonds, while an SPI analysis showed the presence of dominant gliadin aggregates in the size range of 100–200 nm and glutenin networking structures containing fewer but larger porosities. These results provide enough support to suggest using SPI in combination with 1% mono- and diglycerides (MDG) as a substitute foaming agent for eggs in cakes.

2.2. Plant Polymer-Based Foams as Food Packaging Materials

The distribution chain of the food industry is heavily dependent on appropriate packaging materials. When considering the performance required for prolonging shelf life, the properties of most interest in packaging applications, and their relevance in use, can be classified as mechanical and optical properties, chemical stability, and moisture and gas barrier function [69].

For several years, conventional, non-biodegradable plastics have been the principal source of material for food packaging. Many already known reasons are responsible for the wide use of plastics, e.g., they are generally lighter in weight, more easily formed into different shapes, extremely versatile, and have a low cost of production [69]. However, severe concerns related to environmental catastrophe have set suitable conditions for the development of alternative eco-friendly materials derived from natural polymers for packaging [8].

A natural packaging material must fulfil the demands made by the food industry. For example, it should serve as a physical barrier and temperature controller—that is, it must have a high impact resistance and high thermal insulation. It also should act as a biochemical and microbiological preserver (e.g., gas exchange regulator). Equally important is that its production and disposability have a low carbon footprint, as well as low pollution risks and greenhouse gas emissions. Meeting these criteria, plant-derived biopolymers from agro-industrial sources that are renewable, abundant, and inexpensive would present a convenient and attractive alternative [55].

2.2.1. Starch

The low cost, availability, and compostability of starch make it of great significance for the packaging industry [70]. There are hundreds of studies utilizing starch as a replacement for synthetic packaging material. Nonetheless, and regardless of the fabrication method or starch source, all show the same drawback related to water absorption or water vapor permeability [71,72] due to the hydrophilic nature of starch molecules. Therefore, this is one of the biggest problems to solve.

One strategy has been to modify the molecular structure of starch to make it more hydrophobic. In a study by Bergel et al. [44], potato starch was used to fabricate a thermoplastic starch (TPS) foam that was modified by two methods: (i) acetylation and (ii) esterification with maleic anhydride. Their results showed that non-modified TPS foams absorbed 75 g water/100 g solids, while foams with 13% acetylated starch (TPS–Ac) and 20% esterified starch (TPS–Es) presented lower water absorption (42 g and 45 g water/100 g solids, respectively), improving the foam water resistance. Analysis of pure TPS, TPS–Ac, and TPS–Es foam microstructure showed that they have a sandwich-type structure.

This kind of structure is typical in TPS foams produced by a mold compression process or baking process. They consist, roughly speaking, of two sets of layers—outer layers and interior layers. Outer layers have a denser structure, smaller cell size, and less voids than the interior layers which have larger cells and more expanded structures. Additionally, in this study, TPS–Ac presented more or less dense outer layers, depending on the acetylation degree, and a more compact cellular structure than pure TPS foam. Differences in viscosity values of foams might explain distinctive microstructures, as acetyl groups are related to a decline of intermolecular bonds between water and unmodified starch due to their hydrophobicity. If the pastes have low viscosity, they cannot hold vapor bubbles as effectively as more viscous ones during the baking process. Therefore, the lower the viscosity of the paste, the greater the paste expansion, which generates foams with a thinner outer layer and large inner cells.

Similarly, in another study by Bergel et al. [46], two silanes were used for potato starch silylation: 3-chloropropyl trimethoxysilane and methyltrimethoxysilane. The foams were made using modified starch, gelatinized starch, polyvinyl alcohol, and water. Microstructure analyses showed the typical sandwich structure with denser outer layers of small cells and an inner layer of larger and more expanded cells. This microstructure translates into a more compact structure and thicker outer layers, which can be explained by the higher viscosity of the silylated starch pastes applied to make these foams. High viscosity is caused by silane cross-linking. Additionally, mechanical tests showed that foams become more resistant to cracking and fracture with the addition of silylated starch. This may also be due to the cross-linking of silanes which make starch pastes more viscous. Meanwhile, the silylation modification yielded foams with less water absorption. The improved foam performances make them a potential packaging material for use in the food industry.

Cruz-Tirado et al. [47] utilized sweet potato starch and oregano (OEO) or thyme (TEO) essential oils to produce bioactive foams by thermopressing. The essential oils were used at two concentrations (7.5 and 10%). The foams were characterized according to microstructure, mechanical properties, antimicrobial properties, and structural properties. In terms of structure, SEM micrographs revealed that foams presented a sandwich-type structure with two well-defined layers and the presence of air cells. The foam thickness was not significantly affected by the essential oil type and concentration at any level, but the starch–lipid interactions resulted in the formation of amylose–essential oil complexes with lipids localized in the first layer. This structure of the foam may have prevented the essential oil from degrading under the thermoforming temperature. Regarding the solubility and mechanical properties, essential oil addition yielded starch foams with low water solubility but also lower mechanical resistance, especially for 10% OEO. Transversal section microstructure analysis showed that TEO-foams and OEO-foams have more compact structures and fewer porosities, which may have decreased water absorption, especially at the surface. Additionally, strong interactions between OEO and sweet potato starch molecules limited the interactions between chains of amylose–amylose, amylopectin–amylopectin, or amylose–amylopectin, possibly weakening and destabilizing the starch structure. In addition, sweet potato starch and essential oil foams were more effective against Salmonella (Gram-negative bacteria) and *L. monocytogenes* (Gram-positive bacteria) as the essential oil diffuses from inside the foams to the surface. According to the authors, the foam structure might influence essential oil diffusion strongly. The SEM micrographs showed that the essential oil was in the first layer of the foam and was later displaced by water vapor during thermoforming. The foams with 10% essential oil exerted a greater antimicrobial effect due to a greater amount of essential oil that diffused to the environment. The phenolic compounds present in the foam and probably responsible for microbial inhibition are carvacrol, thymol, therpinene, and p-cymene. Therefore, these foams showed good properties to be applied as bioactive food containers.

Another approach by Uslu and Polat [51] and Polat et al. [52], was to prepare glyoxal cross-linked baked corn starch foams with the addition of corn husk fiber, kaolin, and beeswax. Cross-linked starch foams had a more expanded structure, as shown by SEM micrographs. This is likely caused by a quicker gelatinization of the cross-linked starches at a lower temperature, and faster water evaporation during the baking process. In addition, the cell size increased with the cross-linkage addition amount, while cell walls of the cross-linked starch foams were thinner than those of the native foams. Both the tensile and flexural properties of the foams were significantly affected by cross-linking. Foams made from cross-linked starches were more flexible. Inclusion of the corn husk fiber resulted in increased water resistance of cross-linked corn starch foams. Addition of beeswax or kaolin increased the cell size in the center of the foams and decreased the tensile and flexural strength; however, these additives also reduced the water absorption of the foam trays. It is likely that both the physical and chemical properties of fibre contributed to the improvement of the tensile properties of the trays. For example, the long size of the fibre permitted the formation of hydrogen bonds with beeswax and a spreading of the fibre in the direction of tension.

A similar study was developed by Pornsuksomboon et al. [63] in which they obtained very similar results, though they used cassava starch and citric acid as a cross-linker. The citric acid-modified cassava starch foam (CNS) had a higher density, lower thickness, and denser structure than native cassava starch (NS). These differences in morphology are probably due to different viscosity values between the batters. As the viscosity of CNS batter was high compared to NS batter, NS foam was more expandable than CNS foam. On the other hand, the 50/50 NS/CNS ratio foam exhibited a uniform distribution of cell sizes with thinner cell walls than both the NS and CNS foams, also because of the different viscosity of the blended starch batters. In addition, the thermal stability of the blended starch foam was lower than NS foam, probably due to the presence of ester bonds with low thermal stability, while the stabilizing effect of the higher degree of cross-linking and strong hydrogen bonds in the citric acid-modified starch might explain the significantly slower water evaporation and decomposition rate of NS/CNS blend chains.

In the same vein, the morphology and the physical, flexural, and thermal properties of cassava starch foams for packaging applications were researched as a function of cotton fiber and concentrated natural rubber latex (CNRL) content [53]. The main objectives were to solve their two main weaknesses, i.e., lack of flexibility and sensitivity to moisture. Cotton fiber was principally added as a reinforcing material. A comparison among SEM micrographs of starch biofoams, both with and without cotton fiber, showed a sandwich-type structure. However, after the addition of cotton fibers, the foam exhibited denser structures, thicker cell walls, and a lower area porosity (43.37% compared to 52.60%). It seems that cotton fiber presence decreased the chain mobility of starch via hydrogen bonding, resulting in a high viscosity of the starch batter and less expansion of the foam. CNRL helped to control moisture into cassava starch foam. As CNRL content rose, the moisture adsorption capacity of the foam declined (−73.4% and −41.78% at 0 and 100% RH, respectively). This may be due to the hydrophobicity increment of the foam. Foam flexural properties were also tuned by regulating CNRL content. For example, with an amount of 2.5 phr of CNRL, the elongation of the biofoam improved by 24%, while the bending modulus decreased by 2.2%. An interesting study carried out by the same research group involved a soil burial test that assessed the biodegradability of the cotton-fiber-reinforced cassava starch foam. They found that the degradation mainly progresses by hydrolysis and is delayed by the addition of CNRL.

Sunflower proteins and cellulose fibers were also added to cassava starch to produce biodegradable food packaging trays through a baking process [55]. The study was focused on the relationship between the proportions of these three components and their effect on microstructure, physicochemical and mechanical properties of the trays. The results showed that increasing the fiber concentration from 10% to 20% (w/w) raised the water absorption capacity of the material by at least 15%, while mechanical properties were improved. On the contrary, an increase in sunflower proteins up to 20% (w/w) reduced the water absorption capacity and the relative deformation of the trays to 43% and 21%, respectively. The formulation that exhibited a more compact, homogeneous, and dense microstructure, with maximal resistance (6.57 MPa) and 38% reduction in water absorption capacity, contained 20% fiber and 10% protein isolate. This optimized material presented the best mechanical properties, lower water absorption, a lower thickness, and a higher density. Likewise, Mello and Mali [56] used the baking process to produce biodegradable foam trays by mixing malt bagasse with cassava starch. The concentration of malt bagasse varied from 0–20% (w/w) and the microstructural, physical and mechanical properties of foams were assessed. The trays had an amorphous structure as a result of a good distribution of the malt bagasse throughout the polymeric matrix. Foams showed a sandwich-type structure with dense outer skins enclosing small cells. The interior of the foams had large air cells with thin walls. They showed good expansion with large air cells. Their mechanical properties were not affected by variation in the relative humidity (RH) from 33 to 58%. However, when the trays were stored at 90% RH, the stress at break decreased and the strain at break increased. This is likely due to the formation of hydrogen bonds with water favored by the hydrophilicity of starch molecules. Thus, the direct interactions and the proximity between starch chains reduced, while free volume between these molecules increased. Under tensile forces, movements of starch chains were facilitated, and this is reflected in the decrease of the mechanical strength of materials. The sorption isotherm data demonstrated that the inclusion of malt bagasse at 10% (w/w) resulted in a reduction in water absorption of starch foams. Cassava starch trays with malt bagasse might, therefore, be a fitting alternative for packing solid foods.

In another similar study, Machado et al. [57] added sesame cake to cassava starch to produce foams and evaluated the effects on the morphological, physical, and mechanical properties of the materials produced. The content of sesame cake added ranged from 0% to 40% (w/w). Cassava starch-based foams incorporated with sesame cake exhibited improved mechanical properties and reduced density and water capacity absorption when compared to starch control foams. Using sesame cake (SC) concentrations higher than 20% showed better mechanical properties than commercial expanded polystyrene (EPS). Foams produced in this study showed a decrease in flexural stress and modulus of elasticity with the addition of SC. The reduction of these properties correlates with their lower density and larger cells in inner structure in comparison to control foams. Large cells in the foam's inner structure and thinner walls can be associated with water evaporation and leakage through the mold, consequently causing cell rupture. Nevertheless, although enhancements in flexibility and moisture sensibility are still necessary, starch-based foams incorporated with sesame cake might be an alternative for packing solid foods and foods with low moisture content.

Another biodegradable cassava starch-based foam produced by thermal expansion was developed by Engel et al. [58], who incorporated grape stalks and evaluated the morphology (SEM), chemical structure (FTIR), crystallinity (XRD), biodegradability, and applicability for food storage. Foams exhibited sandwich-type structure with denser outer skins that enclose small cells, whereas the inner structure was less dense with large cells. The material also showed good expansion, which might be the result of the occurrence of hydrogen bond-like interactions between the components of the expanded structure during processing of the foam. Biodegradability tests demonstrated neither formation of recalcitrant compounds nor structural alterations that would hinder foam degradation. Foams were completely biodegraded after seven weeks. Additionally, foams made with cassava starch with grape stalks added showed a promising application in the packaging of foods with a low moisture content.

Cassava starch, in combination with pineapple shell, was also utilized as a strengthening material to manufacture biodegradable foam trays by a compression molding process. The starch/fiber ratios were varied to modulate the foam, microstructure and physical and mechanical properties. The foams showed a good distribution of the pineapple shell fiber throughout the polymeric matrix and a semi-crystalline structure. Even though all reinforced foams showed high water absorption, foams produced at a starch/fiber ratio of 95/5 showed the lowest values of thickness and density (2.58 mm and 367 kg m^{-3}, respectively) and the highest crystallinity index value. This starch/fiber ratio also led to foam trays with tensile strengths similar to those of expanded polystyrene samples. This is likely due to the reinforcing effect of the interfacial interaction between the fiber and the starch matrix. However, high proportions of fiber can interfere with the expansibility and produce discontinuity in the starch matrix. An increase in the fiber concentration weakened interactions among starch chains due to a lower proportion of starch in the composites. Based on the results above, the cassava starch-based foams might be a promising biodegradable material to be used for solid food packaging, and future research should focus on the improvement of their physicochemical and structural properties [59].

In the study by Ferreira et al. [60], new biodegradable trays were produced based on the blend of cassava starch with sugarcane bagasse. This mixture was then blended with different fibrous agro-industrial residues, such as cornhusk, malt bagasse, and orange bagasse. Trays produced from those mixtures presented high water sorption during storage under high or medium relative humidity. They were also more rigid and more susceptible to degradation than EPS trays. FTIR analysis revealed that hydrogen bonding between cassava starch and the other biodegradable tray components may have occurred during processing, as well as water interaction with other formulation components (starch, glycerol, and fibers). SEM micrographs showed that fibers of the residues were incorporated into the starch matrix and well distributed, making the material homogeneous, which contributed to good mechanical properties. As a result, the combination of cassava starch, sugarcane bagasse, and cornhusk was shown to be the better mix.

In the works by Matsuda et al. [61] and Vercelheze et al. [62], biodegradable trays were developed based on cassava starch and organically modified montmorillonite, called Cloisite$^{®}$10A and 30B, using a baking process. They studied the changes on the microstructural and physicochemical properties of the trays when using the modified montmorillonite. Foams had the typical sandwich-type structure of the foams made by thermopressing. This structure includes dense outer skins that enclose small cells, similar to other foams made with cassava starch, as seen above. The interior of the foams had large cells with thin walls. Samples produced with the nanoclays showed larger air cells than the control sample. In the samples produced with sugarcane fiber, distribution of these fibers in the foam structure was homogeneous up to a concentration of 20 g fiber/100 g formulation. The density values were not affected by the addition of nanoclays. Probably, the addition of the nanoclays improved the foaming ability of starch pastes, resulting in the greater resistance of cell walls against collapse during the water evaporation that occurred during the baking process, as well as producing more thick trays. Results showed well-shaped foam trays with lower water absorption when using nanoclays in the formulations than using starch alone. The foam densities were between 0.2809 and 0.3075 g/cm^3. There were no dimensional changes during storage in the trays at all RH conditions tested, but no explanation was given to this phenomenon. The trays potentially resulted in an alternative packaging option for foods with low water content.

Oca (*Oxalis tuberosa*) represents a novel starch source. In the work of Cruz-Tirado et al. [64], sugarcane bagasse (SB) and asparagus peel fiber (AP) were mixed with oca starch to produce baked foams. The structure of foams reinforced with SB fiber (starch/fiber ratio of 95/5), AP fiber (95/5) and without addition of fiber (100/0) was heterogeneous. The fiber distribution through the cellulose matrix was dissimilar for both SB and AP fiber. Trays with SB fiber had larger cells arranged in a thinner layer than those with AP fiber, which was probably due to less interference with starch expansion during thermoforming of the tray. Both exhibited the typical sandwich structure. Oca foams mixed with asparagus peel fiber exhibited higher rates of thermal degradation than the control but not to the point of affecting their applicability, while sugarcane bagasse fiber in high concentrations created more dense trays with lower water absorption (WAC) than the control because high SB concentrations decreased starch mass in the mixture, decreasing the foaming of starch, which created a more compact structure, whereas the addition of low SB fiber concentrations probably yielded trays that were more porous with larger diameters of cells that facilitated the entry of water. The density of the oca foams was reduced by lowering the fiber concentrations. Trays were made harder and more deformable by the addition of fiber, though it did not improve the flexural strength of the foams.

2.2.2. Cellulose

Cellulose materials are appropriate for the development of biopolymer-based foams due to their biodegradability and low environmental impact but also because of their low density, high aspect ratio, large surface area, and non-toxicity [7]. In general, cellulose nanofiber-based solid foams can be made using various procedures and these usually comprise three steps: (i) the preparation of a gel, (ii) the creation of the 3-D structure via foaming in the presence of surfactants, and (iii) the removal of the solvent. The subtraction of the solvent can be performed using several techniques, such as, supercritical drying, freeze-drying, oven-drying or ambient conditions. Varying the processing route will impact the nano- or macrostructure of the final product, which subsequently will have an effect on the properties of the solid foam, such as porosity and its mechanical and barrier properties [73].

Cellulose nano- and microfibrils, especially, have been utilized in the production of low-density porous materials that display high specific surface areas, low thermal conductivity, and low dielectric permittivity [70]. Because of their distinctive mechanical and morphological characteristics, the cellulose nano- and microfiber-based foams have attracted industrial interest over the last 20 years [1].

For example, Cervin et al. [74], created a lightweight and strong porous matrix by drying aqueous foams stabilized with surface-modified nanofibrillated cellulose (NFC). The innovation in that study was that they use cellulose as foam-stabilizing particles. As shown by confocal microscopy and high-speed video imaging, NFC nanoparticles stopped the air bubbles from collapsing or coalescing by arranging themselves at the air−liquid interface. Stability was achieved at a solids content around 1% by weight. Careful foam drying resulted in a cellulose-based porous matrix of high porosity (98%), low density (30 mg/cm^3), and with a Young's modulus higher than porous cellulose-based materials made by freeze drying. The size of the pores was in the range of 300 to 500 μm.

Similarly, Ghanbari et al. [75] reported the effect of cellulose nanofibers (CNFs) on thermoplastic starch (TPS) foamed composites. The analyses were focused on the thermal, dynamic mechanical analysis (DMA), density, and water uptake. The results revealed that thermal stability, storage modulus (E'), loss modulus (E''), and damping factor (tan δ) increased for all TPS/CNF samples compared to the pure TPS-foamed composites, while apparent density and water absorption of foams decreased when composed with CNF. Additionally, incorporation of CNFs caused an increase in the glass transition temperature (Tg) of the foams. Moreover, 1.5 (wt.%) CNF concentration gave superior resistance or stability with respect to heat compared to its counterparts. An interesting feature shown by the foams was revealed by SEM images of composite foams containing 1.0 or 1.5 (wt.%) CNF: the size of the cell decreased while density increased as a result of CNF acting as the nucleation agent. CNF favored the formation of the cell nucleation sites and the bubble heterogeneous nucleation during the foaming process.

In the study of Ago et al. [70], various types of isolated lignin-containing cellulosic nanofibrils (LCNF) were used to reinforce waxy corn starch-based biofoams. The addition of LCNF increased the Young's modulus and yielded stress in compression mode by a factor of 44 and 66, respectively. In addition, the water sorption of the foams was decreased by adding LCNF due to relatively lower hydrophilicity of residual lignin. The optimized foams exhibited mechanical properties similar to those of polystyrene foams. Based on the results, cellulose reinforced foams might potentially become a sustainable and biodegradable alternative for packaging and insulation materials.

Using similar components but a different approach, Hassan et al. [76] fabricated biodegradable starch/cellulose composite foams cross-linked with citric acid at 220 °C by compression molding. Increasing the concentration of citric acid made water absorption capacity decrease, while stiffness, tensile strength, flexural strength, and hydrophobicity of the starch/cellulose composite foams increased. For example, tensile strength, flexural modulus, and flexural strength increased from 1.76 MPa, 445 MPa, and 3.76 MPa, for 0 % citric acid to 2.25 MPa, 601.1 MPa, and 7.61 MPa, respectively, for the starch/cellulose composite foam cross-linked with 5% (w/w) citric acid. The foams also showed better thermal stability compared to the non-cross-linked composite foam, indicating that composite foams might be used as biodegradable alternatives to expanded polystyrene packaging.

In another study, lignin from bioethanol production was employed as a reinforcing filler by Luo et al. [77] to fabricate a soy-based polyurethane biofoam (BioPU) from two polyols (soybean oil-derived polyol SOPEP and petrochemical polyol Jeffol A-630) and poly(diphenylmethane diisocyanate) (pMDI). The results are similar to those obtained in other studies where different cellulose sources were used to reinforce lignin-induced cell structure modifications and thereby increase the density and improve the thermal properties of the foam. The mechanical properties were also improved with the presence of lignin, and the samples with 10% concentration had better mechanical properties over other treatments, with values of 0.46 MPa, 11.66 MPa, 0.87 MPa and 26.97 MPa for compressive strength, compressive modulus, flexural strength, and flexural modulus, respectively. According to the authors, the lignin–polyurethane mixtures are characterized by a complex super molecular architecture due to the specific properties of their components. Polyurethanes produced an interpenetrating polymer network (IPN) structure, whereas lignin acted as an emulsifier for polyurethane soft and hard segments because it was subtly dispersed and integrated into the polymer amorphous phase, thereby improving the mechanical properties of foams. The research assessed the potential utilization of lignin in polyurethane applications, such as fillers and coating.

Silva et al. [78], studied the use of different concentrations of cellulose fiber on rigid polyurethane foams (RPFs). Mechanical resistance and thermal stability of the composite foams were not significantly changed by the introduction of cellulose industrial residue fibers, whereas thermal conductivity displayed a minor reduction. Based on those results, cellulose–polyurethane composite foams are potentially useful for applications in thermal insulating areas. Interestingly, the composite foams showed a predisposition to fungal attack in wet environments due to the presence of cellulose fibers. However, in this case, this attribute is appropriate, as it decreases the environmental impact after disposal.

Due to the pressure of environmental concerns over the two last decades, considerable research and development in the area of nanocellulose-based materials have been extensively carried out. As a consequence, new products and applications of nanocellulose are steadily emerging as a range of applications of nanocellulose-based biodegradable polymers, thermoplastic polymers, and porous nanocomposites [1].

3. Conclusions

Applications of plant polymer-based solid foams in the food industry are mainly focused in two areas: edible foams and packaging materials. In these areas, there are many plant polymers that are utilized. However, most of the studies focused on the utilization of starch and cellulose, due to their availability and production costs. Still, it is observed that starch is not more widely studied in the field of edible foams, being a very common by-product of the agriculture and food industry. This is likely because pure starch makes weak and high water absorption foams, so starch must be modified, or other compounds must be incorporated, in order to strengthen the foams and reduce their water absorption. However, these increase the cost of the final product. In addition, a deep knowledge of starch behavior in the presence of other components is required to overcome some disadvantages, such as brittleness and high water absorption capacity.

In this context, most research aims to improve physical characteristics of solid foams, especially the mechanical and thermal properties which are generally impacted by the conditions of the foam process. However, creating a solid knowledge and comprehension of micro- to macrostructure modification will allow for more adequate management and design of processing conditions. Considering the above, there are two kinds of studies to perform. First, those in which researchers evaluate different properties or characteristics and link them to structural changes without further discussion. The second type goes beyond that and explains the changes based on phenomena produced by the several physical and chemical interactions among the components of the foams. This knowledge is crucial in order to tune the structural properties of solid foams through the control of the properties of the liquid foam (bubble size distribution, pore opening, foam density, etc.), which is a big challenge. However, research involving plant polymers in this area is still lacking.

Along with this, the successful application of solid foams depends on other aspects related to foam creation that are very challenging; for instance, the preservation of the liquid foam structure throughout the transition process from liquid to solid foam, and the timescale pairing between the stability of the liquid foam and the solidification. Thus, one of the major drawbacks is rooted in the internal structure of the plant polymer-based foams, which will probably collapse due to poor mechanical properties. Considering this, in the field of edible solid foams, the utilization of functional proteins as foaming agents, e.g., soy protein and lentil protein, helps keep internal structure to an extent. Still, stabilizers are needed, e.g., to act as Pickering particles or to increase viscosity. Other compounds also might help to reinforce the structure, e.g., cellulose, even though this has not been widely explored in this field. In another vein, the incorporation of natural fibers, such as cellulose nanofiber (CNF) or microfibrillated (CMF), has improved the mechanical properties of plant polymer-based solid foams, helping to overcome their natural lack of strength by reinforcing the structure to endure the foam drying process. Other serious limitations arise from the high hydrophilicity of plant polymer compounds in wet conditions and limited thermal resistance. These drawbacks are also addressed by adding cellulose fibers, especially lignin, which may provide the system with better water resistance.

Author Contributions: Conceptualization, M.J.-P. and L.C.; methodology: M.J.-P. and L.C.; investigation, M.J.-P.; resources, M.J.-P.; writing—original draft preparation, M.J.-P.; writing—review and editing, M.J.-P. and L.C.; project administration, M.J.-P.; funding acquisition, M.J.-P. All authors have read and agreed to the published version of the manuscript

Funding: This research and the APC were funded by Agencia Nacional de Investigación (ANID) de Chile, FONDECYT Iniciación, grant number 11180139.

Conflicts of Interest: The authors declare no conflict of interest.

References

1. Kargarzadeh, H.; Huang, J.; Lin, N.; Ahmad, I.; Mariano, M.; Dufresne, A.; Thomas, S.; Gałęski, A. Recent developments in nanocellulose-based biodegradable polymers, thermoplastic polymers, and porous nanocomposites. *Prog. Polym. Sci.* **2018**, *87*, 197–227. [CrossRef]
2. Svagan, A.J.; Samir, M.A.S.A.; Berglund, L.A. Biomimetic Foams of High Mechanical Performance Based on Nanostructured Cell Walls Reinforced by Native Cellulose Nanofibrils. *Adv. Mater.* **2008**, *20*, 1263–1269. [CrossRef]
3. Wu, Q.; Lindh, V.H.; Johansson, E.; Olsson, R.T.; Hedenqvist, M.S. Freeze-dried wheat gluten biofoams; scaling up with water welding. *Ind. Crop. Prod.* **2017**, *97*, 184–190. [CrossRef]

4. Cvrček, L.; Horáková, M. Chapter 14. Non-thermal plasma technology for polymeric materials. In *Non-Thermal Plasma Technology for Polymeric Materials*; Elsevier: Amsterdam, The Netherlands, 2019; pp. 367–407.

5. Fazeli, M.; Keley, M.; Biazar, E. Preparation and characterization of starch-based composite films reinforced by cellulose nanofibers. *Int. J. Biol. Macromol.* **2018**, *116*, 272–280. [CrossRef] [PubMed]

6. Zhong, Y.; Godwin, P.; Jin, Y.; Xiao, H. Biodegradable polymers and green-based antimicrobial packaging materials: A mini-review. *Adv. Ind. Eng. Polym. Res.* **2019**, *3*, 27–35. [CrossRef]

7. Motloung, M.P.; Ojijo, V.; Bandyopadhyay, J.; Ray, S.S. Ray cellulose nanostructure-based biodegradable nanocomposite foams: A brief overview on the recent advancements and perspectives. *Polymers* **2019**, *11*, 1270. [CrossRef]

8. Rydz, J.; Musioł, M.; Zawidlak-Węgrzyńska, B.; Sikorska, W. Present and future of biodegradable polymers for food packaging applications. *Biopolym. Food Des.* **2018**, 431–467. [CrossRef]

9. Kunduru, K.R.; Basu, A.; Domb, A.J. Biodegradable polymers: Medical applications. In *Encyclopedia of Polymer Science and Technology*; Wiley: Hoboken, NJ, USA, 2016; pp. 1–22. [CrossRef]

10. Jin, F.L.; Zhao, M.; Park, M.; Park, S.J. Recent trends of foaming in polymer processing: A review. *Polymers* **2019**, *11*, 6. [CrossRef]

11. Gutiérrez, T.J. Polymers for food applications: News. In *Polymers for Food Applications*; Springer: Cham, Switzerland, 2018; pp. 1–4. [CrossRef]

12. Saiz-Arroyo, C.; Rodríguez-Pérez, M.; Velasco, J.I.; de Saja, J.A. Influence of foaming process on the structure–properties relationship of foamed LDPE/silica nanocomposites. *Compos. Part B Eng.* **2013**, *48*, 40–50. [CrossRef]

13. Bergeron, V.; Walstra, P. Foams (chapter 7). In *Fundamentals of Interface and Colloid Science, Volume V Soft Colloids*; Elsevier Academic Press: Amsterdam, The Netherlands, 2005. Available online: https://www.sciencedirect.com/bookseries/fundamentals-of-interface-and-colloid-science/vol/5/suppl/C (accessed on 28 September 2021).

14. Kumar, S.; Singh, P.; Gupta, S.K.; Ali, J.; Baboota, S. Biodegradable and recyclable packaging materials: A step towards a greener future. In *Reference Module in Materials Science and Materials Engineering*; Elsevier Ltd.: Amsterdam, The Netherlands, 2020; pp. 328–337.

15. Nesic, A.; Castillo, C.; Castaño, P.; Cabrera-Barjas, G.; Serrano, J. Bio-based packaging materials. In *Biobased Products and Industries*; Elsevier: Amsterdam, The Netherlands, 2020; pp. 279–309.

16. Dollet, B.; Raufaste, C. Comptes rendus physique rheology of aqueous foams rhéologie des mousses aqueuses. *C. R. Phys.* **2014**, *15*, 731–747. [CrossRef]

17. Nastaj, M.; Sołowiej, B.G. The effect of various pH values on foaming properties of whey protein preparations. *Int. J. Dairy Technol.* **2020**, *73*, 683–694. [CrossRef]

18. Alavi, F.; Tian, Z.; Chen, L.; Emam-Djomeh, Z. Effect of CaCl2 on the stability and rheological properties of foams and high-sugar aerated systems produced by preheated egg white protein. *Food Hydrocoll.* **2020**, *106*, 105887. [CrossRef]

19. Shogren, R.; Wood, D.; Orts, W.; Glenn, G. Plant-based materials and transitioning to a circular economy. *Sustain. Prod. Consum.* **2019**, *19*, 194–215. [CrossRef]

20. Góral, I.; Wojciechowski, K. Surface activity and foaming properties of saponin-rich plants extracts. *Adv. Colloid Interface Sci.* **2020**, *279*, 102145. [CrossRef] [PubMed]

21. Gonzalez, P.J.; Sörensen, P.M. Characterization of saponin foam from Saponaria officinalis for food applications. *Food Hydrocoll.* **2019**, *101*, 105541. [CrossRef]

22. Cam, I.B.; Topuz, A. Production of soapwort concentrate and soapwort powder and their use in Turkish delight and tahini halvah. *J. Food Process Eng.* **2018**, *41*, e12605. [CrossRef]

23. Schmidt, J.M.; Damgaard, H.; Greve-Poulsen, M.; Larsen, L.B.; Hammershøj, M. Foam and emulsion properties of potato protein isolate and purified fractions. *Food Hydrocoll.* **2018**, *74*, 367–378. [CrossRef]

24. Ozcelik, M.; Ambros, S.; Morais, S.F.; Kulozik, U. Storage stability of dried raspberry foam as a snack product: Effect of foam structure and microwave-assisted freeze drying on the stability of plant bioactives and ascorbic acid. *J. Food Eng.* **2019**, *270*, 109779. [CrossRef]

25. He, Z.; Li, W.; Guo, F.; Li, W.; Zeng, M.; Chen, J. Foaming characteristics of commercial soy protein isolate as influenced by heat-induced aggregation. *Int. J. Food Prop.* **2014**, *18*, 1817–1828. [CrossRef]

26. Zhang, X.; Zhou, J.; Chen, J.; Li, B.; Li, Y.; Liu, S. Edible foam based on pickering effect of bacterial cellulose nanofibrils and soy protein isolates featuring interfacial network stabilization. *Food Hydrocoll.* **2019**, *100*, 105440. [CrossRef]

27. Zhang, X.; Lei, Y.; Luo, X.; Wang, Y.; Li, Y.; Li, B.; Liu, S. Impact of pH on the interaction between soybean protein isolate and oxidized bacterial cellulose at oil-water interface: Dilatational rheological and emulsifying properties. *Food Hydrocoll.* **2021**, *115*, 106609. [CrossRef]

28. Thuwapanichayanan, R.; Prachayawarakorn, S.; Soponronnarit, S. Effects of foaming agents and foam density on drying characteristics and textural property of banana foams. *LWT* **2012**, *47*, 348–357. [CrossRef]

29. Rajkumar, P.; Kailappan, R.; Viswanathan, R.; Raghavan, G.S.V.; Ratti, C. Foam mat drying of alphonso mango pulp. *Dry. Technol.* **2007**, *25*, 357–365. [CrossRef]

30. Zheng, X.-Z.; Liu, C.-H.; Zhou, H. Optimization of Parameters for Microwave-Assisted Foam Mat Drying of Blackcurrant Pulp. *Dry. Technol.* **2011**, *29*, 230–238. [CrossRef]

31. Ceresino, E.B.; Johansson, E.; Sato, H.H.; Plivelic, T.S.; Hall, S.A.; Kuktaite, R. Morphological and structural heterogeneity of solid gliadin food foams modified with transglutaminase and food grade dispersants. *Food Hydrocoll.* **2020**, *108*, 105995. [CrossRef]

32. Ceresino, E.; Johansson, E.; Sato, H.; Plivelic, T.; Hall, S.; Bez, J.; Kuktaite, R. Lupin protein isolate structure diversity in frozen-cast foams: Effects of transglutaminases and edible fats. *Molecules* **2021**, *26*, 1717. [CrossRef] [PubMed]

33. Philipp, C.; Oey, I.; Silcock, P.; Beck, S.M.; Buckow, R. Impact of protein content on physical and microstructural properties of extruded rice starch-pea protein snacks. *J. Food Eng.* **2017**, *212*, 165–173. [CrossRef]

34. Martínez-Sanz, M.; Larsson, E.; Filli, K.B.; Loupiac, C.; Assifaoui, A.; López-Rubio, A.; Lopez-Sanchez, P. Nano-/microstructure of extruded Spirulina/starch foams in relation to their textural properties. *Food Hydrocoll.* **2020**, *103*, 105697. [CrossRef]

35. Mitrus, M.; Moscicki, L. Extrusion-cooking of starch protective loose-fill foams. *Chem. Eng. Res. Des.* **2014**, *92*, 778–783. [CrossRef]

36. Savage, G.P. SAPONINS. In *Encyclopedia of Food Sciences and Nutrition*; Elsevier: Amsterdam, The Netherlands, 2003; pp. 5095–5098.

37. Çelik, I.; Yılmaz, Y.; Işık, F.; Üstün, O. Effect of soapwort extract on physical and sensory properties of sponge cakes and rheological properties of sponge cake batters. *Food Chem.* **2007**, *101*, 907–911. [CrossRef]

38. Alting, A.C.; Pouvreau, L.; Giuseppin, M.L.F.; van Nieuwenhuijzen, N.H. Potato proteins. In *Handbook of Food Proteins*; Elsevier: Amsterdam, The Netherlands, 2011; pp. 316–334.

39. Silva, S.S.; Fernandes, E.; Pina, S.; Silva-Correia, J.; Vieira, S.; Oliveira, J.; Reis, R. 2.11 Polymers of biological origin. *Compr. Biomater.* **2017**, *2*, 228–252.

40. Sun, X.S. Thermal and mechanical properties of soy proteins. In *Bio-Based Polymers and Composites*; Elsevier: Amsterdam, The Netherlands, 2005; pp. 292–326.

41. Lin, M.; Tay, S.H.; Yang, H.; Yang, B.; Li, H. Replacement of eggs with soybean protein isolates and polysaccharides to prepare yellow cakes suitable for vegetarians. *Food Chem.* **2017**, *229*, 663–673. [CrossRef]

42. Jarpa-Parra, M.; Wong, L.; Wismer, W.; Temelli, F.; Han, J.; Huang, W.; Eckhart, E.; Tian, Z.; Shi, K.; Sun, T.; et al. Quality characteristics of angel food cake and muffin using lentil protein as egg/milk replacer. *Int. J. Food Sci. Technol.* **2017**, *52*, 1604–1613. [CrossRef]

43. Jarpa-Parra, M. Lentil protein: A review of functional properties and food application. An overview of lentil protein functionality. *Int. J. Food Sci. Technol.* **2018**, *53*, 892–903. [CrossRef]

44. Bergel, B.F.; Osorio, S.D.; da Luz, L.M.; Santana, R.M.C. Effects of hydrophobized starches on thermoplastic starch foams made from potato starch. *Carbohydr. Polym.* **2018**, *200*, 106–114. [CrossRef] [PubMed]

45. Semeijn, C.; Buwalda, P.L. *Potato Starch*; Elsevier Ltd.: Amsterdam, The Netherlands, 2018.

46. Bergel, B.F.; Araujo, L.L.; Silva, A.L.D.S.D.; Santana, R.M.C. Effects of silylated starch structure on hydrophobization and mechanical properties of thermoplastic starch foams made from potato starch. *Carbohydr. Polym.* **2020**, *241*, 116274. [CrossRef] [PubMed]

47. Cruz-Tirado, J.P.; Ferreira, R.S.B.; Lizárraga, E.; Tapia-Blacido, D.R.; Silva, N.C.C.; Angelats-Silva, L.; Siche, R. Bioactive Andean sweet potato starch-based foam incorporated with oregano or thyme essential oil. *Food Packag. Shelf Life* **2020**, *23*, 100457. [CrossRef]

48. Cinelli, P.; Chiellini, E.; Lawton, J.W.; Imam, S.H. Foamed articles based on potato starch, corn fibers and poly (vinyl alcohol). *Polym. Degrad. Stab.* **2006**, *91*, 1147–1155. [CrossRef]

49. Hamaker, B.R.; Tuncil, Y.E.; Shen, X. *Carbohydrates of the Kernel*, 3rd ed.; Elsevier Inc.: Amsterdam, The Netherlands, 2018.

50. Loy, D.D.; Lundy, E.L. *Nutritional Properties and Feeding Value of Corn and Its Coproducts*, 3rd ed.; Elsevier Inc.: Amsterdam, The Netherlands, 2018.

51. Uslu, M.K.; Polat, S. Effects of glyoxal cross-linking on baked starch foam. *Carbohydr. Polym.* **2012**, *87*, 1994–1999. [CrossRef]

52. Polat, S.; Uslu, M.K.; Aygün, A.; Certel, M. The effects of the addition of corn husk fibre, kaolin and beeswax on cross-linked corn starch foam. *J. Food Eng.* **2013**, *116*, 267–276. [CrossRef]

53. Sanhawong, W.; Banhalee, P.; Boonsang, S.; Kaewpirom, S. Effect of concentrated natural rubber latex on the properties and degradation behavior of cotton-fiber-reinforced cassava starch biofoam. *Ind. Crop. Prod.* **2017**, *108*, 756–766. [CrossRef]

54. Wheatley, C.C.; Chuzel, G.; Zakhia, N. CASSAVA | The Nature of the Tuber. In *Encyclopedia of Food Sciences and Nutrition*, 2nd ed.; Caballero, B., Ed.; Elsevier Academic Press: Amsterdam, The Netherlands, 2003; pp. 964–969.

55. Salgado, P.R.; Schmidt, V.C.; Ortiz, S.E.M.; Mauri, A.N.; Laurindo, J.B. Biodegradable foams based on cassava starch, sunflower proteins and cellulose fibers obtained by a baking process. *J. Food Eng.* **2008**, *85*, 435–443. [CrossRef]

56. Mello, L.R.; Mali, S. Use of malt bagasse to produce biodegradable baked foams made from cassava starch. *Ind. Crop. Prod.* **2014**, *55*, 187–193. [CrossRef]

57. Machado, C.M.; Benelli, P.; Tessaro, I.C. Sesame cake incorporation on cassava starch foams for packaging use. *Ind. Crop. Prod.* **2017**, *102*, 115–121. [CrossRef]

58. Engel, J.B.; Ambrosi, A.; Tessaro, I.C. Development of biodegradable starch-based foams incorporated with grape stalks for food packaging. *Carbohydr. Polym.* **2019**, *225*, 115234. [CrossRef] [PubMed]

59. Cabanillas, A.; Nuñez, J.; Tirado, L.J.P.C.; Vejarano, R.; Tapia-Blácido, D.R.; Arteaga, H.; Siche, R. Pineapple shell fiber as reinforcement in cassava starch foam trays. *Polym. Polym. Compos.* **2019**, *27*, 496–506. [CrossRef]

60. Ferreira, D.C.; Molina, G.; Pelissari, F. Biodegradable trays based on cassava starch blended with agroindustrial residues. *Compos. Part B Eng.* **2020**, *183*. [CrossRef]

61. Matsuda, D.K.; Verceheze, A.E.; Carvalho, G.M.; Yamashita, F.; Mali, S. Baked foams of cassava starch and organically modified nanoclays. *Ind. Crop. Prod.* **2012**, *44*, 705–711. [CrossRef]

62. Vercelheze, A.E.; Fakhouri, F.M.; Dall'Antônia, L.H.; Urbano, A.; Youssef, E.Y.; Yamashita, F.; Mali, S. Properties of baked foams based on cassava starch, sugarcane bagasse fibers and montmorillonite. *Carbohydr. Polym.* **2012**, *87*, 1302–1310. [CrossRef]

63. Pornsuksomboon, K.; Holló, B.B.; Szécsényi, K.M.; Kaewtatip, K. Properties of baked foams from citric acid modified cassava starch and native cassava starch blends. *Carbohydr. Polym.* **2016**, *136*, 107–112. [CrossRef]

64. Cruz-Tirado, J.P.; Siche, R.; Cabanillas, A.; Díaz-Sánchez, L.; Vejarano, R.; Tapia-Blácido, D.R. Properties of baked foams from oca (Oxalis tuberosa) starch reinforced with sugarcane bagasse and asparagus peel fiber. *Procedia Eng.* **2017**, *200*, 178–185. [CrossRef]

65. Zhu, F.; Cui, R. Comparison of molecular structure of oca (Oxalis tuberosa), potato, and maize starches. *Food Chem.* **2019**, *296*, 116–122. [CrossRef] [PubMed]

66. Ramos Diaz, J.M.; Kirjoranta, S.; Tenitz, S.; Penttilä, P.A.; Serimaa, R.; Lampi, A.M.; Jouppila, K. Use of amaranth, quinoa and kañiwa in extruded corn-based snacks. *J. Cereal Sci.* **2013**, *58*, 59–67. [CrossRef]

67. Anton, A.A.; Gary Fulcher, R.R.; Arntfield, S.D. Physical and nutritional impact of fortification of corn starch-based extruded snacks with common bean (Phaseolus vulgaris L.) flour: Effects of bean addition and extrusion cooking. *Food Chem.* **2009**, *113*, 989–996. [CrossRef]

68. Zhang, W.; Li, S.; Zhang, B.; Drago, S.R.; Zhang, J. Relationships between the gelatinization of starches and the textural properties of extruded texturized soybean protein-starch systems. *J. Food Eng.* **2016**, *174*, 29–36. [CrossRef]

69. Nastaj, M.; Sołowiej, B.G.; Terpiłowski, K.; Mleko, S. Effect of erythritol on physicochemical properties of reformulated high protein meringues obtained from whey protein isolate. *Int. Dairy J.* **2020**, *105*, 104672. [CrossRef]

70. Nastaj, M.; Mleko, S.; Terpiłowski, K.; Tomczyńska-Mleko, M. Effect of sucrose on physicochemical properties of high-protein meringues obtained from whey protein isolate. *Appl. Sci.* **2021**, *11*, 4764. [CrossRef]

71. Emblem, A. Plastics properties for packaging materials. In *Packaging Technology: Fundamentals, Materials and Processes*; Elsevier: Amsterdam, The Netherlands, 2012; pp. 287–309.

72. Ago, M.; Ferrer, A.; Rojas, O.J. Starch-based biofoams reinforced with lignocellulose nanofibrils from residual palm empty fruit bunches: Water sorption and mechanical strength. *ACS Sustain. Chem. Eng.* **2016**, *4*, 5546–5552. [CrossRef]

73. Löbmann, K.; Svagan, A.J. Cellulose nanofibers as excipient for the delivery of poorly soluble drugs. *Int. J. Pharm.* **2017**, *533*, 285–297. [CrossRef]

74. Cervin, N.T.; Andersson, L.; Ng, J.B.S.; Olin, P.; Bergström, L.; Wågberg, L. Lightweight and strong cellulose materials made from aqueous foams stabilized by nanofibrillated cellulose. *Biomacromolecules* **2013**, *14*, 503–511. [CrossRef]

75. Ghanbari, A.; Tabarsa, T.; Ashori, A.; Shakeri, A.; Mashkour, M. Thermoplastic starch foamed composites reinforced with cellulose nanofibers: Thermal and mechanical properties. *Carbohydr. Polym.* **2018**, *197*, 305–311. [CrossRef]

76. Hassan, M.; Tucker, N.; Le Guen, M. Thermal, mechanical and viscoelastic properties of citric acid-crosslinked starch/cellulose composite foams. *Carbohydr. Polym.* **2019**, *230*, 115675. [CrossRef]

77. Luo, X.; Mohanty, A.; Misra, M. Lignin as a reactive reinforcing filler for water-blown rigid biofoam composites from soy oil-based polyurethane. *Ind. Crop. Prod.* **2013**, *47*, 13–19. [CrossRef]

78. Silva, M.C.; Takahashi, J.A.; Chaussy, D.; Belgacem, M.N.; Silva, G.G. Composites of rigid polyurethane foam and cellulose fiber residue. *J. Appl. Polym. Sci.* **2010**, *117*, 3665–3672. [CrossRef]

applied sciences

Article

Design of a New Fermented Beverage from Medicinal Plants and Organic Sugarcane Molasses via Lactic Fermentation

Hamza Gadhoumi [1,2,*], Maria Gullo [3], Luciana De Vero [3], Enriqueta Martinez-Rojas [4], Moufida Saidani Tounsi [2] and El Akrem Hayouni [2]

[1] Faculty of Sciences of Tunis, University of Tunis El Manar, El Manar, Tunis 2092, Tunisia
[2] Laboratory of Aromatic and Medicinal Plants, Center of Biotechnology at the Ecopark of Borj-Cédria, BP 901, Hammam-Lif 2050, Tunisia; moufidatounsi@yahoo.fr (M.S.T.); a.hayouni@gmail.com (E.A.H.)
[3] Department of Life Sciences, University of Modena and Reggio Emilia, Partner of the JRU MIRRI-IT, 42122 Reggio Emilia, Italy; maria.gullo@unimore.it (M.G.); luciana.devero@unimore.it (L.D.V.)
[4] Department of Agriculture and Food Sciences, Neubrandenburg University of Applied Sciences, 17033 Neubrandenburg, Germany; martinez-rojas@hs-nb.de
* Correspondence: gadhoumih@gmail.com

Abstract: Functional beverages obtained using medicinal plants and fermented with lactic acid bacteria are gaining much interest from the scientific community, driven by the growing demand for food and beverages with beneficial properties. In this work, three different batches of medicinal plants and organic sugarcane molasses, named FB-lc, FB-sp and FB-lcsp, were prepared and fermented by using *Lactobacillus acidophilus* ATCC 43121, *Bifidobacterium breve* B632 and a mix of both strains' culture, respectively. The three fermented beverages revealed a high level of polyphenols (expressed as gallic acid equivalent), ranging from 182.50 to 315.62 µg/mL. The highest content of flavonoids (152.13 µg quercetin equivalent/mL) and tannins (93.602 µg catechin equivalent/mL) was detected in FB-lcsp trial. The IR spectroscopy analysis showed a decrease in sugar (pyranose forms, D-glucopyranose and rhamnosides). In addition, the aromatic compounds of the fermented beverages, detected by GC-MS headspace analysis, showed twenty-four interesting volatile compounds, which could give positive aroma attributes to the flavor of the beverages. The highest antioxidant activity was observed in the beverage obtained by the mix culture strains. Accordingly, the production of these beverages can be further investigated for considering their well-being effects on human health.

Keywords: lactic fermentation; functional beverages; volatile compounds; antioxidant activity

Citation: Gadhoumi, H.; Gullo, M.; De Vero, L.; Martinez-Rojas, E.; Saidani Tounsi, M.; Hayouni, E.A. Design of a New Fermented Beverage from Medicinal Plants and Organic Sugarcane Molasses via Lactic Fermentation. *Appl. Sci.* **2021**, *11*, 6089. https://doi.org/10.3390/app11136089

Academic Editor: Maria Kanellaki

Received: 26 May 2021
Accepted: 28 June 2021
Published: 30 June 2021

Publisher's Note: MDPI stays neutral with regard to jurisdictional claims in published maps and institutional affiliations.

1. Introduction

Since ancient times, medicinal plants have been used for human well-being as they are a source of healthy metabolic compounds. They still are widely used for the production of food, medicines and fuel. The improvement of the biological properties of plants, vegetables and herbs can be achieved through fermentation technology, which can transform complex substrates and change the quantity of some bioactive compounds [1]. Additionally, the fermentation process allows the preservation of food and beverages while improving the nutritional and organoleptic quality of the products [2]. Selected microorganisms are widely applied to obtain fermented foods and beverages. Among them, yeasts, lactic acid bacteria (LAB) and acetic acid bacteria (AAB) are the main organisms with beneficial technological applications and for which a wide knowledge of use is available [3,4]. In some cases, mixed cultures are preferred over single cultures since they can offer improved processes and higher flavor and bioactive compound content [5], as well as health benefits [6]. Applying selected microbial cultures to suitable raw materials, the fermentation process can be modulated, also enhancing the content of bioactive metabolites and other compounds with antioxidant activity [7–10]. Since the production of specific metabolites is a strain-specific trait, the choice of microorganisms for the intended use is fundamental to obtain fermented products with enhanced properties [11–13].

A comprehensive inventory of microorganisms and their status as GRAS (generally recognized as safe) and/or QPS (qualified presumption of safety) organisms for the intended use is provided by Bourdichon and Casaregola [14].

In particular, LAB strains can produce key volatile compounds, active amino acids, peptides and sugars that contribute to the final flavor of fermented food products [15–17].

Moreover, due to the probiotic activity of some LAB, they can confer further attributes to fermented foods and beverages [18,19].

Plants are also suitable substrates for the growth of LAB [20–22]. Aromatic and medicinal plants are particularly valuable for selective bioprocesses as they already contain many bioactive compounds, including phenolic compounds, carotenoids, anthocyanins and tocopherols [23,24]. Moreover, the content of these compounds can be increased by the metabolic activity of the microorganisms involved in the fermentation process [25–27]. The metabolism of phenolic compounds correlated with increased antioxidant activity was observed during lactic acid fermentation [23,28].

Several authors have successfully obtained fermented products with improved antioxidant activity from vegetables, fruits, sugarcane molasses or plants [29,30]. Functional beverages that contain significant amounts of bioactive compounds can offer several health benefits. For instance, they can prevent some damaging physiological activities including metabolic and cardiovascular diseases [21–33].

Several LAB strains have been proposed as candidate probiotics due to their ability to provide functional attributes to foods and beverages.

Within the *Bifidobacterium* genus, some species are reported as potential probiotics for the treatment of enteric disorders in newborns such as infantile colics or as preventive agents for infantile diarrhea of bacterial origin. This function was also associated with strong antimicrobial activity against coliforms and other pathogenic bacteria [18]. Moreover, *Lactobacillus* species show several probiotic traits and are widely used in dairy and no dairy foods and beverages production. Among the functional attributes, those contributing to lower cholesterol levels are reported to be related to the action of strains belonging to the *Lactobacillus acidophilus* species [19]. The selection of appropriate organisms is the first step in designing and implementing bioprocesses for obtaining functional foods and beverages, as well as the availability of strain cultures in culture collections with known and validated functions [34,35].

This study aims to explore the effect of single and mixed culture of LAB strains, previously indicated as candidate probiotics, for the production of functional beverages starting with plants and organic sugarcane molasses as the raw material. The main fermentation parameters, the biochemical composition (carbohydrate, polyphenols, proteins), the volatile content and the antioxidant activity of the obtained beverages is assessed.

2. Materials and Methods

2.1. Plant Material and Fermentation

The plant material used for the formulation of three fermented plant beverages is shown in Table 1. Some plants were collected from Orbata National Park Mountain (Gafsa, Tunisia) with coordinates: N 34°22′49.8″ and E 9°3′23.4″ and others were purchased from the commercial central market of Tunisia. A voucher specimen was deposited in the Laboratory of Extremophile plants, Center of Biotechnology at the Ecopark of Borj-Cédria, and was identified by Prof. Abderrazak Smaoui. The selected plant materials were harvested and mixed with 30 g of organic sugarcane molasses solubilized in 1 L of distilled water. After the adjustment of the pH at 7.8, the solution was sterilized by pasteurization (80 °C in 15 min). Three equivalent batches were prepared and fermented using the selected strains *Lactobacillus acidophilus* ATCC 43121 and *Bifidobacterium breve* B632, provided by the Microbiology laboratory at the Centre of Biotechnology of Borj–Cédria.

Table 1. Plant material used for the fermentation process.

Plants	Botanical Family	Weight (g/L)	Used Organ	Condition
Nigella sativa	*Ranunculaceae*	5	seeds	Dry
Foeniculum Vulgare	*Apiaceae*	28	seeds	Dry
Ficus indica	*Cactaceae*	90	fruits	Fresh
Linum usitatissimum	*Linaceae*	10	seeds	Dry
Vitis vinifera	*Vitaceae*	5	seeds	Dry
Lavandula multifida	*Lamiaceae*	18	leaves	Fresh
Periploca laevigata	*Apocynaceae*	10	root	Fresh
Thymus hirtus sp. algeriensis.	*Lamiaceae*	33	leaves	Fresh
Zingiber officinalis	*Zingiberaceae*	20	root	Fresh

Specifically, a single culture of each strain, *L. acidophilus* ATCC 43121 and *B. breve* B632, was used to ferment batches named FB-lc and FB-sp, respectively. In addition, a mixed culture from both the strains was prepared and used to ferment the batch named FB-lcsp. The LAB cultures were previously grown on de Man Rogosa Sharpe (MRS broth, Oxoid, UK) at 37 °C for 24 h. Then, 10 mL of each culture inoculum with a concentration of 6 log cells/mL was added to each batch. After 30 days of fermentation, the three beverages were sterilized by filtration (0.2-μm filters) and preserved at +4 °C until analysis.

2.2. Fermentative Parameters

Titratable acidity (TA) was measured by titration using 0.1 mol/L NaOH solution and phenolphthalein as an indicator. The pH was detected by a digital potentiometer. The evaluation of reducing sugars was assessed using the dinitrosalicylic acid (DNS) method described by Warwick et al. [36]. Total proteins were measured using Bradford methods and the free amino acid according to the method described by Hermosín et al. [37]. Viable cell counts were made by inoculating 0.1 mL of samples, subjected to serial dilution in phosphate buffered saline (PBS), on MRS agar plates. Bacteria colonies were counted after 2 days of incubation at 30 °C.

2.3. Total Phenolic Content

According to the Folin–Ciocalteu method described by Tlili et al. [38], the total phenolic content was detected. A volume of 0.5 mL of Folin–Ciocalteu reagent and 1.25 mL of Na_2CO_3 (7% *w/v*) were added to 0.125 mL of each fermented beverage. The absorbance of each sample was measured at 765 nm after incubation of the tubes for 90 min in the dark. The total polyphenols content was expressed as μg gallic acid equivalents per mL of fermented beverage (μg GAE/mL).

2.4. Total Flavonoid Content

The total flavonoid content was assessed according to the method reported by Dewanto et al. [39]. Total flavonoids, expressed as μg of quercetin equivalent per mL of fermented beverage (μg QE/mL), were estimated concerning the quercetin standard curve (concentration range: 100–750 μg/mL).

2.5. Condensed Tannins Contents

The condensed tannins were assessed according to the method reported by Sun et al. [40]. The number of condensed tannins, expressed as μg catechin equivalent per mL of fermented beverage (μg CE/mL), was measured using a standard calibration curve.

2.6. Polysaccharides Analyses

2.6.1. Extraction

The polysaccharides dissolved in the filtrate obtained from the fermented beverages were precipitated by adding absolute ethanol (four times the volume of the filtrate) and were incubated at 4 °C for 24 h. The solution was then centrifuged at 4500× *g* rpm for

10 min. The precipitate obtained was collected and again dissolved in 20 mL of distilled water deproteinized with 2 mL of Sevag reagent (chloroform/butanol 4: 1, *v/v*) according to the method described by Navarini et al. [41]. Finally, the solution obtained was dialyzed using a dialysis membrane (2 µm) against bi-distilled water.

2.6.2. Infrared Spectrum Analysis

The organic functional groups were investigated by Fourier transform infrared (FTIR) analysis, which was performed using an FTIR spectrophotometer (Bruker, Ettlingen, Germany) with a spectral range of 4000–400 cm^{-1} [42]. Briefly, 2 mg of each sample of polysaccharides was diluted with 200 mg spectroscopic grade potassium bromide (KBr) powder, ground and pressed into 1 mm pellets.

2.7. Headspace Solid-Phase Microextraction of Volatile Compounds

The volatile compounds were analyzed by Headspace Solid-Phase Microextraction (HS-SPME) coupled with Gas Chromatography–Mass Spectrometry (GC–MS) as previously described by Tian et al. [43]. The GC–MS headspace analysis was performed with an iron sources temperature of 240 °C and an ionization voltage of 70 eV. The mass spectrometer was operated in scan mode from m/z 50 to 350. Peak areas were determined for each compound by integrating a selected ion unique to that compound. The Kovats retention index (RI) was calculated with a homologous series of n-alkanes (C6-C28) under the same conditions applied for the sample analyses. The volatile compounds (OAV > 1) were considered to contribute to the aroma of fermented beverages [44,45].

2.8. Antioxidant Activity Determination

The antioxidant activities of different fermented beverages were determined by the following tests. The α, α-diphenyl-β-picrylhydrazyl (DPPH) free radical scavenging method was performed according to Tlili et al. [38]. The reducing power was determined using the method described by Oyaizu [46]. The appearance of the blue green color was measured at 700 nm, vitamin C was used as the control. The total antioxidant capacity of each extract was estimated using the method described by Prieto et al. [47]; the absorbance was detected at 695 nm. Moreover, a 2,2-azino-bis diammonium (ABTS$^{\bullet+}$) radical scavenging assay was assessed using the method described by Hayouni et al. [48]. A total of 25 µL of the fermented beverages or the standard trolox was added to 2 mL of diluted ABTS$^{\bullet+}$ solution and the absorbance was measured at 734 nm. The scavenging ability was expressed as IC$_{50}$ (µg/mL), which is defined as the concentration inhibitive at 50% of free radical scavenging.

2.9. Statistical Analyses

All the experiments were performed in triplicate. The results were reported as mean values of three replicates ± the standard deviation (sd). The differences between the average variables were considered significant at $p < 0.05$. XLSTAT statistical software version 2016 (Addinsoft, New York, USA) was used for data processing.

3. Results and Discussion

3.1. Fermentation Trials

The colony counts of the selected LAB cultures, starting from the value of 6.5 log cfu/mL, increased during the 30 days of the fermentation trials. The highest value of 12 log cfu/mL was reached in the samples fermented with the mixed culture (FB-lcsp), after 20 days of fermentation. Later, the LAB population decreased, and the final values detected were 11, 10 and 8 log cfu/mL for FB-lcsp, FB-lc and FB-sp, respectively (Figure 1a). A substantial decrease of the pH and a corresponding increase of the titratable acidity was observed during the fermentation process (Figure 1b). The lowest pH value (3.7) and the highest TA value (3.4 mL NaOH 0.1N) were assessed for the FB-lcsp samples.

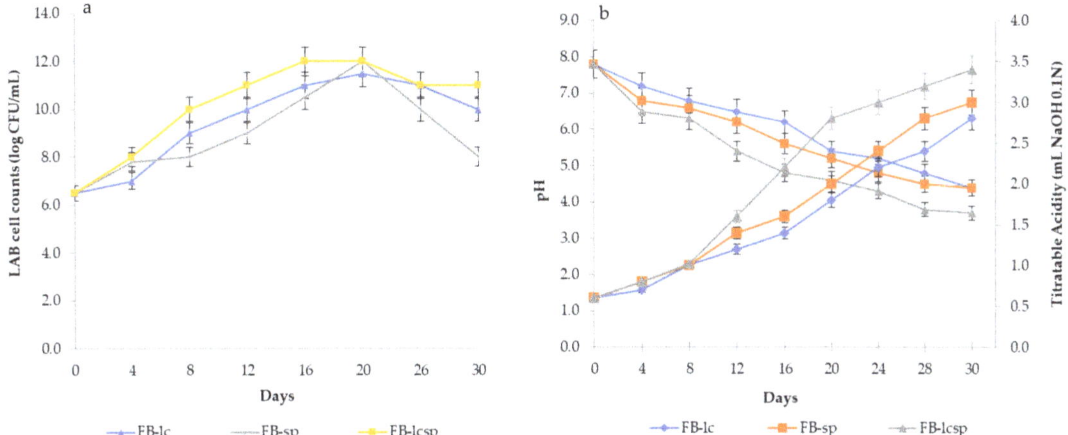

Figure 1. LAB cells count, expressed as log of colony forming unit/mL (**a**), pH and titratable acidity (**b**) of the three different products during the lactic acid fermentation process. Results are reported as mean values of 3 replicates.

A decrease of total and reducing sugars was observed, probably due to the acid production by the LAB. In addition, a degradation of the total proteins and a decrease of the free amino acid content occurred (Table 2). These results demonstrate the effectiveness of the fermentation, confirming the degradation of sugars and proteins by the LAB and the production of organic acids and new metabolites. It is interesting to note that the variation of all the considered parameters was higher for the trial obtained from FB-Lcsp, due to the combined effect of *B. breve* B632 and *L. acidophilus* ATCC 43121 cultures. During the fermentation process, the soluble components of proteins, carbohydrates and free sugars were readily available, and the consumption of these nutrients by LAB led to a drop in pH levels [49]. Moreover, the beverage was obtained by selecting hetero-fermentative LAB results that had a sparkling attribute, due to CO_2 produced, which is desired by consumers [49].

Table 2. Total sugars and proteins, total phenolic compounds, flavonoids and tannins contents of obtained beverages.

Chemical Parameters	Batch		
	FB-lc	**FB-sp**	**FB-lcsp**
Total sugars (mg/L)	23.7 ± 2.2	22.1 ± 3.3	20.3 ± 0.5
Reducing sugars (mg/L)	12.5 ± 1.4	10.3 ± 0.8	9.5 ± 0.4
Total Protein (mg/L)	17.8 ± 0.5	16.63 ± 1.4	15.6 ± 0.3
Total phenolics (μg GAE/mL) *	182.5 ± 12	289.7 ± 11	315.6 ± 21
Flavonoides (μg RE/mL) **	95.2 ± 8.2	119.3 ± 2	152.1 ± 5.3
Tannins (μg CE/mL) ***	87.6 ± 7.11	73.1 ± 3	93.6 ± 9.1

FB-sp: obtained beverage fermented by *Bifidobacterium breve B632*. FB-lcsp: the obtained beverage fermented by a mix of both strains. FB-lc: the obtained beverage fermented by *L. acidophilus* ATCC 43121. * μg of gallic acid equivalents per mL of beverages. ** μg of rutin equivalent per mL of beverages. *** μg of catechin equivalent per mL of beverages. Results are expressed as the mean of 3 replicates.

3.2. Polyphenol, Flavanoids and Condensed Tannins Content

The polyphenols represent the major group of antioxidants of plant materials [50,51]. They include a wide variety of bioactive metabolites that contain at least one aromatic-hydroxyl group in addition to other substituents with large spectra of biological activities [45]. They are divided into different groups such as flavonoids, tannins and phenolic acids [45]. In this study, the amount of total phenolics, flavonoids and condensed tannins varied in the three beverages (Table 2). A high level of polyphenols, with values of 315.6;

289.7 and 182.5 µg GAE/mL, were observed for FB-lcsp, FB-sp and FB-lc, respectively. The highest content of flavonoids (152.1 µg RE/mL) and tannins (93.6 µg CE/mL) was found in the FB-lcsp trial. It is well known that the secondary metabolites possess a wide range of biochemical and pharmacological properties [48]. These results suggest that the mixed culture improves the extraction of secondary metabolites such as flavonoids, tannins and phenolic acids. During fermentation, many changes of composition occur, leading to a modified ratio of nutrients and anti-nutrients and, therefore, the properties of the product, such as bioactivity and digestibility, are modified [52]. Lactic fermentation of plants has been shown to increase the concentration of several phenolic compounds [53]. So, the richness of our fermented plant beverages with phenolic compounds, especially FB-lcsp, could be an important starting point to explore their biological activities.

3.3. Infrared Spectrum

The infrared spectrum was used to identify the polysaccharides by detecting the stretching vibration of related groups, which allows identification of a variety of functional groups, the type of sugar and glycosides. Figure 2 shows the FTIR spectra with the characteristic absorptions of the beverages before and after fermentation.

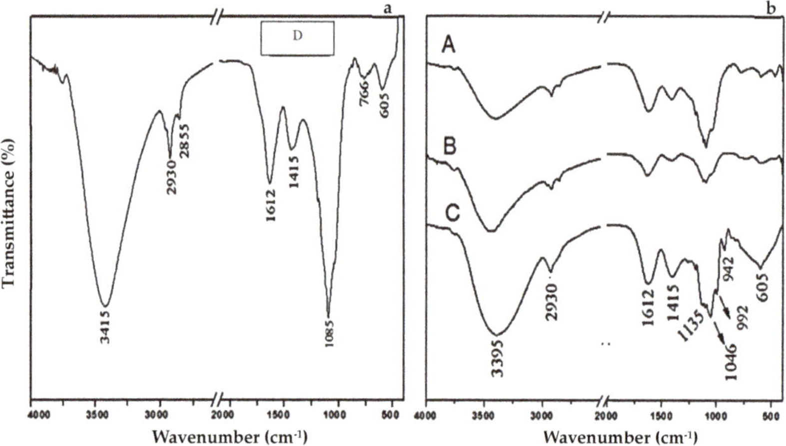

Figure 2. Infrared spectrum of polysaccharides, recorded for wavelengths 500–4000 cm^{-1}. D refers to the polysaccharides detected in not fermented beverage (**a**). A, B, and C refer to the polysaccharides detected in the beverages fermented with the selected cultures FB-sp, FB-lcsp, and FB-lc, respectively (**b**).

Substantial differences between the spectra were, indeed, observed, and the intensity of the peaks was higher before the fermentation (Figure 2a) and decreased in the fermented beverages (Figure 2b), indicating an almost total degradation of sugars.

In particular, the infrared spectrum analysis of the polysaccharides, recorded for wavelengths 500–4000 cm^{-1}, showed an intense peak between (3415–3395 cm^{-1}), representative of the vibrations of the hydroxyl groups (-OH), followed by a weak vibration of the C-H band at 2930 cm^{-1} [54]. Peaks with vibration bands between (3415–2930 cm^{-1}) are unidentified organic groups (Figure 2a). The peak at 1612 cm^{-1} is attributed to the vibration of the C = O group in its protonated carboxylic form. The peak at 1415 cm^{-1} is attributed to the vibration of the COO- group in its deprotonated carboxylic form polysaccharide extract [55]. Moreover, the peaks detected at 1200–1000 cm^{-1} indicate that the fractions are pyranose forms of carbohydrate [56]. The peaks with vibration bands at 992 cm^{-1} and 942 cm^{-1} (Figure 2b) derive from the antisymmetric stretching of tetrahydropyran, suggesting the presence of an arrangement in the carbohydrate units [57]. The peak at 605 cm^{-1} is related to D-glucopyranose. Additionally, the vibration at 766 cm^{-1} is related to rhamnoside present in the polysaccharide. The functional groups, such as –OH, –COOH and C=O, largely found in carbohydrate in addition to anionic and cationic groups, have a good ability to counteract oxidative stress, which is expressed by excessive production of reactive oxygen species [58].

3.4. Volatile Compounds

Twenty-four volatile compounds were identified with remarkable differences between the three beverages (Table 3). Among the volatile compounds detected, some are typically produced by plants and other compounds are generated by the LAB metabolism [59]. The data showed high variations in aroma compounds. Some compounds were present only in FB-sp (1-pentanol 2-methyl, and 2-butyl acetate), and others were detected only in FB-lcsp (isoamyl alcohol, propanoic acid- 1-methylpropyl ester, α-pinene, camphene, l-phellandrene, and α-terpinene). Lactic acid, acetic acid, propyl ester, propanoic acid, propyl ester; 1,8-cineole and fenchone were also found in the three beverages. During fermentation, many changes of composition occur, leading to a modified ratio of nutrients and anti-nutrients and, therefore, the properties of the product, such as bioactivity and digestibility, are modified [60]. Among volatile compounds, alcohols are considered the principal compounds in the aromatic profile of fermented plant beverages and are a common result of the metabolism of carbohydrate and amino acids [61]. Alcohols can represent essential preservative and aromatic components in fermented plant beverages. Moreover, they can also be used as a solvent for other volatile compounds and, thus, can contribute more to the global flavor [62]. The presence of ester compounds in the fermented beverages due to the esterification of free acids with alcohol was also observed. Several volatile components in fermented plant beverages can contribute to sweet aromatic notes [63]. Our results suggest that the mixed culture of LAB could provide a suitable aroma to the fermented plants beverages.

Table 3. Aromatic compounds identified from fermented beverage extracts of by GC–MS head space.

Compounds	Ret Time	Type	FB-lc		FB-sp		FB-lcsp	
			Area	Area %	Area	Area %	Area	Area %
acetic acid, ethyl ester	3.179	BV	NF	NF	3,714,514	2.94	42,368,725	7.472
butyraldehyde, 2-methyl-	3.631	BB	3,166,922	2.372	3,165,840	2.50	3,259,781	0.574
2-amino-1,3-propanediol	4.085	BV	NF	NF	3,512,197	2.78	24,311,405	4.287
n-propyl acetate	4.282	PV	2,752,332	2.062	3,065,189	2.42	112,465,982	19.83
1-butanol, 3-methyl-, formate	4.676	BV	2,205,104	1.652	2,365,918	1.87	8,500,143	1.499
propanoic acid	4.975	PV	1,216,092	0.911	1,305,607	1.03	9,172,563	1.617
propanoic acid, propyl ester	5.857	PV	724,850	0.543	832,916	0.66	5,365,518	0.946
2-oxopentanedioic acid	5.996	BV	1,244,074	0.932	2,245,910	1.78	NF	NF
lactic acid	7.841	BV	566,468	0.424	9,635,316	7.62	2,134,730	0.376
alpha-pinene	8.138	VV	480,350	0.359	526,813	0.42	12,589,365	2.220
camphene	8.414	VB	449,670	0.336	563,489	0.45	6,925,660	1.221
α-phellandrene	9.404	PV	NF	NF	448,923	0.00	13,479,075	2.377
alpha-terpinene	9.633	BV	841,209	0.630	1,935,872	0.36	9,501,040	1.676
p-cimene	9.790	VB	28,89,327	2.164	2,889,327	0.00	18,702,987	3.298
D-limonene	9.879	BV	NF	NF	125,683	1.53	10,850,613	1.914
eucalyptol	9.950	VB	82,174,283	61.569	45,227,359	2.29	8,296,759	1.463
gamma-terpinene	10.522	PV	887,853	0.665	889,853	0.10	2,486,838	0.438
terpinolene	11.298	VV	2,302,629	1.725	2,302,629	35.77	159,646,370	28.158
fenchone	11.300	VV	NF	NF	NF	0.70	29,477,277	5.199
linalol	11.554	BB	604,976	0.453	20,304,384	1.82	36,435,120	6.426
camphor	12.844	BB	12,301,015	9.216	12,303,015	9.73	8,640,472	1.523
endo-borneol	13.341	PV	3,391,528	2.541	4,485,728	16.06	28,204,267	4.975
terpinen-4-ol	13.556	VB	15,049,231	11.275	3,263,980	9.73	14,150,705	2.496
α-terpineol	13.885	VB	218,307	0.163	1,318,397	3.55	NF	NF

FB-sp: beverage fermented by *Bifidobacterium breve B632*. FB-lcsp: beverage fermented by a mix of both strains. FB-lc: beverage fermented by *Lactobacillus acidophilus* ATCC 43121. NF: not found. BV: baseline and valley points (start at a baseline point, end at a dropline from a valley point). VB: Valley dropline and baseline points. PV: penetration point and valley point. BB: start at a baseline point, end at a dropline baseline point. VV: start and dropline at a valley point.

3.5. Antioxidant Activity

The metabolization and depolymerization of phenolic compounds, correlated with increased antioxidant activity, have been observed during lactic acid fermentation [64]. In this study, the results of total antioxidant capacity (TAC) showed a similar motif to secondary metabolite contents. The TAC differed significantly among the strains used (Table 4). The three beverages showed high antioxidant activity, in particular, FB-lcsp (82.69 µg GAE/mL). Lower values were detected in samples FB-sp and FB-lc with 73.2 and 64.83 µg GAE/mL, respectively.

Table 4. Total antioxidant capacity and the free radical scavenging activity of DPPH, ABTS and the reducing power essay of the three fermented beverages.

Antioxidant Activity	FB-lc	FB-sp	FB-lcsp	BHT	Trolox	Vitamin C
Total antioxidant capacity (µg EGA/mL)	64.8 ± 0.6	73.2 ± 2.4	82.6 ± 6.3			
DPPH (µg/mL)	17.8 ± 1.6	16.5 ± 3.1	13.4 ± 8.3	11.8 ± 2.6		
ABTS (µg/mL sample)	24.6 ± 1.6	19.3 ± 1.1	17.4 ± 0.96		10.2 ± 1.5	
Reducing power (µg/mL)	55.9 ± 4.8	48.1 ± 4.1	36.1 ± 3.1			12.5 ± 2

The scavenging ability was expressed as IC_{50} (µg/mL). The reducing power activity (FRPA) was expressed as EC_{50} (µg/mL). Results are expressed as mean of 3 replicates $\pm$ standard deviation.

The DPPH scavenging activity of the fermented beverages also varied among the fermentation trials (Table 4). Samples of FB-lcsp showed the highest antioxidant activity, compared to FB-sp and FB-lc. The free radical scavenging activity $ABTS^{\bullet+}$ of each sample was also different (Table 4). This activity was higher in samples FB-lcsp (IC_{50}; 17.4 ± 0.96 µg/mL) compared to those detected in FB-sp and FB-lc samples (IC_{50}; 19.3 ± 1.1 and 24.6 ± 1.6 µg/mL, respectively).

Regarding the reducing power activity (FRPA) of fermented beverages, expressed as EC_{50} (µg/mL), all the beverages were able to lower Fe^{3+} to Fe^{2+}, but the results differed among the three samples. Specifically, the highest FRPA was detected in FB-lcsp samples (Table 4). A significantly lower activity was found in FB-sp and FB-lc samples, indicating that the double culture was able to enhance and improve the antioxidant activity in the fermented beverages more than the monoculture. Indeed, it is well known that the total antioxidant capacity is essentially due to the presence of phenolic and bioactive compounds. Several studies reported that the polyphenols present in fermented beverages possess high scavenging activity against oxidative stress [65–67]. Recently, fungal fermentation has been used to increase the antioxidant and anti-cholesterolemic activities of food and medical compounds as well as for food preservation [68]. Moreover, many bioactive compounds, formed during seaweed fermentation, have been reported to scavenge free radicals and reactive oxygen species [69].

4. Conclusions

In this study, medicinal plants and organic sugarcane molasses were tested as raw materials to produce functional beverages via lactic fermentation. *L. acidophilus* ATCC 43121 and *B. breve* B632 were selected and used in single and multiple cultures. In particular, the strain *B. breve* B632 is widely described in the literature as a potential probiotic that is useful for the treatment of enteric disorders in newborns. Strains belonging to the *L. acidophilus* species are also valuable for functional attributes among which is the ability to lower cholesterol level.

In our work, the three fermented beverages revealed an enhanced level of polyphenols, flavonoids and tannins, especially those obtained by using the mixed culture FB-lcsp. In addition, the detection of aromatic compounds in the fermented beverages showed twenty-four interesting volatile compounds, which could give pleasant-positive aroma attributes to the flavor of the beverages. The highest antioxidant activity was observed in the beverage obtained by the mix culture strains. This innovative approach, based on lactic fermentation performed by a mixed LAB culture, can generally be used to ameliorate secondary metabolites production and extraction from plant materials.

Accordingly, the assessment of selected strains that are able to efficiently ferment medicinal plants and organic sugarcane molasses is a crucial step for designing new functional beverages with well-being effects on human health.

Author Contributions: H.G.: conceptualization, methodology, original draft; E.A.H., E.M.-R., and M.S.T. designed the study and performed the experiments. H.G., M.G., and L.D.V., contributed to drafting the manuscript, validation, data curation, visualization, writing. All authors have read and agreed to the published version of the manuscript.

Funding: This study was supported by University of Tunis EL Manar; University of Applied Sciences, Neubrandenburg, Germany; University of Modena and Reggio Emilia, Italy; Institute of Pasteur of Tunis for the infrastructure and facilities, and Laboratory of Aromatic and Medicinal Plants, Center of Biotechnology (LR15CBBC06) at the Ecopark of Borj-cédria. BP-901, 2050 Hammam-Lif. Tunisia.

Institutional Review Board Statement: Not applicable.

Informed Consent Statement: Not applicable.

Data Availability Statement: Not applicable.

Acknowledgments: We are grateful, and provide sincere thanks, to the honorable Vice Chancellor of the University of Applied Sciences. Neubrandenburg, Germany, and the University of Modena and Reggio Emilia, Italy for the infrastructure and facilities. This study was supported by a grant from the Ministry of Tunisia, Laboratory of Aromatic and Medicinal Plants, Center of Biotechnology (LR15CBBC06) at the Ecopark of Borj-Cédria. BP-901, 2050 Hammam-Lif. Tunisia.

Conflicts of Interest: The authors declare no conflict of interest.

References

1. Hussein, H.J.; Hameed, I.H.; Hadi, M.Y. Using gas chromatography-mass spectrometry (GC-MS) technique for analysis of bioactive compounds of methanolic leaves extract of *Lepidium sativum*. *Res. J. Pharm. Tech.* **2017**, *10*, 3981–3989. [CrossRef]
2. Das, A.; Raychaudhuri, U.; Chakraborty, R. Cereal based functional food of Indian subcontinent: A review. *J. Food Sci. Technol.* **2012**, *49*, 665–672. [CrossRef] [PubMed]
3. Gullo, M.; Zanichelli, G.; Verzelloni, E.; Lemmetti, F.; Giudici, P. Feasible acetic acid fermentations of alcoholic and sugary substrates in combined operation mode. *Process. Biochem.* **2016**, *51*, 1129–1139. [CrossRef]
4. Narvhus, J.A.; Gadaga, T.H. The role of interaction between yeasts and lactic acid bacteria in African fermented milks: A review. *Int. J. Food Microbiol.* **2003**, *86*, 51–60. [CrossRef]
5. Smit, G.; Smit, B.A.; Engels, W.J.M. Flavour formation by lactic acid bacteria and biochemical flavour profiling of cheese products. *FEMS Microbiol. Rev.* **2005**, *29*, 591–610. [CrossRef]
6. Chen, Y.; Huang, Y.; Bai, Y.; Fu, C.; Zhou, M.; Gao, B.; Wang, C.; Li, D.; Hu, Y.; Xu, N. Effects of mixed cultures of Saccharomyces cerevisiae and *Lactobacillus plantarum* in alcoholic fermentation on the physicochemical and sensory properties of citrus vinegar. *LWT Food Sci. Technol.* **2017**, *84*, 753–763. [CrossRef]
7. Rice-Evans, C.; Miller, N.; Paganga, G. Antioxidant properties of phenolic compounds. *Trends Plant Sci.* **1997**, *2*, 152–159. [CrossRef]
8. Mathew, S.; Abraham, T.E. In vitro antioxidant activity and scavenging effects of *Cinnamomum verum* leaf extract assayed by different methodologies. *Food Chem. Toxicol.* **2006**, *44*, 198–206. [CrossRef]
9. De Vero, L.; Bonciani, T.; Verspohl, A.; Mezzetti, F.; Giudici, P. High-glutathione producing yeasts obtained by genetic improvement strategies: A focus on adaptive evolution approaches for novel wine strains. *AIMS Microbiol.* **2017**, *3*, 155. [CrossRef]
10. La China, S.; Zanichelli, G.; De Vero, L.; Gullo, M. Oxidative fermentations and exopolysaccharides production by acetic acid bacteria: A mini review. *Biotechnol. Lett.* **2018**, *40*, 1289–1302. [CrossRef]
11. Bonciani, T.; De Vero, L.; Giannuzzi, E.; Verspohl, A.; Giudici, P. Qualitative and quantitative screening of the β-glucosidase activity in *Saccharomyces cerevisiae* and *Saccharomyces uvarum* strains isolated from refrigerated must. *Lett. Appl. Microbiol.* **2018**, *67*, 72–78. [CrossRef]
12. Kieronczyk, A.; Skeie, S.; Langsrud, T.; Yvon, M. Cooperation between *Lactococcus lactis* and nonstarter lactobacilli in the formation of cheese aroma from amino acids. *Appl. Environ. Microbiol.* **2003**, *69*, 734–739. [CrossRef]
13. Bonciani, T.; De Vero, L.; Mezzetti, F.; Fay, J.C.; Giudici, P. A multi-phase approach to select new wine yeast strains with enhanced fermentative fitness and glutathione production. *Appl. Microbiol. Biotechnol.* **2018**, *102*, 2269–2278. [CrossRef]
14. Bourdichon, F.; Casaregola, S.; Farrokh, C.; Frisvad, J.C.; Gerds, M.L.; Hammes, W.P.; Harnett, J.; Huys, G.; Laulund, S.; Ouwehand, A.; et al. Food fermentations: Microorganisms with technological beneficial use. *Int. J. Food Microbiol.* **2012**, *154*, 87–97. [CrossRef]

15. Schindler, S.; Zelena, K.; Krings, U.; Bez, J.; Eisner, P.; Berger, R.G. Improvement of the aroma of pea (*Pisum sativum*) protein extracts by lactic acid fermentation. *Food Biotechnol.* **2012**, *26*, 58–74. [CrossRef]

16. Harada, R.; Yuzuki, M.; Ito, K.; Shiga, K.; Bamba, T.; Fukusaki, E. Influence of yeast and lactic acid bacterium on the constituent profile of soy sauce during fermentation. *J. Biosci. Bioeng.* **2016**, *123*, 203–208. [CrossRef]

17. Zhao, C.J.; Schieber, A.; Gänzle, M.G. Formation of taste-active amino acids, amino acid derivatives and peptides in food fermentations—A review. *Food Res. Int.* **2016**, *89*, 39–47. [CrossRef]

18. Aloisio, I.; Santini, C.; Biavati, B.; Dinelli, G.; Cencič, A.; Chingwaru, W.; Mogna, L.; Di Gioia, D. Characterization of *Bifidobacterium* spp. strains for the treatment of enteric disorders in newborns. *Appl. Microbiol. Biotechnol.* **2012**, *96*, 1561–1576. [CrossRef]

19. Gilliland, S.E.; Nelson, C.R.; Maxwell, C. Assimilation of cholesterol by *Lactobacillus acidophilus*. *Appl. Environ. Microbiol.* **1985**, *49*, 377–381. [CrossRef]

20. Yoon, K.Y.; Woodams, E.E.; Hang, Y.D. Production of probiotic cabbage juice by lactic acid bacteria. *Bioresour. Technol.* **2006**, *97*, 1427–1430. [CrossRef]

21. Andersen, J.M.; Barrangou, R.; Abou Hachem, M.; Lahtinen, S.J.; Goh, Y.J.; Svensson, B.; Klaenhammer, T.R. Transcriptional analysis of prebiotic uptake and catabolism by *Lactobacillus acidophilus* NCFM. *PLoS ONE* **2012**, *7*, e44409. [CrossRef]

22. Filho, E.; Pierini, D.; Robazza, C.; Tenenbaum, G.; Bertollo, M. Shared mental models and intra-team psychophysiological patterns: A test of the juggling paradigm. *J. Sports Sci.* **2017**, *35*, 112–123. [CrossRef]

23. Mousavi, Z.E.; Mousavi, S.M.; Razavi, S.H.; Hadinejad, M.; Emam-Djomeh, Z.; Mirzapour, M. Effect of fermentation of pomegranate juice by *Lactobacillus plantarum* and *Lactobacillus acidophilus* on the antioxidant activity and metabolism of sugars, organic acids and phenolic compounds. *Food Biotechnol.* **2013**, *27*, 1–13. [CrossRef]

24. Naczk, M.; Shahidi, F. Phenolics in cereals, fruits and vegetables: Occurrence, extraction and analysis. *J. Pharm. Biomed. Anal.* **2006**, *41*, 1523–1542. [CrossRef]

25. Katina, K.; Laitila, A.; Juvonen, R.; Liukkonen, K.H.; Kariluoto, S.; Piironen, V.; Landberg, R.; Åman, P.; Poutanen, K. Bran fermentation as a means to enhance technological properties and bioactivity of rye. *Food Microbiol.* **2007**, *24*, 175–186. [CrossRef]

26. Đorđević, T.M.; Šiler-Marinković, S.S.; Dimitrijević-Branković, S.I. Effect of fermentation on antioxidant properties of some cereals and pseudo cereals. *Food Chem.* **2010**, *119*, 957–963. [CrossRef]

27. Ng, C.C.; Wang, C.Y.; Wang, Y.P.; Tzeng, W.S.; Shyu, Y.T. Lactic acid bacterial fermentation on the production of functional antioxidant herbal *Anoectochilus formosanus* Hayata. *J. Biosci. Bioeng.* **2011**, *111*, 289–293. [CrossRef]

28. Bounaix, M.S.; Gabriel, V.; Morel, S.; Robert, H.; Rabier, P.; Remaud-Simeon, M.; Gabriel, B.; Fontagne-Faucher, C. Biodiversity of exopolysaccharides produced from sucrose by sourdough lactic acid bacteria. *J. Agric. Food Chem.* **2009**, *57*, 10889–10897. [CrossRef]

29. Martins, S.; Mussatto, S.I.; Martínez-Avila, G.; Montañez-Saenz, J.; Aguilar, C.N.; Teixeira, J.A. Bioactive phenolic compounds: Production and extraction by solid-state fermentation. A review. *Biotechnol. Adv.* **2011**, *29*, 365–373. [CrossRef]

30. Torino, M.I.; Limón, R.I.; Martínez-Villaluenga, C.; Mäkinen, S.; Pihlanto, A.; Vidal-Valverde, C.; Frias, J. Antioxidant and antihypertensive properties of liquid and solid-state fermented lentils. *Food Chem.* **2013**, *136*, 1030–1037. [CrossRef]

31. Mishra, S.S.; Behera, P.K.; Kar, B.; Ray, R.C. Advances in probiotics, prebiotics and nutraceuticals. In *Innovations in Technologies for Fermented Food and Beverage Industries*; Springer: Cham, Switzerland, 2018; pp. 121–141.

32. Bartkiene, E.; Mozuriene, E.; Lele, V.; Zokaityte, E.; Gruzauskas, R.; Jakobsone, I.; Juodeikiene, G.; Ruibys, R.; Bartkevics, V. Changes of bioactive compounds in barley industry by-products during submerged and solid-state fermentation with antimicrobial *Pediococcus acidilactici* strain LUHS29. *Nutr. Food Sci.* **2020**, *8*, 340–350. [CrossRef] [PubMed]

33. Peng, M.; Tabashsum, Z.; Anderson, M.; Truong, A.; Houser, A.K.; Padilla, J.; Akmel, A.; Bhatti, J.; Rahaman, S.O.; Biswas, D. Effectiveness of probiotics, prebiotics, and prebiotic-like components in common functional foods. *Compr. Rev. Food Sci. Food Saf.* **2020**, *19*, 1908–1933. [CrossRef] [PubMed]

34. De Vero, L.; Boniotti, M.B.; Budroni, M.; Buzzini, P.; Cassanelli, S.; Comunian, R.; Gullo, M.; Logrieco, A.F.; Mannazzu, I.; Musumeci, R.; et al. Preservation, Characterization and Exploitation of Microbial Biodiversity: The Perspective of the Italian Network of Culture Collections. *Microorganisms* **2019**, *7*, 685. [CrossRef] [PubMed]

35. De Vero, L.; Iosca, G.; Gullo, M.; Pulvirenti, A. Functional and Healthy Features of Conventional and Non-Conventional Sourdoughs. *Appl. Sci.* **2021**, *11*, 3694. [CrossRef]

36. Warwick, L.; Marsden, P.; Gray, G.J.; Nippard, M.R. Quinlan. Evaluation of the DNS method for analysing lignocellulosic hydrolysates. *J. Chem. Technol. Biotechnol.* **1982**, *32*, 1016–1022.

37. Hermosín, I.; Chicón, R.M.; Cabezudo, M.D. Free amino acid composition and botanical origin of honey. *Food Chem.* **2003**, *83*, 263–268. [CrossRef]

38. Tlili, N.; Elfalleh, W.; Hannachi, H.; Yahia, Y.; Khaldi, A.; Ferchichi, A.; Nasri, N. Screening of natural antioxidants from selected medicinal plants. *Int. J. Food Prop.* **2013**, *16*, 1117–1126. [CrossRef]

39. Dewanto, V.; Wu, X.; Adom, K.K.; Liu, R.H. Thermal processing enhances the nutritional value of tomatoes by increasing total antioxidant activity. *J. Agric. Food Chem.* **2002**, *50*, 3010–3014. [CrossRef]

40. Sun, B.; Ricardo-da-Silva, J.M.; Spranger, I. Critical factors of vanillin assay for catechins and proanthocyanidins. *J. Agric. Food Chem.* **1998**, *46*, 4267–4274. [CrossRef]

41. Navarini, L.; Gilli, R.; Gombac, V.; Abatangelo, A.; Bosco, M.; Toffanin, R. Polysaccharides from hot water extracts of roasted *Coffea arabica* beans: Isolation and characterization. *Carbohydr. Polym.* **1999**, *40*, 71–81. [CrossRef]

42. You, L.; Gao, Q.; Feng, M.; Yang, B.; Ren, J.; Gu, L.; Cui, C.; Zhao, M. Structural characterisation of polysaccharides from *Tricholoma matsutake* and their antioxidant and antitumour activities. *Food Chem.* **2013**, *138*, 2242–2249. [CrossRef]
43. Tian, H.; Shen, Y.; Yu, H.; He, Y.; Chen, C. Effects of 4 probiotic strains in coculture with traditional starters on the flavor profile of yogurt. *J. Food Sci.* **2017**, *82*, 1693–1701. [CrossRef]
44. Guth, H. Quantitation and sensory studies of character impact odorants of different white wine varieties. *J. Agric. Food Chem.* **1997**, *45*, 3027–3032. [CrossRef]
45. Xi, W.; Zheng, H.; Zhang, Q.; Li, W. Profiling taste and aroma compound metabolism during apricot fruit development and ripening. *Int. J. Mol. Sci.* **2016**, *17*, 998. [CrossRef]
46. Oyaizu, M. Studies on products of browning reaction antioxidative activities of products of browning reaction prepared from glucosamine. *Jpn. J. Nutr. Diet.* **1986**, *44*, 307–315. [CrossRef]
47. Prieto, P.; Pineda, M.; Aguilar, M. Spectrophotometric quantitation of antioxidant capacity through the formation of a phospho-molybdenum complex: Specific application to the determination of vitamin E. *Anal. Biochem.* **1999**, *269*, 337–341. [CrossRef]
48. Hayouni, E.A.; Abedrabba, M.; Bouix, M.; Hamdi, M. The effects of solvents and extraction method on the phenolic contents and biological activities in vitro of Tunisian *Quercus coccifera* L. and *Juniperus phoenicea* L. fruit extracts. *Food Chem.* **2007**, *105*, 1126–1134. [CrossRef]
49. Sõukand, R.; Hajdari, A.; Pieroni, A.; Biró, M.; Dénes, A.; Doğan, Y.; Quave, C.L.; Kalle, R.; Reade, B.; Mustafa, B.; et al. An ethnobotanical perspective on traditional fermented plant foods and beverages in Eastern Europe. *J. Ethnopharmacol.* **2015**, *170*, 284–296. [CrossRef]
50. Di Cagno, R.; Surico, R.F.; Paradiso, A.; De Angelis, M.; Salmon, J.C.; Buchin, S.; De Garaand, L.; Gobbetti, M. Effect of autochthonous lactic acid bacteria starters on health-promoting and sensory properties of tomato juices. *Int. J. Food Microbiol.* **2009**, *128*, 473–483. [CrossRef]
51. Di Cagno, R.; Minervini, G.; Rizzello, C.G.; De Angelis, M.; Gobbetti, M. Effect of lactic acid fermentation on antioxidant, texture, color and sensory properties of red and green smoothies. *Food Microbiol.* **2011**, *28*, 1062–1071. [CrossRef]
52. Zhang, Z.; Lv, G.; Pan, H.; Fan, L.; Soccol, C.R.; Pandey, A. Production of powerful antioxidant supplements via solid-state fermentation of wheat (*Triticum aestivum Linn.*) by *Cordyceps militaris. Food Technol. Biotech.* **2012**, *50*, 32–39.
53. Kang, J.; Li, Z.; Wu, T.; Jensen, G.S.; Schauss, A.G.; Wu, X. Anti-oxidant capacities of flavonoid compounds isolated from acai pulp (*Euterpe oleracea Mart.*). *Food Chem.* **2010**, *122*, 610–617. [CrossRef]
54. Xu, P.; Duong, D.M.; Seyfried, N.T.; Cheng, D.; Xie, Y.; Robert, J.; Rush, J.; Hochstrasser, M.; Finley, D.; Peng, J. Quantitative proteomics reveals the function of unconventional ubiquitin chains in proteasomal degradation. *Cell* **2009**, *137*, 133–145. [CrossRef] [PubMed]
55. Manrique, G.D.; Lajolo, F.M. FT-IR spectroscopy as a tool for measuring degree of methyl esterification in pectins isolated from ripening papaya fruit. *Postharvest Biol. Technol.* **2002**, *25*, 99–107. [CrossRef]
56. Karmakar, P.; Pujol, C.A.; Damonte, E.B.; Ghosh, T.; Ray, B. Polysaccharides from *Padina tetrastromatica*: Structural features, chemical modification and antiviral activity. *Carbohydr. Polym.* **2010**, *80*, 513–520. [CrossRef]
57. Li, Q.; Xie, Y.; Su, J.; Ye, Q.; Jia, Z. Isolation and structural characterization of a neutral polysaccharide from the stems of *Dendrobium densiflorum. Int. J. Biol. Macromol.* **2012**, *50*, 1207–1211. [CrossRef]
58. Wang, J.; Li, X.; Han, T.; Yang, Y.; Jiang, Y.; Yang, M.; Xu, Y.; Harpaz, S. Effects of different dietary carbohydrate levels on growth, feed utilization and body composition of juvenile grouper *Epinephelus akaara. Aquaculture* **2016**, *459*, 143–147. [CrossRef]
59. Phillip, A.J. Breeding for improved sugar content in sugarcane. *Field Crops Res.* **2005**, *92*, 277–290.
60. Braga, C.M.; Zielinski, A.A.F.; da Silva, K.M.; de Souza, F.K.F.; Pietrowski, G.D.A.M.; Couto, M.; Granato, D.; Wosiacki, G.; Nogueira, A. Classification of juices and fermented beverages made from unripe, ripe and senescent apples based on the aromatic profile using chemometrics. *Food Chem.* **2013**, *141*, 967–974. [CrossRef]
61. Cheng, H. Volatile flavor compounds in yogurt: A review. *Crit. Rev. Food Sci. Nutr.* **2010**, *50*, 938–950.
62. Di Cagno, R.; Filannino, P.; Gobbetti, M. Lactic acid fermentation drives the optimal volatile flavor-aroma profile of pomegranate juice. *Int. J. Food Microbiol.* **2017**, *248*, 56–62. [CrossRef]
63. Rita, R.D.; Zanda, K.; Daina, K.; Dalija, S. Composition of aroma compounds in fermented apple juice: Effect of apple variety, fermentation temperature and inoculated yeast concentration. *Procedia Food Sci.* **2011**, *1*, 1709–1716. [CrossRef]
64. Hartmann, G.; Krieg, A.M. Mechanism and function of a newly identified CpG DNA motif in human primary B cells. *J. Immunol.* **2000**, *164*, 944–953. [CrossRef]
65. Hur, S.J.; Lee, S.Y.; Kim, Y.C.; Choi, I.; Kim, G.B. Effect of fermentation on the antioxidant activity in plant-based foods. *Food Chem.* **2014**, *160*, 346–356. [CrossRef]
66. James, N.D.; Sydes, M.R.; Mason, M.D.; Clarke, N.W.; Anderson, J.; Dearnaley, D.P.; Dwyer, J.; Jovic, G.; Ritchie, A.W.S.; Russell, J.M.; et al. Celecoxib plus hormone therapy versus hormone therapy alone for hormone-sensitive prostate cancer: First results from the STAMPEDE multiarm, multistage, randomised controlled trial. *Lancet Oncol.* **2012**, *13*, 549–558. [CrossRef]
67. Pacifico, S.; Gallicchio, M.; Stintzing, F.C.; Fiorentino, A.; Fischer, A.; Meyer, U. Antioxidant properties and cytotoxic effects on human cancer cell lines of aqueous fermented and lipophilic quince (*Cydonia oblonga Mill.*) preparations. *Food Chem. Toxicol.* **2012**, *50*, 4130–4135. [CrossRef]

68. Xiao, Y.; Xing, G.; Rui, X.; Li, W.; Chen, X.; Jiang, M.; Dong, M. Enhancement of the antioxidant capacity of chickpeas by solid state fermentation with *Cordyceps militaris* SN-18. *J. Funct. Foods* **2014**, *10*, 210–222. [CrossRef]
69. Uchida, M.; Kurushima, H.; Ishihara, K.; Murata, Y.; Touhata, K.; Ishida, N.; Niwa, K.; Araki, T. Characterization of fermented seaweed sauce prepared from nori (*Pyropia yezoensis*). *J. Biosci. Bioeng.* **2017**, *123*, 327–332. [CrossRef]

Review

Functional and Healthy Features of Conventional and Non-Conventional Sourdoughs

Luciana De Vero *, Giovanna Iosca, Maria Gullo and Andrea Pulvirenti *

Unimore Microbial Culture Collection (UMCC), Department of Life Sciences, University of Modena and Reggio Emilia, Via Amendola 2, 42122 Reggio Emilia, Italy; giovanna.iosca@unimore.it (G.I.); maria.gullo@unimore.it (M.G.)
* Correspondence: luciana.devero@unimore.it (L.D.V.); andrea.pulvirenti@unimore.it (A.P.)

Abstract: Sourdough is a composite ecosystem largely characterized by yeasts and lactic acid bacteria which are the main players in the fermentation process. The specific strains involved are influenced by several factors including the chemical and enzyme composition of the flour and the sourdough production technology. For many decades the scientific community has explored the microbiological, biochemical, technological and nutritional potential of sourdoughs. Traditionally, sourdoughs have been used to improve the organoleptic properties, texture, digestibility, palatability, and safety of bread and other kinds of baked products. Recently, novel sourdough-based biotechnological applications have been proposed to meet the demand of consumers for healthier and more natural food and offer new inputs for the food industry. Many researchers have focused on the beneficial effects of specific enzymatic activities or compounds, such as exopolysaccharides, with both technological and functional roles. Additionally, many studies have explored the ability of sourdough lactic acid bacteria to produce antifungal compounds for use as bio-preservatives. This review provides an overview of the fundamental features of sourdoughs and their exploitation to develop high value-added products with beneficial microorganisms and/or their metabolites, which can positively impact human health.

Keywords: sourdough; yeasts; lactic acid bacteria; bioactive compounds; exopolysaccharides; antifungal activity

Citation: De Vero, L.; Iosca, G.; Gullo, M.; Pulvirenti, A. Functional and Healthy Features of Conventional and Non-Conventional Sourdoughs. *Appl. Sci.* **2021**, *11*, 3694. https://doi.org/10.3390/app11083694

Academic Editor: Antonio Valero

Received: 17 March 2021
Accepted: 18 April 2021
Published: 20 April 2021

Publisher's Note: MDPI stays neutral with regard to jurisdictional claims in published maps and institutional affiliations.

1. Introduction

Sourdoughs, in all their different types produced worldwide, represent an awesome ecosystem which can offer several opportunities for conventional and non-conventional microbial exploitation to sustain the ecological and nutritional needs of new consumers [1].

Sourdough is a mixture of water and flour that is fermented by cultures of indigenous yeasts and lactic acid bacteria (LAB) [2–4]. In addition to these microorganisms, *Proteobacteria* may also be present, specially at the beginning of fermentation [5,6]. Among them, acetic acid bacteria (AAB), such as those belonging to *Gluconobacter* sp., *Acetobacter* sp., and *Komagataeibacter* sp., allow faster acidification of the dough and influence the volatile attributes of the final product [7–10]. The indigenous microflora of sourdough is the result of the microbial interaction among microorganisms coming from the flour, the bakery environment, and the vegetable matrices, such as fruits, must, or vinegar, which can be added to the original mixture to accelerate the start-up of fermentation [11].

For many decades the scientific community has explored the microbiological, biochemical, technological, and nutritional potential of sourdoughs and the overall literature, produced in the last 30 years, has been recently reviewed by Arora et al. [12]. What stands out are the novel sourdough-based biotechnological applications proposed, in the last decade, to meet the demand of consumers for healthier and more natural food [13,14].

This review highlights the fundamental features of sourdoughs and their exploitation to develop high value-added products and offer new inputs for the food industry.

In addition to the sourdough technology, this review deals with meaningful studies on the enzymatic activities that have positive effects on human health and the production of compounds with functional properties. Moreover, it reports the potential exploitation of sourdough lactic acid bacteria as bio-preservatives against fungal growth.

2. Sourdough Technology

Based on the technology applied for their production, three main types of sourdough can be distinguished: type I, which is the artisan bakery firm sourdough; type II, referring to industrial liquid sourdoughs; and type III, which indicates industrial dried sourdoughs [15,16] (Figure 1).

Usually, mature type I sourdough contains a mixture of typical yeast and mesophilic LAB strains that characterize sourdough-based products [6]. The stability of the dough depends on the type of flour, the quality and nutritional value of the cereals, the temperature and humidity during processes, and the microbial composition of the inoculum [17,18].

The common yeast species are mainly *Saccharomyces cerevisiae*, *Kazachstania humilis* (formerly *Candida humilis*), *Kazachstania exigua*, *Pichia kudriavzevii*, and *Torulaspora delbrueckii* [19,20]. Moreover, almost 95% of the traditional sourdough population is dominated by heterofermentative LAB alone or in association with homofermentative lactobacilli [15,21]. *Lactobacillus sanfranciscensis* (currently *Fructilactobacillus sanfranciscensis*) was most frequently detected in association with yeasts belonging to *Kazachstania* species, predominantly *K. humilis* [22–24]. Other frequently representative LAB include *Lactobacillus plantarum* (currently *Lactiplantibacillus plantarum*), *Lactobacillus brevis* (currently *Levilactobacillus brevis*), *Leuconostoc* spp., and *Weissella* spp. [1,5,25,26].

To avoid any confusion for the *Lactobacillus* strains reported in previous studies with the former classification of the genus, the old name of the species will be maintained in the present review; the following link is suggested for the appropriate name conversion: http://lactobacillus.ualberta.ca, accessed on 17 March 2021.

The type I sourdough is prepared according to the traditional method, conducted by daily refreshments, also called "back-slopping", which keeps it metabolically active. Accordingly, a selection of sourdough microbiota occurs, due to the back-slopping, which is repeated five to ten times [8,27]. These sourdoughs are typical of various traditional Italian sweet baked products, including those that are commercialized and consumed during holiday seasons, such as *Panettone* and *Colomba* [3,18,28].

Type II and III sourdoughs (the latter is made by dehydrating the stabilized form of type II sourdough) were developed for industries with the aim of providing a more standardized process [6]. Both are made with selected cultures added at a ratio of 100:1 (LAB to yeasts) to obtain specific features of the baked products and inhibit the growth of unwanted microbiota [8].

The selection of microbial strains useful as starter cultures is fundamental for sourdough industries and relies on various metabolic traits that have both technological and functional interest [29,30]. In this context, qualified microbial culture collections, constitute a fundamental cornerstone for the investigation of sourdoughs' microbiota to select strains with desired features [31].

Among LAB, the selection of potential starters is generally made within the *Lactobacillaceae* and the most often used are acid-tolerant strains such as *L. amylovorus*, *L. panis*, *L. pontis*, and *L. reuteri* [27,32,33].

Different non-lactobacillus strains have also been tested as suitable starters because of the positive effect of their compounds on the sourdough flavor, which can open new prospects in the sourdough industries. As reported by Montemurro et al. [34], *Pediococcus pentosaceus* OA1 and S3N3 and *Leuconostoc citreum* PRO17 were selected on the basis of optimal acidification and growth performance, as well as the intense proteolytic activity in whole wheat flour doughs.

Recently, AAB have also been considered useful starters for the production of desired metabolites [7]. Accordingly, *Acetobacter pasteurianus* IMDO 386B and *Gluconobacter oxydans*

IMDO A845 strains were tested for type II sourdough production processes and the latter, in particular, had an attractive impact on the production of volatile organic compounds [7].

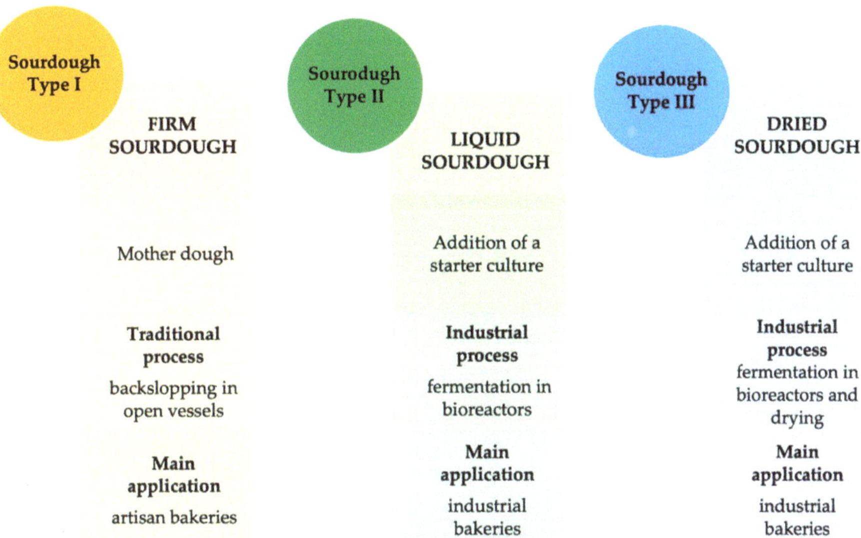

Figure 1. Schematic description of the three main types of sourdough.

3. Nutritional and Functional Features of Sourdough

Sourdough fermentation is certainly the most conventional and efficient tool for guaranteeing rheology, sensory, hygiene and shelf-life features [14]. Additionally, it has the potentiality to enhance the nutritional and functional features of wheat flours [35–38].

The positive effects provided by sourdough technology can be summarized as:

- preservation of food through acetic acid, lactic acid, alcoholic, and alkaline fermentations;
- food enrichment with compounds that originate either from biochemical reactions (e.g., essential amino acids, proteins and essential fatty acids), or biosynthesis (e.g., vitamins);
- development of aromas, flavors and textures in food substrates;
- detoxification during food fermentation processing.

Moreover, it has been reported that sourdough fermentation can lower glycemic index, increase mineral bioavailability, reduce gluten content and reduce starch digestibility, mainly through the organic acids production and other complementary mechanisms [12,39]. The main outcomes of sourdough fermentation are shown in Figure 2.

In recent years the development of high-quality gluten-free (GF) products has become an important socio-economic issue and a new approach in this framework is represented by the application of sourdough fermentation [40].

Common products for celiacs available on the market are often characterized by poor palatability and lack of minerals and fibers [41]. Moreover, they usually have a low amount of health promoting nutrients, such as B vitamins groups, which are essential in the human diet; for instance, folate (B11) is involved in fundamental metabolic reaction, biosynthesis of nucleotides, building blocks of DNA and RNA and prevent neural tube defects in newborns [40]. Consequently, fermentation with LAB and yeasts can be considered an effective methodology in food production suitable for the development of new kinds of GF products with nutraceutical and health-promoting features [42]. Applying microbial selected cultures to suitable raw materials, the fermentation process can be modulated, also enhancing the content of bioactive metabolites and other compounds, such as glutathione, which exert an antioxidant activity [22,43–45].

Moreover, the sourdough fermentation can be useful in preserving some good sensory characteristics, such as products' structure and/or softness maintenance during storage, which are usually compromised by the lack of gluten [46,47]. Therefore, the selection of specific sourdough cultures has been considered as a new tool for GF food processing [14,48,49].

Various ecological studies have provided useful information on the presence of competitive LAB and yeasts strains, which can be used as candidates for starter development [50]. In rice, maize, teff, and amaranth sourdough, for instance, is frequently used to isolate microorganisms such as: *L. fermentum*, *L. plantarum*, and *L. paralimentarius* [51]. Above them, strains of *L. helveticus*, *L. pontis*, and *S. cerevisiae* are most competitive in different kind of cereals, pseudo-cereals, and cassava sourdoughs [10,52]. Positive cooking performance, obtained with the use of sourdough in GF products, in terms of definite volume, flavor, texture and mouthfeel encourages further studies and the development of an industrial production [40].

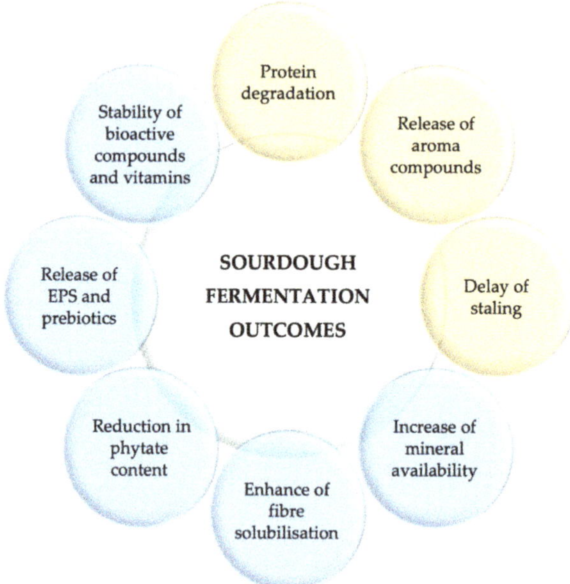

Figure 2. Main technological effects of sourdough on rheology, flavor and shelf life (in yellow), and main health-promoting features (in light blue).

In particular, the flavor profile of the product can be greatly affected by the use of sourdough, depending on type of starter cultures, fermentation conditions, baking conditions and raw material. Recently, researchers have proposed the use of sourdough-fermented ingredients to reach new sensorial profile and enhance nutritional value in pasta [13]. The use of sourdough for pasta fortification has been explored under several aspects and has also been applied with success to mixed flour composed by pseudo-cereals and legumes, allowing an increasing of γ-aminobutyric acid (GABA) levels [53]. GABA is a non-protein amino acid with several physiological functions, such as induction of hypotension and prevention of diabetes as well as diuretic and tranquilizer effects [54,55]. A systematic review on its production from LAB has been recently reported by Cui et al. [56]. The use of non-conventional flours to obtain food products characterized by peculiar flavor, abundance of proteins with high nutritional value, dietary fibers, polyphenols, and minerals, represent an attractive feature to be explored by food industries for new applications of sourdough microbiota [26,42].

3.1. Enzymatic Activities with Beneficial Effects on Human Health

It is known that cereal grains are significant sources of minerals such as magnesium, potassium, iron and zinc. However, they also contain phytic acid or myo-inositol hexakiphosphate (IP6) (1–4% of dry weight) [57,58]. This compound is an anti-nutritional factor for humans and animals, in fact, the central hexaphosphate ring is highly charged due to six anionic groups and acts as a chelator of dietary minerals reducing their bioavailability [59,60]. Sourdough fermentation, owing to the pH reduction, provides suitable conditions for cereal endogenous phytase activities [36,61]. The phytases that take part in the process are also exogenous and they are produced by a large number of microorganisms among which are sourdough yeasts and LAB. Their activity reduces to less than a half the phytate content of whole wheat bread and allows to increase bioavailability of minerals, free amino acids, and proteins. A recent screening on 152 LAB, isolated from cereal-based substrates, revealed a widespread capacity of the isolates (95%) for degrading phytic acid. Among the isolates, strains *L. brevis* LD65 and *L. plantarum* PB241, showed the highest phytase activity; on the contrary, *Weissella confusa* strains showed low or no phytase activity [62].

Allergies, intolerances, and sensitive individuals are also positively influenced by the LAB present in the sourdough, as they can show specific enzymatic activities toward gluten proteins. The proteolytic activity carried out by endogenous and exogenous proteases during the fermentation process with sourdough seems to lead to a complete hydrolysis of gluten, meeting the needs of individuals affected by celiac disease [32].

3.2. Effects of Organic Acids

Most of the valuable properties attributed to sourdough are due to the acidification activity, due to LAB and AAB. In fact, during fermentation, several organic acids, e.g., lactic, acetic, citric, pyruvic and succinic acid are produced. Among them, lactic and acetic acid are the most important as they can greatly affect the aroma profile and rheological properties of sourdoughs [58]. Their production depends upon several factor including flour type, starter used, metabolic activity, technological performance, and acidification properties of the wheat sourdoughs [63]. Organic acids produced during sourdough fermentation can have preservative and antimicrobial effects, improving storability and safety [64], as well as have positive health effects. For instance, acetic acid, propionic and lactic acid have the capacity of lowering the insulin response [65]. The mechanisms of these acids seem to vary; lactic acid acts to lower the rate of starch digestion in bread while propionic and acetic acids seem to extend the gastric emptying rate [66].

As suggested by the work of Östman et al. [67], lactic acid, in particular, is designed to lower blood sugar and promote an interaction between starch and gluten favoring the reduction of starch bioavailability.

4. Exopolysaccharides

Exopolysaccharides (EPS) are biopolymers of high molecular weight produced by several microorganisms, such as LAB, AAB and microalgae [68–70].

Based on the chemical composition and biosynthesis mechanisms, microbial EPS are classified into two distinct groups: (1) homopolysaccharides (HoPS), such as glucans and fructans, (2) heteropolysaccharides (HePS), e.g., gellan and xanthan [71]. Among the HoPS, dextran, levan, and cellulose are mostl important in the food industry for their significant features [71,72]. In particular, EPS-producing LAB are attractive for application in bakery products thanks to their ability as viscosifiers, texturizers, emulsifiers, and syneresis-lowering agents [62,70]. In addition to polysaccharides naturally occurring in cereal grains flour and dough, microbial EPS from sucrose can be produced in sourdough through the activity of glycosyltransferases [73]. It has been reported that fructan from *L.*

sanfransiscensis can positively affect the dough rheology and bread texture more than the external addition of the same polysaccharides [74]. Furthermore, it has been shown that dextran produced in situ can improve texture and cover unpleasant flavors of wholegrain bread suggesting microbial EPS as a possible substitute of sweeteners [75].

Sourdough technology using EPS-producing LAB strains seem to be a valid solution to improve GF baked products [70]. For instance, microbial EPS can replace hydrocolloids, which are fundamental components in GF products to get acceptable quality levels in terms of texture, volume, and shelf life [76].

Other evidence has been recently reported by Franco et al. [77], who described EPS, produced by *Pediococcus*, *Leuconostoc*, and *Weisella* strains in quinoa sourdough, able to improve the organoleptic and rheological attributes in GF-free doughs. Consequently, the development of EPS-producing starter cultures for different types of flour sourdoughs has been receiving a growing-interest in the last years, also due to the potential health benefits associated with EPS themselves as prebiotics, which can be exploited to make added-value functional products [78]

5. Bioactive Compounds

Cereals contain different phytochemicals, such as phenolic acids, phytosterols, alkylresorcinols, tocols, lignans, and folate [79]. Among other processing conditions, e.g., milling and malting, the sourdough fermentation is the one that most affects the levels and bioavailability of phytochemicals and increases the level of extractable phenolic compounds [14,80]. Furthermore, bioactive compounds are synthesized during fermentation, while other components involved in grain-related digestion problems or pathologies, such as gluten sensitivity or gastrointestinal syndromes, are reduced [81].

As reported by Katina et al. [82], folate and other free phenolic acids increased up to seven and ten times in germinated rye during sourdough fermentation. Moreover, comparing the capability of different yeasts and LAB to affect the folate content in a rye sourdough, it was demonstrated that the synthesis of folate by bacteria was minimal, while the yeasts were able to increase its content over three-fold in the best case [83].

Regarding the vitamin E, tocopherol and tocotrienol, a reduction during the sourdough preparation and dough making has been reported, probably due to the sensitivity of the compounds to the air [14].

Several authors have stated LAB as the microbes most suitable in bioactive peptides enrichment [55,84]. According to Rizzello et al. [49,55], selected sourdough lactobacilli with specific proteinase and peptidase activities toward cereal proteins, were effectively used for releasing Angiotensin I-Converting Enzyme (ACE)-inhibitory peptides during a long-time sourdough fermentation. These peptides are of great interest for functional foods as they may be used for avoiding hypertension and for other therapeutic purposes [55]. Another study showed that sourdough LAB can increase the concentration of lunasin, a cancer-preventing peptide, during fermentation of various flours, including those of wheat, barley, amaranth, soybean, or rye [85]. Specifically, *L. curvatus* SAL33 and *L. brevis* AM7 strains used as sourdough starters were able to synthesize this compound, increasing its concentration up to 2–4 times during fermentation.

6. Antifungal Compounds

One of the big issues in the bakery product industries are moulds – the primary cause of spoilage, off-flavors and potentially producing harmful secondary metabolites commonly called mycotoxins. Aflatoxins are the most common hazardous mycotoxins, causing both chronic and acute toxicity to humans and cattle. Usually, chemical preservatives are successfully used against fungal growth, however, the exploitation of sourdough LAB, as bio-preservatives, has gained great interest among researchers and industries driven by the growing demand for clean label products in which chemically derived ingredients are replaced by natural alternatives [86].

Generally, LAB can be considered protective microorganisms not only for their production of lactate and acetate, which act as effective preservatives, but also for releasing other active compounds during fermentation. Indeed, a synergistic effect between pH and antifungal metabolites seems to be responsible of the LAB protective activity [87,88]. The majority of antifungal substances produced from LAB include organic acids, hydrogen peroxide, reuterin, proteinaceous and phenolic compounds, hydroxyl fatty acids, and other low-molecular-weight compounds [89,90].

Several studies have investigated the ability of sourdough LAB to prevent moulding events on leavened goods; some examples referred to specific strains are reported in Table 1. A comprehensive review on the ability of LAB to serve as antifungal and anti-mycotoxigenic agents have been recently provided by Sadiq et al. [86].

Table 1. Some lactic acid bacteria strains from sourdoughs tested for their antifungal activity.

LAB Strains	Compounds with Antifungal Activity	Fungal Target Tested	Reference
L. sanfrancisencis CB1	Acetic, caproic, formic, propionic, butyric and n-valeric acids	*Fusarium graminearum* 623	[91]
L. plantarum 21B	Phenyllactic acid and 4-hydroxyphenyllactic acid	*Aspergillus niger* FTDC3227	[92]
L. plantarum CRL 778 *L. reuteri* CRL 1100 *L. brevis* CRL 772 *L. brevis* CRL 796	Lactic, acetic, and phenyllactic acids	*Penicillium* sp. *Aspergillus niger* *Fusarium graminearum*	[93]
L. buchneri FUA 3525 *L. diolovorans* DSM 14421	Propionate and acetate	*Aspergillum clavatus* *Cladisporium* spp. *Mortierella* spp. *Penicillium Roquefort*	[94]
L. rossiae LD108, *L. paralimentarius* PB127	Lactic acid, acetic acid, phenyllactic acid and diacetyl	*Aspergillum japonicus*	[95]
L. amylovorus DSM 19280	Lactic acid, acetic acid, 3-phenylpropanoic acid, p-coumaric, (E)-2-methylcinnamic acid, 3-phenyllactic acid and cyclic dipetides	*Fusarium culmorum* FST 4.05 *Aspergillus niger* FST4.21 *Penicillium expansum* FST 4.22 *Penicillium roqueforti* FST 4.11	[96,97]
L. paracasei subsp. *tolerans* L17	Cell-wall binding and enzyme-mediated degradation	*Fusarium proliferatum* M 5991 *Fusarium proliferatum* M 5689 *Fusarium graminearum* R 4053	[98,99]

7. Sourdough Effectiveness on Rheology, Shelf-Life and Safety

The impact of sourdough on properties and rheological behavior of dough has been widely investigated. Generally, changes may be attributed to several intrinsically related factors, including variations in the rate or amount of acid produced [100].

For instance, a low proteolytic degradation of wheat proteins affects the physical properties of gluten and, thus, influences the firmness and staling of the final baked product [101]. The specific proteolytic activity of sourdough microorganisms also has a great impact [102]. As proved by Clarke et al. [103], the use of sourdough, prepared either from a single strain or a mixed strain starter culture, significantly influenced the rheological properties of wheat flour dough; sourdough prepared with starter cultures increased the softening level of the dough.

Regarding gas production in sourdoughs, Hammes and Gänzle [27] proved that the contribution of yeasts and LAB changes according to the type of starter and the dough technology. This aspect is particularly important in the GF products where the lack of the viscoelastic gluten system is responsible for low expansion and gas retention during leavening [104]. The efficiency of the addition of sourdough on GF bread quality was demonstrated in various formulations made of a different kind of flour by Picozzi et al. [105], which used a Type I GF-sourdough with a stable association between *L. sanfranciscensis* and *C. humilis*. The positive effects of sourdough on rheology and texture, in terms of volume and softness, led to an extended shelf-life of the baked products. This aspect is certainly of great interest for the bakery industry which has recently reevaluated the traditional sourdough fermentation to contrast the short shelf-life of baked products, like bread, mainly caused by spoilage microorganisms.

In fact, sourdough can act as a natural preservative able to replace the use of chemical preservatives. Above the antifungal activity described in the previous section, antibacterial activity has been also scientifically proven [106]. Bread quality and safety can be affected by spore-forming bacteria, such as *Bacillus subtilis*, which mainly occur on the outer parts of grains, and consequently can contaminate also the other ingredients and/or bakery environment [107,108]. The antibacterial effect of sourdough is generally attributed to the synergistic activity of several compounds produced by yeasts and LAB, which includes the synthesis of organic acids, EPS, antimicrobial compounds, bioactive peptides as well as the conversion of phenolic compounds and lipids [106,109,110].

Therefore, sourdough technology combined with the use of high-quality flours represent a tool for improving both organoleptic and healthy features of the baked goods. Optimized sourdoughs obtained with non-conventional flours are the new input for food companies which want to satisfy the needs of consumers affected by allergies and food intolerance.

Legumes and pseudocereals, such as amaranth and quinoa, which have very different chemical composition and technological properties compared to wheat, can be a valid alternative useful for the development of new food products included in different kinds of bread, pasta, or snacks [13]. Regarding the allergies issue, the European Regulation No. 1169/2011 have been adopted with the aim to improve the labeling of foods with the clear indication of ingredients and nutritional values. Accurate food labeling can, in fact, allow consumers suffering from allergies or intolerances of knowing the specific ingredients present in food products and help them to make healthier choices [111].

8. Conclusions

Sourdough fermentation has emerged in human history and since then, it has been empirically used for the improvement of the organoleptic properties, texture, digestibility, palatability, and safety of different food matrices. Nowadays, sourdough fermentation is widely employed to enrich food with beneficial microorganisms and/or their metabolites, which positively impact human health. This result can be achieved either through the exploitation of the wild microbiota naturally associated to raw materials or as the result of the inoculation of selected starters.

The wide and successful use of sourdough, maintained over time for its peculiar and unique features, confirms this ancient biotechnology as an effective answer to the modern world demand for natural, healthy, and eco-friendly food.

Author Contributions: L.D.V., conceptualization, writing—review and editing; G.I., writing partially the original draft and editing; M.G., writing, review and editing; A.P., supervision and review. All authors have read and agreed to the published version of the manuscript.

Funding: This work was supported by the EU Project "Implementation and Sustainability of Microbial Resource Research Infrastructure for the 21st Century" (IS_MIRRI21, Grant Agreement Number 871129).

Institutional Review Board Statement: Not applicable.

Informed Consent Statement: Not applicable.

Acknowledgments: The JRU MIRRI-IT (http://www.mirri-it.it/) is greatly acknowledged for scientific support.

Conflicts of Interest: The authors declare no conflict of interest.

References

1. De Vuyst, L.; Van Kerrebroeck, S.; Harth, H.; Huys, G.; Daniel, H.M.; Weckx, S. Microbial ecology of sourdough fermentations: Diverse or uniform? *Food Microbiol.* **2014**, *37*, 11–29. [CrossRef]
2. Gobbetti, M. The sourdough microflora: Interactions of lactic acid bacteria and yeasts. *Trends Food Sci. Technol.* **1998**, *9*, 267–274. [CrossRef]
3. Garofalo, C.; Silvestri, G.; Aquilanti, L.; Clementi, F. PCR-DGGE analysis of lactic acid bacteria and yeast dynamics during the production processes of three varieties of Panettone. *J. Appl. Microbiol.* **2008**, *105*, 243–254. [CrossRef]
4. Pulvirenti, A.; Solieri, L.; Gullo, M.; De Vero, L.; Giudici, P. Occurence and dominance of yeast species in sourdough. *Lett. Appl. Microbiol.* **2004**, *38*, 113–117. [CrossRef]
5. Celano, G.; De Angelis, M.; Minervini, F.; Gobbetti, M. Different flour microbial communities drive to sourdoughs characterized by diverse bacterial strains and free amino acid profiles. *Front. Microbiol.* **2016**, *7*, 1770. [CrossRef] [PubMed]
6. De Vuyst, L.; Van Kerrebroeck, S.; Leroy, F. Microbial ecology and process technology of sourdough fermentation. *Adv. Appl. Microbiol.* **2017**, *100*, 49–160. [PubMed]
7. Comasio, A.; Van Kerrebroeck, S.; Harth, H.; Verté, F.; De Vuyst, L. Potential of bacteria from alternative fermented foods as starter cultures for the production of wheat sourdoughs. *Microorganisms* **2020**, *8*, 1534. [CrossRef] [PubMed]
8. Siepmann, F.B.; Ripari, V.; Waszczynskyj, N.; Spier, M.R. Overview of Sourdough Technology: From Production to Marketing. *Food Bioprocess Technol.* **2018**, *11*, 242–270. [CrossRef]
9. Chaves-Lopez, C.; Serio, A.; Delgado-Ospina, J.; Rossi, C.; Grande-Tovar, C.D.; Paparella, A. Exploring the bacterial microbiota of colombian fermented maize dough "Masa agria" (Maiz añejo). *Front. Microbiol.* **2016**, *7*, 1168. [CrossRef] [PubMed]
10. Vogelmann, S.A.; Seitter, M.; Singer, U.; Brandt, M.J.; Hertel, C. Adaptability of lactic acid bacteria and yeasts to sourdoughs prepared from cereals, pseudocereals and cassava and use of competitive strains as starters. *Int. J. Food Microbiol.* **2009**, *130*, 205–212. [CrossRef]
11. Catzeddu, P.; Mura, E.; Parente, E.; Sanna, M.; Farris, G.A. Molecular characterization of lactic acid bacteria from sourdough breads produced in Sardinia (Italy) and multivariate statistical analyses of results. *Syst. Appl. Microbiol.* **2006**, *29*, 138–144. [CrossRef] [PubMed]
12. Arora, K.; Ameur, H.; Polo, A.; Di Cagno, R.; Rizzello, C.G.; Gobbetti, M. Thirty years of knowledge on sourdough fermentation: A systematic review. *Trends Food Sci. Technol.* **2021**, *108*, 71–83. [CrossRef]
13. Montemurro, M.; Coda, R.; Rizzello, C.G. Recent advances in the use of sourdough biotechnology in pasta making. *Foods* **2019**, *8*, 129. [CrossRef] [PubMed]
14. Gobbetti, M.; De Angelis, M.; Di Cagno, R.; Calasso, M.; Archetti, G.; Rizzello, C.G. Novel insights on the functional/nutritional features of the sourdough fermentation. *Int. J. Food Microbiol.* **2019**, *302*, 103–113. [CrossRef]
15. De Vuyst, L.; Neysens, P. The sourdough microflora: Biodiversity and metabolic interactions. *Trends Food Sci. Technol.* **2005**, *16*, 43–56. [CrossRef]
16. Chavan, R.S.; Chavan, S.R. Sourdough Technology-A Traditional Way for Wholesome Foods: A Review. *Compr. Rev. Food Sci. Food Saf.* **2011**, *10*, 169–182. [CrossRef]
17. Gullo, M.; Romano, A.D.; Pulvirenti, A.; Giudici, P. *Candida humilis*—Dominant species in sourdoughs for the production of durum wheat bran flour bread. *Int. J. Food Microbiol.* **2003**, *80*, 55–59. [CrossRef]
18. Foschino, R.; Gallina, S.; Andrighetto, C.; Rossetti, L.; Galli, A. Comparison of cultural methods for the identification and molecular investigation of yeasts from sourdoughs for Italian sweet baked products. *FEMS Yeast Res.* **2004**, *4*, 609–618. [CrossRef]
19. Pulvirenti, A.; Caggia, C.; Restuccia, C.; Gullo, M.; Giudici, P. DNA fingerprinting methods used for identification of yeasts isolated from Sicilian sourdoughs. *Ann. Microbiol.* **2001**, *51*, 107–120.
20. Carbonetto, B.; Nidelet, T.; Guezenec, S.; Perez, M.; Segond, D.; Sicard, D. Interactions between *Kazachstania humilis* yeast species and lactic acid bacteria in Sourdough. *Microorganisms* **2020**, *8*, 240. [CrossRef]
21. Hammes, W.P.; Vogel, R.F. The genus *Lactobacillus*. In *The Genera of Lactic Acid Bacteria*, 1st ed.; Springer: Boston, MA, USA, 1995; pp. 19–54.
22. Xu, D.; Zhang, Y.; Tang, K.; Hu, Y.; Xu, X.; Gänzle, M.G. Effect of Mixed Cultures of Yeast and Lactobacilli on the Quality of Wheat Sourdough Bread. *Front. Microbiol.* **2019**, *10*, 2113. [CrossRef]
23. De Vuyst, L.; Harth, H.; Van Kerrebroeck, S.; Leroy, F. Yeast diversity of sourdoughs and associated metabolic properties and functionalities. *Int. J. Food Microbiol.* **2016**, *239*, 26–34. [CrossRef]
24. Foschino, R.; Picozzi, C.; Galli, A. Comparative study of nine *Lactobacillus fermentum* bacteriophages. *J. Appl. Microbiol.* **2001**, *91*, 394–403. [CrossRef] [PubMed]
25. Zheng, J.; Wittouck, S.; Salvetti, E.; Franz, C.M.A.P.; Harris, H.M.B.; Mattarelli, P.; O'toole, P.W.; Pot, B.; Vandamme, P.; Walter, J.; et al. A taxonomic note on the genus *Lactobacillus*: Description of 23 novel genera, emended description of the genus *Lactobacillus* beijerinck 1901, and union of *Lactobacillaceae* and *Leuconostocaceae*. *Int. J. Syst. Evol. Microbiol.* **2020**, *70*, 2782–2858. [CrossRef] [PubMed]
26. Gänzle, M.G.; Zheng, J. Lifestyles of sourdough lactobacilli—Do they matter for microbial ecology and bread quality? *Int. J. Food Microbiol.* **2019**, *302*, 15–23. [CrossRef]
27. Hammes, W.P.; Gänzle, M.G. Sourdough breads and related products. In *Microbiology of Fermented Foods*; Wood, B.J.B., Ed.; Springer: Boston, MA, USA, 1998; pp. 199–216.

28. Vernocchi, P.; Valmorri, S.; Gatto, V.; Torriani, S.; Gianotti, A.; Suzzi, G.; Guerzoni, M.E.; Gardini, F. A survey on yeast microbiota associated with an Italian traditional sweet-leavened baked good fermentation. *Food Res. Int.* **2004**, *37*, 469–476. [CrossRef]
29. Pulvirenti, A.; Rainieri, S.; Boveri, S.; Giudici, P. Optimizing the selection process of yeast starter cultures by preselecting strains dominating spontaneous fermentations. *Can. J. Microbiol.* **2009**, *55*, 326–332. [CrossRef]
30. Corsetti, A. *Handbook on Sourdough Biotechnology*, 1st ed.; Springer: Boston, MA, USA, 2013; pp. 1–298.
31. De Vero, L.; Boniotti, M.B.; Budroni, M.; Buzzini, P.; Cassanelli, S.; Comunian, R.; Gullo, M.; Logrieco, A.F.; Mannazzu, I.; Musumeci, R.; et al. Preservation, characterization and exploitation of microbial biodiversity: The perspective of the Italian network of culture collections. *Microorganisms* **2019**, *7*, 685. [CrossRef]
32. Clarke, C.I.; Arendt, E.K. A Review of the Application of Sourdough Technology to Wheat Breads. *Adv. Food Nutr. Res.* **2005**, *49*, 137–161. [PubMed]
33. Huys, G.; Daniel, H.M.; De Vuyst, L. Taxonomy and biodiversity of sourdough yeasts and lactic acid bacteria. In *Handbook on Sourdough Biotechnology*; Gobbetti, M., Gänzle, M., Eds.; Springer: Boston, MA, USA, 2013.
34. Montemurro, M.; Celano, G.; De Angelis, M.; Gobbetti, M.; Rizzello, C.G.; Pontonio, E. Selection of non-*Lactobacillus* strains to be used as starters for sourdough fermentation. *Food Microbiol.* **2020**, *90*, 103–491. [CrossRef]
35. Coda, R.; Di Cagno, R.; Gobbetti, M.; Rizzello, C.G. Sourdough lactic acid bacteria: Exploration of non-wheat cereal-based fermentation. *Food Microbiol.* **2014**, *37*, 51–58. [CrossRef] [PubMed]
36. Gobbetti, M.; Rizzello, C.G.; Di Cagno, R.; De Angelis, M. How the sourdough may affect the functional features of leavened baked goods. *Food Microbiol.* **2014**, *37*, 30–40. [CrossRef] [PubMed]
37. Facco Stefanello, R.; Nabeshima, E.H.; de Oliveira Garcia, A.; Heck, R.T.; Valle Garcia, M.; Martins Fries, L.L.; Venturini Copetti, M. Stability, sensory attributes and acceptance of panettones elaborated with *Lactobacillus fermentum* IAL 4541 and *Wickerhamomyces anomallus* IAL 4533. *Food Res. Int.* **2019**, *116*, 973–984. [CrossRef]
38. Çakır, E.; Arıcı, M.; Durak, M.Z. Biodiversity and techno-functional properties of lactic acid bacteria in fermented hull-less barley sourdough. *J. Biosci. Bioeng.* **2020**, *130*, 450–456. [CrossRef]
39. Siepmann, F.B.; Sousa de Almeida, B.; Waszczynskyj, N.; Spier, M.R. Influence of temperature and of starter culture on biochemical characteristics and the aromatic compounds evolution on type II sourdough and wheat bread. *LWT* **2019**, *108*, 199–206. [CrossRef]
40. Zannini, E.; Pontonio, E.; Waters, D.M.; Arendt, E.K. Applications of microbial fermentations for production of gluten-free products and perspectives. *Appl. Microbiol. Biotechnol.* **2012**, *93*, 473–485. [CrossRef]
41. Zoumpopoulou, G.; Tsakalidou, E. Gluten-free products. In *The Role of Alternative and Innovative Food Ingredients and Products in Consumer Wellness*; Galanakis, C., Ed.; Academic Press: Cambridge, MA, USA, 2019; pp. 213–237.
42. Rizzello, C.G.; Coda, R.; Gobbetti, M. Use of sourdough fermentation and non wheat flours for enhancing nutritional and healthy properties of wheat-based foods. In *Fermented Foods in Health and Disease Prevention*; Frías, J., Martínez-Villaluenga, C., Peñas, E., Eds.; Academic Press: Cambridge, MA, USA, 2016; pp. 433–452.
43. Pophaly, S.D.S.D.; Singh, R.; Pophaly, S.D.S.D.; Kaushik, J.K.; Tomar, S.K. Current status and emerging role of glutathione in food grade lactic acid bacteria. *Microb. Cell Fact.* **2012**, *11*, 1–14. [CrossRef]
44. De Vero, L.; Bonciani, T.; Verspohl, A.; Mezzetti, F.; Giudici, P. High-glutathione producing yeasts obtained by genetic improvement strategies: A focus on adaptive evolution approaches for novel wine strains. *AIMS Microbiol.* **2017**, *3*, 155–170. [CrossRef]
45. Bonciani, T.; De Vero, L.; Mezzetti, F.; Fay, J.C.; Giudici, P. A multi-phase approach to select new wine yeast strains with enhanced fermentative fitness and glutathione production. *Appl. Microbiol. Biotechnol.* **2018**, *102*, 2269–2278. [CrossRef]
46. Moore, M.M.; Heinbockel, M.; Dockery, P.; Ulmer, H.M.; Arendt, E.K. Network formation in gluten-free bread with application of transglutaminase. *Cereal Chem.* **2006**, *83*, 28–36. [CrossRef]
47. Moore, M.M.; Juga, B.; Schober, T.J.; Arendt, E.K. Effect of lactic acid bacteria on properties of gluten-free sourdoughs, batters, and quality and ultrastructure of gluten-free bread. *Cereal Chem.* **2007**, *84*, 357–364. [CrossRef]
48. Di Cagno, R.; Rizzello, C.G.; De Angelis, M.; Cassone, A.; Giuliani, G.; Benedusi, A.; Limitone, A.; Surico, R.F.; Gobbetti, M. Use of selected sourdough strains of Lactobacillus for removing gluten and enhancing the nutritional properties of gluten-free bread. *J. Food Prot.* **2008**, *71*, 1491–1495. [CrossRef] [PubMed]
49. Rizzello, C.G.; De Angelis, M.; Di Cagno, R.; Camarca, A.; Silano, M.; Losito, I.; De Vincenzi, M.; De Bari, M.D.; Palmisano, F.; Maurano, F.; et al. Highly efficient gluten degradation by lactobacilli and fungal proteases during food processing: New perspectives for celiac disease. *Appl. Environ. Microbiol.* **2007**, *73*, 4499–4507. [CrossRef] [PubMed]
50. Hüttner, E.K.; Dal Bello, F.; Arendt, E.K. Identification of lactic acid bacteria isolated from oat sourdoughs and investigation into their potential for the improvement of oat bread quality. *Eur. Food Res. Technol.* **2010**, *230*, 849–857. [CrossRef]
51. Moroni, A.V.; Dal Bello, F.; Arendt, E.K. Sourdough in gluten-free bread-making: An ancient technology to solve a novel issue? *Food Microbiol.* **2009**, *26*, 676–684. [CrossRef] [PubMed]
52. Moroni, A.V.; Arendt, E.K.; Dal Bello, F. Biodiversity of lactic acid bacteria and yeasts in spontaneously-fermented buckwheat and teff sourdoughs. *Food Microbiol.* **2011**, *28*, 497–502. [CrossRef] [PubMed]
53. Coda, R.; Rizzello, C.G.; Gobbetti, M. Use of sourdough fermentation and pseudo-cereals and leguminous flours for the making of a functional bread enriched of γ-aminobutyric acid (GABA). *Int. J. Food Microbiol.* **2010**, *137*, 236–245. [CrossRef]
54. Siragusa, S.; De Angelis, M.; Di Cagno, R.; Rizzello, C.G.; Coda, R.; Gobbetti, M. Synthesis of γ-aminobutyric acid by lactic acid bacteria isolated from a variety of Italian cheeses. *Appl. Environ. Microbiol.* **2007**, *73*, 7283–7290. [CrossRef]

55. Rizzello, C.G.; Cassone, A.; Di Cagno, R.; Gobbetti, M. Synthesis of angiotensin I-converting enzyme (ACE)-inhibitory peptides and γ-aminobutyric acid (GABA) during sourdough fermentation by selected lactic acid bacteria. *J. Agric. Food Chem.* **2008**, *56*, 6936–6943. [CrossRef]

56. Cui, Y.; Miao, K.; Niyaphorn, S.; Qu, X. Production of gamma-aminobutyric acid from lactic acid bacteria: A systematic review. *Int. J. Mol. Sci.* **2020**, *21*, 995. [CrossRef]

57. Pandey, A.; Szakacs, G.; Soccol, C.R.; Rodriguez-Leon, J.A.; Soccol, V.T. Production, purification and properties of microbial phytases. *Bioresour. Technol.* **2001**, *77*, 203–214. [CrossRef]

58. Corsetti, A.; Settanni, L. Lactobacilli in sourdough fermentation. *Food Res. Int.* **2007**, *40*, 539–558. [CrossRef]

59. De Angelis, M.; Gallo, G.; Corbo, M.R.; McSweeney, P.L.H.; Faccia, M.; Giovine, M.; Gobbetti, M. Phytase activity in sourdough lactic acid bacteria: Purification and characterization of a phytase from *Lactobacillus sanfranciscensis* CB1. *Int. J. Food Microbiol.* **2003**, *87*, 259–270. [CrossRef]

60. Zotta, T.; Ricciardi, A.; Parente, E. Enzymatic activities of lactic acid bacteria isolated from Cornetto di Matera sourdoughs. *Int. J. Food Microbiol.* **2007**, *115*, 165–172. [CrossRef] [PubMed]

61. Palla, M.; Blandino, M.; Grassi, A.; Giordano, D.; Sgherri, C.; Quartacci, M.F.; Reyneri, A.; Agnolucci, M.; Giovannetti, M. Characterization and selection of functional yeast strains during sourdough fermentation of different cereal wholegrain flours. *Sci. Rep.* **2020**, *10*, 1–15. [CrossRef]

62. Milanović, V.; Osimani, A.; Garofalo, C.; Belleggia, L.; Maoloni, A.; Cardinali, F.; Mozzon, M.; Foligni, R.; Aquilanti, L.; Clementi, F. Selection of cereal-sourced lactic acid bacteria as candidate starters for the baking industry. *PLoS ONE* **2020**, *15*, e0236190.

63. Robert, H.; Gabriel, V.; Lefebvre, D.; Rabier, P.; Vayssier, Y.; Fontagné-Faucher, C. Study of the behaviour of *Lactobacillus plantarum* and *Leuconostoc* starters during a complete wheat sourdough breadmaking process. *LWT Food Sci. Technol.* **2006**, *39*, 256–265. [CrossRef]

64. Rosenquist, H.; Hansen, Å. The antimicrobial effect of organic acids, sourdough and nisin against *Bacillus subtilis* and *B. licheniformis* isolated from wheat bread. *J. Appl. Microbiol.* **1998**, *85*, 621–631. [CrossRef]

65. Katina, K.; Arendt, E.; Liukkonen, K.H.; Autio, K.; Flander, L.; Poutanen, K. Potential of sourdough for healthier cereal products. *Trends Food Sci. Technol.* **2005**, *16*, 104–112. [CrossRef]

66. Poutanen, K.; Flander, L.; Katina, K. Sourdough and cereal fermentation in a nutritional perspective. *Food Microbiol.* **2009**, *26*, 693–699. [CrossRef]

67. Östman, E.M.; Nilsson, M.; Liljeberg Elmståhl, H.G.M.; Molin, G.; Björck, I.M.E. On the effect of lactic acid on blood glucose and insulin responses to cereal products: Mechanistic studies in healthy subjects and in vitro. *J. Cereal Sci.* **2002**, *36*, 339–346. [CrossRef]

68. Mamlouk, D.; Gullo, M. Acetic Acid Bacteria: Physiology and Carbon Sources Oxidation. *Indian J. Microbiol.* **2013**, *53*, 377–384. [CrossRef] [PubMed]

69. La China, S.; Zanichelli, G.; De Vero, L.; Gullo, M. Oxidative fermentations and exopolysaccharides production by acetic acid bacteria: A mini review. *Biotechnol. Lett.* **2018**, *40*, 1289–1302. [CrossRef] [PubMed]

70. Vigentini, I.; Fabrizio, V.; Dellacà, F.; Rossi, S.; Azario, I.; Mondin, C.; Benaglia, M.; Foschino, R. Set-up of bacterial cellulose production from the genus *Komagataeibacter* and its use in a gluten-free bakery product as a case study. *Front. Microbiol.* **2019**, *10*, 1953. [CrossRef] [PubMed]

71. Zannini, E.; Waters, D.M.; Coffey, A.; Arendt, E.K. Production, properties, and industrial food application of lactic acid bacteria-derived exopolysaccharides. *Appl. Microbiol. Biotechnol.* **2016**, *100*, 1121–1135. [CrossRef]

72. La China, S.; De Vero, L.; Anguluri, K.; Brugnoli, M.; Mamlouk, D.; Gullo, M. Kombucha tea as a reservoir of cellulose producing bacteria: Assessing diversity among *Komagataeibacter* isolates. *Appl. Sci.* **2021**, *11*, 1595. [CrossRef]

73. Arena, M.P.; Russo, P.; Spano, G.; Capozzi, V. From Microbial Ecology to Innovative Applications in Food Quality Improvements: The Case of Sourdough as a Model Matrix. *J* **2020**, *3*, 9–19. [CrossRef]

74. Kaditzky, S.; Seitter, M.; Hertel, C.; Vogel, R.F. Performance of *Lactobacillus sanfranciscensis* TMW 1.392 and its levansucrase deletion mutant in wheat dough and comparison of their impact on bread quality. *Eur. Food Res. Technol.* **2008**, *227*, 433–442. [CrossRef]

75. Falasconi, I.; Fontana, A.; Patrone, V.; Rebecchi, A.; Garrido, G.D.; Principato, L.; Callegari, M.L.; Spigno, G.; Morelli, L. Genome-assisted characterization of *Lactobacillus fermentum*, *Weissella cibaria*, and *Weissella confusa* strains isolated from sorghum as starters for sourdough fermentation. *Microorganisms* **2020**, *8*, 1388. [CrossRef] [PubMed]

76. Galle, S.; Schwab, C.; Arendt, E.; Gänzle, M.G. Exopolysaccharide-forming *Weissella* strains as starter cultures for sorghum and wheat sourdoughs. *J. Agric. Food Chem.* **2010**, *58*, 5834–5841. [CrossRef]

77. Franco, W.; Pérez-Díaz, I.M.; Connelly, L.; Diaz, J.T. Isolation of exopolysaccharide-producing yeast and lactic acid bacteria from quinoa (*Chenopodium quinoa*) sourdough fermentation. *Foods* **2020**, *9*, 337. [CrossRef] [PubMed]

78. Lynch, K.M.; Coffey, A.; Arendt, E.K. Exopolysaccharide producing lactic acid bacteria: Their techno-functional role and potential application in gluten-free bread products. *Food Res. Int.* **2018**, *110*, 52–61. [CrossRef] [PubMed]

79. Mattila, P.; Pihlava, J.M.; Hellström, J. Contents of phenolic acids, alkyl- and alkenylresorcinols, and avenanthramides in commercial grain products. *J. Agric. Food Chem.* **2005**, *53*, 8290–8295. [CrossRef] [PubMed]

80. Filannino, P.; Di Cagno, R.; Gobbetti, M. Metabolic and functional paths of lactic acid bacteria in plant foods: Get out of the labyrinth. *Curr. Opin. Biotechnol.* **2018**, *49*, 64–72. [CrossRef]

81. Fernández-Peláez, J.; Paesani, C.; Gómez, M. Sourdough Technology as a Tool for the Development of Healthier Grain-Based Products: An Update. *Agronomy* **2020**, *10*, 1962. [CrossRef]
82. Katina, K.; Laitila, A.; Juvonen, R.; Liukkonen, K.H.; Kariluoto, S.; Piironen, V.; Landberg, R.; Åman, P.; Poutanen, K. Bran fermentation as a means to enhance technological properties and bioactivity of rye. *Food Microbiol.* **2007**, *24*, 175–186. [CrossRef]
83. Kariluoto, S.; Aittamaa, M.; Korhola, M.; Salovaara, H.; Vahteristo, L.; Piironen, V. Effects of yeasts and bacteria on the levels of folates in rye sourdoughs. *Int. J. Food Microbiol.* **2006**, *106*, 137–143. [CrossRef]
84. Coda, R.; Cassone, A.; Rizzello, C.G.; Nionelli, L.; Cardinali, G.; Gobbetti, M. Antifungal Activity of *Wickerhamomyces anomalus* and *Lactobacillus plantarum* during Sourdough Fermentation: Identification of Novel Compounds and Long-Term Effect during Storage of Wheat Bread. *Appl. Environ. Microbiol.* **2011**, *77*, 3484–3492. [CrossRef]
85. Rizzello, C.G.; Nionelli, L.; Coda, R.; Gobbetti, M. Synthesis of the cancer preventive peptide lunasin by lactic acid bacteria during sourdough fermentation. *Nutr. Cancer* **2012**, *64*, 111–120. [CrossRef]
86. Sadiq, F.A.; Yan, B.; Tian, F.; Zhao, J.; Zhang, H.; Chen, W. Lactic Acid Bacteria as Antifungal and Anti-Mycotoxigenic Agents: A Comprehensive Review. *Compr. Rev. Food Sci. Food Saf.* **2019**, *18*, 1403–1436. [CrossRef] [PubMed]
87. Cortés-Zavaleta, O.; López-Malo, A.; Hernández-Mendoza, A.; García, H.S. Antifungal activity of lactobacilli and its relationship with 3-phenyllactic acid production. *Int. J. Food Microbiol.* **2014**, *173*, 30–35. [CrossRef] [PubMed]
88. Ndagano, D.; Lamoureux, T.; Dortu, C.; Vandermoten, S.; Thonart, P. Antifungal activity of 2 lactic acid bacteria of the *Weissella* genus isolated from food. *J. Food Sci.* **2011**, *76*, M305–M311. [CrossRef] [PubMed]
89. Dalié, D.K.D.; Deschamps, A.M.; Richard-Forget, F. Lactic acid bacteria—Potential for control of mould growth and mycotoxins: A review. *Food Control* **2010**, *21*, 370–380. [CrossRef]
90. Axel, C.; Brosnan, B.; Zannini, E.; Peyer, L.C.; Furey, A.; Coffey, A.; Arendt, E.K. Antifungal activities of three different *Lactobacillus* species and their production of antifungal carboxylic acids in wheat sourdough. *Appl. Microbiol. Biotechnol.* **2016**, *100*, 1701–1711. [CrossRef] [PubMed]
91. Corsetti, A.; Gobbetti, M.; Rossi, J.; Damiani, P. Antimould activity of sourdough lactic acid bacteria: Identification of a mixture of organic acids produced by *Lactobacillus sanfrancisco* CB1. *Appl. Microbiol. Biotechnol.* **1998**, *50*, 253–256. [CrossRef] [PubMed]
92. Lavermicocca, P.; Valerio, F.; Evidente, A.; Lazzaroni, S.; Corsetti, A.; Gobbetti, M. Purification and characterization of novel antifungal compounds from the sourdough *Lactobacillus plantarum* strain 21B. *Appl. Environ. Microbiol.* **2000**, *66*, 4084–4090. [CrossRef]
93. Gerez, C.L.; Torino, M.I.; Obregozo, M.D.; De Font Valdez, G. A ready-to-use antifungal starter culture improves the shelf life of packaged bread. *J. Food Prot.* **2010**, *73*, 758–762. [CrossRef]
94. Zhang, C.; Brandt, M.J.; Schwab, C.; Gänzle, M.G. Propionic acid production by cofermentation of *Lactobacillus buchneri* and *Lactobacillus diolivorans* in sourdough. *Food Microbiol.* **2010**, *27*, 390–395. [CrossRef]
95. Garofalo, C.; Zannini, E.; Aquilanti, L.; Silvestri, G.; Fierro, O.; Picariello, G.; Clementi, F. Selection of sourdough lactobacilli with antifungal activity for use as biopreservatives in bakery products. *J. Agric. Food Chem.* **2012**, *60*, 7719–7728. [CrossRef]
96. Ryan, L.A.M.; Zannini, E.; Dal Bello, F.; Pawlowska, A.; Koehler, P.; Arendt, E.K. *Lactobacillus amylovorus* DSM 19280 as a novel food-grade antifungal agent for bakery products. *Int. J. Food Microbiol.* **2011**, *146*, 276–283. [CrossRef]
97. Axel, C.; Röcker, B.; Brosnan, B.; Zannini, E.; Furey, A.; Coffey, A.; Arendt, E.K. Application of *Lactobacillus amylovorus* DSM19280 in gluten-free sourdough bread to improve the microbial shelf life. *Food Microbiol.* **2015**, *47*, 36–44. [CrossRef]
98. Hassan, Y.I.; Bullerman, L.B. Antifungal activity of *Lactobacillus paracasei* subsp. tolerans against Fusarium proliferatum and Fusarium graminearum in a liquid culture setting. *J. Food Prot.* **2008**, *71*, 2213–2216.
99. Hassan, Y.I.; Zhou, T.; Bullerman, L.B. Sourdough lactic acid bacteria as antifungal and mycotoxin-controlling agents. *Food Sci. Technol. Int.* **2016**, *22*, 79–90. [CrossRef]
100. Angioloni, A.; Romani, S.; Pinnavaia, G.G.; Dalla Rosa, M. Characteristics of bread making doughs: Influence of sourdough fermentation on the fundamental rheological properties. *Eur. Food Res. Technol.* **2006**, *222*, 54–57. [CrossRef]
101. Corsetti, A.; Gobbetti, M.; De Marco, B.; Balestrieri, F.; Paoletti, F.; Russi, L.; Rossi, J. Combined effect of sourdough lactic acid bacteria and additives on bread firmness and staling. *J. Agric. Food Chem.* **2000**, *48*, 3044–3051. [CrossRef] [PubMed]
102. Gobbetti, M.; Corsetti, A.; Rossi, J. Interaction between lactic acid bacteria and yeasts in sour-dough using a rheofermentometer. *World J. Microbiol. Biotechnol.* **1995**, *11*, 625–630. [CrossRef] [PubMed]
103. Clarke, C.I.; Schober, T.J.; Arendt, E.K. Effect of single strain and traditional mixed strain starter cultures on rheological properties of wheat dough and on bread quality. *Cereal Chem.* **2002**, *79*, 640–647. [CrossRef]
104. Cappa, C.; Lucisano, M.; Raineri, A.; Fongaro, L.; Foschino, R.; Mariotti, M. Gluten-Free Bread: Influence of Sourdough and Compressed Yeast on Proofing and Baking Properties. *Foods* **2016**, *5*, 69. [CrossRef] [PubMed]
105. Picozzi, C.; Mariotti, M.; Cappa, C.; Tedesco, B.; Vigentini, I.; Foschino, R.; Lucisano, M. Development of a Type I gluten-free sourdough. *Lett. Appl. Microbiol.* **2016**, *62*, 119–125. [CrossRef]
106. Melini, V.; Melini, F. Strategies to extend bread and GF bread shelf-life: From Sourdough to antimicrobial active packaging and nanotechnology. *Fermentation* **2018**, *4*, 5–10.
107. Lavermicocca, P.; Valerio, F.; De Bellis, P.; Sisto, A. Leguérinel, Sporeforming bacteria associated with bread production: Spoilage and toxigenic potential. In *Food Hygiene and Toxicology in Ready-to-Eat Foods*; Kotzekidou, P., Ed.; Academic Press: San Diego, CA, USA, 2016; pp. 275–293.

108. Capozzi, V.; Fragasso, M.; Romaniello, R.; Berbegal, C.; Russo, P.; Spano, G. Spontaneous Food Fermentations and Potential Risks for Human Health. *Fermentation* **2017**, *3*, 49. [CrossRef]
109. Gobbetti, M.; De Angelis, M.; Corsetti, A.; Di Cagno, R. Biochemistry and physiology of sourdough lactic acid bacteria. *Trends Food Sci. Technol.* **2005**, *16*, 57–69. [CrossRef]
110. Luti, S.; Galli, V.; Venturi, M.; Granchi, L.; Paoli, P.; Pazzagli, L. Bioactive Properties of Breads Made with Sourdough of Hull-Less Barley or Conventional and Pigmented Wheat Flours. *Appl. Sci.* **2021**, *11*, 3291. [CrossRef]
111. Marcotrigiano, V.; Lanzilotti, C.; Rondinone, D.; De Giglio, O.; Caggiano, G.; Diella, G.; Orsi, G.B.; Montagna, M.T.; Napoli, C. Food labelling: Regulations and Public Health implications. *Ann. Ig* **2018**, *30*, 220–228.

 *applied sciences*

Article

Modelling of the Acetification Stage in the Production of Wine Vinegar by Use of Two Serial Bioreactors

Carmen M. Álvarez-Cáliz [1], Inés María Santos-Dueñas [1,*], Jorge E. Jiménez-Hornero [2] and Isidoro García-García [1]

[1] Department of Chemical Engineering, Campus de Rabanales, University of Cordoba, 14071 Cordoba, Spain; q42alcac@yahoo.es (C.M.Á.-C.); isidoro.garcia@uco.es (I.G.-G.)

[2] Department of Electrical Engineering and Automatic Control, Campus de Rabanales, University of Cordoba, 14071 Cordoba, Spain; jjimenez@uco.es

* Correspondence: ines.santos@uco.es; Tel.: +34-957-218-658

Received: 28 November 2020; Accepted: 16 December 2020; Published: 18 December 2020

Abstract: In the scope of a broader study about modelling wine acetification, the use of polynomial black-box models seems to be the best choice. Additionally, the use of two serially arranged bioreactors was expected to result in increased overall acetic acid productivity. This paper describes the experiments needed to obtain enough data for modelling the process and the use of second-order polynomials for this task. A fractional experimental design with central points was used with the ethanol concentrations during loading of the bioreactors, their operation temperatures, the ethanol concentrations at unloading time, and the unloaded volume in the first one as factors. Because using two serial reactors imposed some constraints on the operating ranges for the process, an exhaustive combinatorial analysis was used to identify a working combination of such ranges. The obtained models provided highly accurate predictions of the mean overall rate of acetic acid formation, the mean total production of acetic acid of the two-reactor system, and ethanol concentration at the time the second reactor is unloaded. The operational variables associated with the first bioreactor were the more strongly influential to the process, particularly the ethanol concentration at the time the first reactor was unloaded, the unloaded volume, and the ethanol concentration when loading.

Keywords: vinegar; wine; acetification; bioprocesses; experimental design; polynomial modelling; black-box models

1. Introduction

Vinegar production is a biotechnological process essentially involving the biological conversion of ethanol from a given source into acetic acid. Vinegar can be obtained from alcohol, wine, cereals or fruits, among other sources [1–4]. The key step of the process is possibly that by which a complex microbiota of acetic acid bacteria (AAB) convert ethanol into acetic acid in a bioreactor. The bacterial mixture affecting the conversion arises from the natural microbial selection in the acetification medium. In practice, only AAB can exist in an environment containing medium concentrations of ethanol and acetic acid at the beginning but low levels of the former and high levels of the latter at the end [5]. These conditions make it unnecessary to sterilize containers or keep aseptic conditions during operation.

Acetification bioreactors usually operate in a semi-continuous mode. Thus, once the reactors are in full operation, the ethanol concentration is allowed to decrease to a preset level and then an also preset fraction of the reactor contents is unloaded, the remainder being allowed to stand in it in order to act as an inoculum in the next conversion cycle [6–10]. After the bioreactor is unloaded, it is slowly replenished with a fresh alcoholic substrate to start a new ethanol depletion cycle.

Because the temperature and airflow rate are usually fixed, the operational variables that can be altered include the ethanol concentration at the time the reactor is unloaded, the proportion of broth that is unloaded, the loading mode and/or rate, and the total concentration of the culture medium—which is the combination of the ethanol concentration in % *v/v* and the acidity in % *w/v* [11–16]. These variables influence the mean ethanol concentration and acidity in each cycle [8–10], which in turn affect the AAB concentration and cell activity [17,18]. As a result, the acetification conditions will be more or less stressful for the bacteria affecting the process (AAB).

Usually, an acetification bioreactor is operated in an automated manner in order to not alter the spontaneous dynamics of the system. In practice, this allows repeated cycling to be easily and rapidly achieved [19]. Also, the characteristics of AAB [20–23] make cultivation and selection outside their typical natural or industrial environment rather difficult. In addition, their complex identification, behaviour, and interactions, and their potential synergistic effects, require determining their optimum conditions of operation in an empirical manner.

Notwithstanding the previous difficulties, there is a wealth of technical experience and knowledge about the most suitable working methods and operating conditions for acetification. The use of non-segregated non-structured models [24] could be appropriated to model this kind of system; working with this type of approach allowed us to reach most of the knowledge on this particular process, which has allowed quantitative relations between operational variables and diverse industrial objective functions such as productivity to be established [8–10,25–30].

The advent of massive methods of analysis, such as several omic techniques, has considerably expanded available knowledge about the acetification process at the molecular level, and is bound to help improve the stability and food safety of the end-product [31]. There is, however, an ongoing search for more or less structured modelling approaches to relating the variables of the overall process.

Previous modelling studies [7,10,30] led to proposing the use of two serially arranged bioreactors to optimize the outcome of the acetification process under operationally restricted conditions. Thus, vinegar production is most often subjected to strict regulations with regards to the properties of the end-product, which, for example, should contain very little or no ethanol. In practice, however, ethanol in the acetification medium should never be depleted before the reactor is unloaded since that would place AAB under extremely stressful conditions (viz., a high acidity and a lack of substrate) and render them virtually useless in subsequent biomass conversion cycles. As a result, many industrial acetification plants use additional bioreactors to deplete ethanol present in the vinegar following the unloading of the production bioreactors.

In this scenario, modelling of the two serial bioreactors system requires the use of an appropriate experimental design. Previous experience with modelling of the biotransformation stage in the vinegar production process suggests that black-box models based on second-order generalized polynomials [7,19,32–34] provide a more accurate depiction of the experimental results than do existing alternatives, allowing it to describe potential interactions between independent variables to be considered—and with added advantages like ease of development and statistical validation. The number of experiments needed for the accurate fitting of a polynomial equation depends on whether the polynomial is linear or non-linear [35]. In any case, the greater the number of polynomial terms is (increasing the accuracy of the model), the more experiments will be required to calculate their coefficients. All testing should be conducted in the framework of an experimental design using the minimum possible number of runs to identify interactions between variables and allowing representative equations for the target process to be established as possible.

Based on the previous comments, the main aim of this work was to obtain polynomial models for several key variables of the two serially arranged bioreactors system and to determine an appropriate experimental design for gathering the experimental data needed to estimate such models. A fractional factorial design with central points including the six major operational variables involved on the acetification process has been used, considering the fulfilment of several restrictions arising from the operation of the two bioreactors, which required identifying and examining the impact of such variables (specifically, establishing their lower and upper limits in the framework of the experimental design). Once the experiments were conducted, polynomial models of relevant target variables of the process such as productivity, acetification rate, etc. were fitted using the gathered experimental data. As far as we know, this thorough study where a detailed analysis considering the constraints for the operational variables, the so many experiments, as well as the replications carried out for each experimental set of variables' values has not been done before; additionally, the modelling approach for this acetification set-up has not been reported in any other previous work.

2. Materials and Methods

2.1. Operating Mode

A diagram showing the operation of two bioreactors working serially is depicted in Figure 1, where the process involved the following operations:

1. Once the ethanol concentration in the first reactor was reduced to E_{u1}, a volume fraction of culture medium V_{u1} was unloaded, whereas the remainder was used as inoculum for the fresh medium to be loaded in order to replace the unloaded fraction. Immediately before the first reactor was partially unloaded, the second was completely unloaded in order to receive the portion withdrawn from the first.
2. Both reactors were loaded with fresh substrate (wine) controlling the ethanol concentrations to E_{l1} in the first and E_{l2} in the second. This allowed drastic changes in the ethanol and acetic acid concentrations to be avoided. In fact, intermittently adding appropriate amounts of fresh substrate allowed a constant concentration of ethanol to be maintained throughout the loading stage.
3. Once loading was finished, both reactors continued to operate until the second was again unloaded prior to the first.

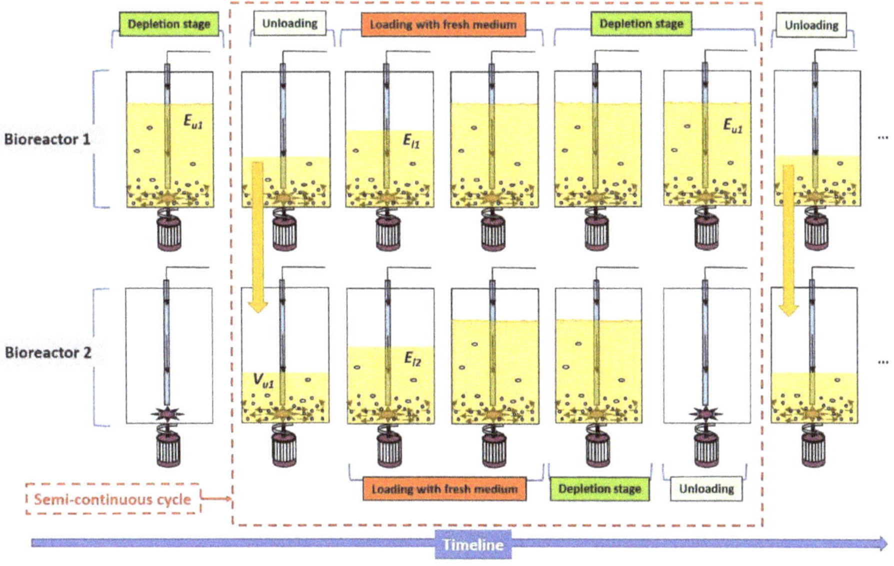

Figure 1. The two serial reactors operating in a semi-continuous mode.

This scheme allowed the following operational variables to be changed: ethanol concentration at the time the first reactor was unloaded (E_{u1}), volume unloaded from the first reactor (V_{u1}), the ethanol concentration while the first and second reactor were loaded (E_{l1} and E_{l2}, respectively), and the temperature in each reactor (T_1 and T_2, respectively). A total of six operational variables were thus involved. Although the alcoholic strength of the wine and the airflow rate could also have been used as variables, the feed wine usually contains 12% (*v/v*) ethanol, and the airflow rate is usually as high as possible (especially when productivity is to be maximized). Also, using gas condensers minimized the sweeping of volatiles.

Except for the temperatures, the working range for the previous variables could not be freely chosen because each variable was subjected to physical limitations, especially because the serial operation of the two reactors imposed constraints arising from the mutual relationships between the variables. This required careful planning of the experimental work in order to define the number of experiments needed to obtain accurate information with a view to predicting the influence of each variable on the resulting system behavior. As described in Section 3.1, a fractional factorial design for this purpose was used [36].

2.2. Experimental Set-Up

Figure 2 shows the two serially arranged bioreactors used to conduct the experiments. The reactors were two 8 L Acetators from Heinrich Frings GmbH & Co. KG (Rheinbach, Germany) equipped with self-aspirating turbines in the bottom in order to obtain a dispersion of fine air bubbles. The reactors were highly efficient in transferring oxygen between the gas phase and the substrate [11]. The ensemble was interfaced to a computer for automated operation, especially with regards to loading and unloading of reactors, and monitoring of the process. This resulted in highly reproducible loading and unloading, and in efficient acquisition of data [6,11,37].

Figure 2. The experimental set-up for the bioconversion of ethanol into acetic acid.

2.3. Raw Material

The substrate used was white wine from the Montilla–Moriles D.O. (Córdoba, Spain) containing (11.5 ± 0.5)% (*v/v*) ethanol and an initial acidity of (0.4 ± 0.1)% (*w/v*) as acetic acid.

2.4. Microorganisms

The inoculum used was a mixed culture of acetic acid bacteria (AAB) obtained from bioreactors previously operating in repeated acetification cycles with the same type of wine as the substrate. This is the usual mode of operation for the vinegar industry.

2.5. Analytical Methods

The only variable not determined automatically was acidity, which was measured by acid—base titration with an NaOH solution approximately 0.5 N that was previously standardized with potassium hydrogen phthalate. The volume and ethanol concentration were measured in a continuous manner by using an EJA 110 differential pressure probe from Yokogawa Electric Corp. (Tokyo, Japan) and an Alkosens probe equipped with an Acetomat transducer from Heinrich Frings, respectively.

2.6. Mathematical Methods

2.6.1. Polynomial Models and Fitting Methods

The models used here were based on second-order non-linear polynomial equations of the following type:

$$Y = b_0 + \sum_{i=1}^{n} b_i \cdot X_i + \sum_{\substack{i=1 \\ i<j}}^{n} b_{ij} \cdot X_i \cdot X_j + \sum_{i=1}^{n} b_{ii} \cdot X_i^2, \tag{1}$$

where Y denotes the dependent variable, b_0 the independent regression term, b_i the coefficients of the linear (first-order) terms, b_{ij} those of non-linear (interaction) terms, b_{ii} those of quadratic terms, X_i the independent variables, X_j the independent variables with $i < j$, and n the number of independent variables. The Y variable is the response or output of the polynomial model and X variables are the inputs.

The equations were fitted by using one of several specific methods along with experimental data to calculate the previous coefficients. Such methods allow identifying those terms in Equation (1) that are significant and those that are not. In practice, they solve a least-squares problem by minimizing the sum of squares of the residuals (prediction errors). Second-order polynomial models can be fitted by using several methods [38–43]. In this work, the Forward Stepwise Regression method has been used, which successively incorporates one by one those independent variables that contribute to predicting a dependent variable from the most to the least predictive. The sequence in which terms are added to or removed from a polynomial in each step of the process is established according to the F-to-enter, F-to-remove (reference values set by the modeler for the F-values associated to each term of the model), R and R^2 criteria. Then, the F-statistics also give a measure of the model sensitivity to each term of the polynomial relating the input variables.

2.6.2. ANalysis Of VAriance (ANOVA)

Before the model equations were fitted, the experimental data to be used for this purpose were checked for statistical differences by analysis of variance (ANOVA), which compares the means for two or more data samples in terms of their variances. ANOVA's null hypothesis is that all means are identical and the alternative hypothesis that at least one mean will be different from all others [36]. ANOVA is used to identify sources of differences among data samples and to assess whether the differences among sample media are too large to be assigned to random errors alone [44].

3. Results and Discussion

3.1. Experimental Design

The experimental design that was initially intended to use was a Box-Behnken or Doehlert central composite factorial design. With two levels per factor, the total number of experiments needed was $2^n + 2n + 1$, where the first term represents the overall factorial design, the second the central points of the faces, and the third the central point. The factors (operational variables) used were the ethanol concentration at the loading stage in the first bioreactor (E_{l1}), the ethanol concentration at unloading time in the first bioreactor (E_{u1}), the volume unloaded from the first bioreactor into the second one

(V_{u1}), the working temperature in the first bioreactor (T_1), the ethanol concentration at loading time in the second bioreactor (E_{l2}), and the working temperature in the second bioreactor (T_2). A total of 77 different experiments were in theory needed for to model these operational variables. In practice, however, this would have involved too many runs—probably more than needed to fit the model. Also, each experiment would have to be replicated many times to obtain reproducible results and lag phases would be needed each time the operational variables were modified to investigate a new case.

It was therefore necessary to reduce the number of experiments without sacrificing experimental information. This was accomplished by using a fractional factorial design requiring only 2^{n-p} experiments, where n is the total number of variables and p the number of those variables obtained from some interactions of the others. Thus, considering $p = 2$ (variables T_1 and T_2) and design generators $I_1 = V_{u1} \cdot E_{u1} \cdot E_{l1}$ and $I_2 = V_{u1} \cdot E_{u1} \cdot E_{l2}$ for a 1/4 fraction design, the experimental setup is shown in Table 1. Normalized values −1 and +1 have been used for V_{u1}, E_{u1}, E_{l1}, E_{l2} and columns for T_1, T_2 have been obtained using I_1, I_2 generators, respectively. For example, normalized values for T_1 were the product of normalized values of V_{u1}, E_{u1} and E_{l1}; similarly, normalized values for T_2 were the product of normalized values of V_{u1}, E_{u1} and E_{l2}.

Table 1. Normalized values of the operational variables for the proposed 2^{6-2} fractional factorial experimental design.

Experiment No.	V_{u1}	E_{u1}	E_{l1}	E_{l2}	T_1	T_2
1	−1	−1	−1	−1	−1	−1
2	+1	−1	−1	−1	+1	+1
3	−1	+1	−1	−1	+1	+1
4	+1	+1	−1	−1	−1	−1
5	−1	−1	+1	−1	+1	−1
6	+1	−1	+1	−1	−1	+1
7	−1	+1	+1	−1	−1	+1
8	+1	+1	+1	−1	+1	−1
9	−1	−1	−1	+1	−1	+1
10	+1	−1	−1	+1	+1	−1
11	−1	+1	−1	+1	+1	−1
12	+1	+1	−1	+1	−1	+1
13	−1	−1	+1	+1	+1	+1
14	+1	−1	+1	+1	−1	−1
15	−1	+1	+1	+1	−1	−1
16	+1	+1	+1	+1	+1	+1

It was thought essential to use more than one central experiment (level 0 of all operational variables) at different times in order to examine the variance of the response variables. In this work, we used two central experiments (see Table 2). We would also use far-spaced experiments (levels −2 and +2) symmetrically placed at a distance α from center, α being the fourth root of the number of fractional factorial design experiments excluding central points (i.e., $\alpha = \sqrt[4]{16} = 2$). Therefore, the distance of levels (−2) and (+2) from level (0) was twice that from (0) to (−1) and (+1). This required adding two experiments per operational variable (see Table 3). In summary, a total of 30 experiments would be required with the proposed design (i.e., less than one-half those required by a pure factorial design), with the added advantage that the ranges of operating conditions would be extended to far extreme values.

Table 2. Normalized values of the operational variables for the central points of the proposed 2^{6-2} experimental design.

Experiment No.	V_{u1}	E_{u1}	E_{l1}	E_{l2}	T_1	T_2
17	0	0	0	0	0	0
18	0	0	0	0	0	0

Table 3. Normalized values of the operational variables for the extended points of the proposed 2^{6-2} experimental design.

Experiment No.	V_{u1}	E_{u1}	E_{l1}	E_{l2}	T_1	T_2
19	+2	0	0	0	0	0
20	−2	0	0	0	0	0
21	0	+2	0	0	0	0
22	0	−2	0	0	0	0
23	0	0	+2	0	0	0
24	0	0	−2	0	0	0
25	0	0	0	+2	0	0
26	0	0	0	−2	0	0
27	0	0	0	0	+2	0
28	0	0	0	0	−2	0
29	0	0	0	0	0	+2
30	0	0	0	0	0	−2

Because of the serial operation of the two bioreactors, several specific constraints must be considered for previous variables, since they cannot take certain values in practice:

Constraints on E_{l1} :

1. The ethanol concentration of the feed wine would invariably be 12% (*v/v*).
2. $E_{l1} \geq E_{u1}$ as it would make no sense to have an ethanol concentration at the reactor loading stage lower than that at unloading time.
3. E_{l1} must be less than the maximum ethanol concentration in the first bioreactor $(E_{l1})_{max}$ at any time. The latter concentration depends on E_{u1}, V_{u1} and the ethanol concentration of the feed wine, and can be calculated from the following mass balance:

$$(E_{l1})_{max} = \frac{(8 - V_{u1}) \cdot E_{u1} + V_{u1} \cdot 12}{8},$$ (2)

Constraints on E_{u1}:

1. E_{u1} must be greater than 1% (*v/v*) in order to prevent ethanol depletion from happening too early in the first bioreactor—the second bioreactor was used for this purpose. Also, too low a value would slow down acetification and decrease productivity as a result [15].
2. E_{u1} must not be too high. Otherwise, ethanol not consumed in the first bioreactor would have to be used in the second and the probability of complete depletion before unloading would be diminished. Also, ethanol concentrations higher than 5–6% (*v/v*) could detract from microbial activity [15,45]. Therefore, E_{u1} must not exceed 5% (*v/v*).
3. $E_{u1} < (E_{l1})_{max}$ (Equation (2)).

Constraints on V_{u1}:

This variable must range from 1 to 7 L if the self-aspirating turbine with which the bioreactors were equipped is to operate properly. The turbine was used to feed oxygen to the microorganisms and help homogenize the culture medium.

Constraints on T_1 and T_2:

The values of these variables must be compatible with the activity of acetic acid bacteria (AAB), which are the microorganisms affecting the process. The optimum temperature for AAB to oxidize ethanol into acetic acid is 25–35 °C. Because AAB activity drops outside this range [46,47], temperatures from 26 to 34 °C have been used here.

Constraints on E_{l2}:

1. The ethanol concentration in the feed wine is typically 12% (*v/v*).

2. E_{l2} must be less than the maximum ethanol concentration in the medium of the second bioreactor at any time. Such a concentration depends on E_{u1}, V_{u1} and the ethanol concentration of feed wine, and can be easily obtained from a simple mass balance:

$$(E_{l2})_{max} = \frac{V_{u1} \cdot E_{u1} + (8 - V_{u1}) \cdot 12}{8},$$

(3)

Based on the previous constraints, E_{u1}, V_{u1}, E_{l1} and E_{l2} are mutually related; also, feasible E_{u1} and V_{u1} values span the range 1–5% (*v/v*) and 1–7 L, respectively. However, not all potential combinations of the values of the operational variables can be used in practice. Thus, the initial volume of the second bioreactor—unloaded from the first—V_{u1}, together with E_{u1}, imposes some additional constraints on E_{l1} and E_{l2}. As a result, the experimental design is subjected to the following constraints:

- Based on constraint 2 on E_{l1} and the experiments of Table 1, E_{l1} level (−1) must not be less than E_{u1} level (+1). Likewise, E_{l1} level (−2) must be greater than or at least equal to E_{u1} level (0) (see Table 3).
- Based on Tables 2 and 3, E_{l1} level (0) can only be used in conjunction with E_{u1} levels (0), (−2) and (+2). Therefore, E_{l1} level (0) must be greater than or equal to E_{u1} level (+2), unless no experiments under the extended conditions are to be conducted—in which case E_{l1} level (0) must be greater than or equal to E_{u1} level (0).
- Based on Equation (2), E_{l1} level (+1) must be less than the smallest $(E_{l1})_{max}$ value. Also, based on Equation (3), E_{l2} level (+1) must be less than the smallest $(E_{l2})_{max}$ value. Such values are obtained by using all possible combinations of levels (−1) and (+1) of V_{u1} and E_{u1} (see Table 1).
- E_{l1} and E_{l2} level (+2) must be less than the smallest value of $(E_{l1})_{max}$ (Equation (2)) and $(E_{l2})_{max}$ (Equation (3)), respectively, as obtained by using level (0) of V_{u1} and E_{u1} (see Table 3).
- E_{l1} and E_{l2} level (0) must be less than the smallest values of $(E_{l1})_{max}$ (Equation (2)) and $(E_{l2})_{max}$ (Equation (3)), respectively, as obtained by using all possible combinations of levels (0), (−2) and (+2) of V_{u1} and E_{u1} (see Tables 2 and 3).

Based on the foregoing, the constraints on E_{l1} and E_{l2} can be summarized as follows:
Constraints on E_{l1}:

- Level (+1) must be less than the smallest $(E_{l1})_{max}$ value (Equation (2)) when V_{u1} and E_{u1} levels (−1) and (+1) are to be used; also, level (−1) must be greater than or equal to E_{u1} level (+1).
- Level (0) must be less than the smallest $(E_{l1})_{max}$ value (Equation (2)) when V_{u1} and E_{u1} levels (0), (−2) and (+2) are to be used, but greater than or equal to E_{u1} level (+2).
- Level (+2) must be less than $(E_{l1})_{max}$ (Equation (2) when V_{u1} and E_{u1} are used at level (0) and level (−2) must be greater than or equal to E_{u1} level (0).

Constraints on E_{l2}:

- Level (+1) must be less than the smallest $(E_{l2})_{max}$ value (Equation (3)) when V_{u1} and E_{u1} levels (−1) and (+1) are to be used.
- Level (0) must be less than the smallest $(E_{l2})_{max}$ value (Equation (3)) when V_{u1} and E_{u1} are to be used at levels (0), (−2) and (+2).
- Level (+2) must be less than $(E_{l2})_{max}$ (Equation (3)) with level (0) of V_{u1} and E_{u1}.

A systematic analysis of all possible combinations of the values of the operational variables with provision for the above-described constraints was done by using a self-developed script in MATLAB [48] (see MATLAB file "Get_feasible_combinations.m" in Supplementary Materials). Each combination corresponded to a complete experimental design, as depicted in Tables 1–3. The software used different values for level (0) of each variable within a preset range in combination with different values for levels (−2) and (+2)—levels (−1) and (+1) were calculated automatically in each case. The specific ranges considered for level (0) were 1–5% (v/v) for E_{u1}, E_{l1} and E_{l2}; and 1–7 L for V_{u1}, all in 0.5 steps. A 0.5 step over the range 0.5–5 was used for iteration of levels (−2) and (+2) in each case.

The script initially identified 148611 feasible combinations. Such a large number required additional constraints to be imposed based on the following practical arguments:

A1. There must be a simultaneous maximum difference between levels (−1) and (+1) in E_{u1} and V_{u1} (see Table 1). This constraint expanded the feasible range of operational conditions and allowed the influence of each variable and mutual relations to be more accurately detected and examined as a result.

A2. Because in the semi-continuous operation mode it is a common practice to unload at least half of the working volume from the first bioreactor, V_{u1} level (−1) must be greater than or equal to 4 L.

A3. Because the initial ethanol concentration in the second bioreactor must not be too high in order to facilitate appropriate performance (see constraint 2 on E_{u1}), E_{u1} level (+2) must be less than 5% (v/v).

Introducing these additional constraints in the MATLAB script reduced the number of feasible combinations to 310 (see tab "Pass 1" in Excel file "Results.xlsx" in Supplementary Materials). Subsequently, in line with the previous argument A1, the MATLAB script was used with those combinations maximizing the difference between E_{l1} levels (−2) and (+2), which further reduced the number of combinations to 31 (see tab "Pass 2" in Excel file "Results.xlsx" in Supplementary Materials). Finally, selecting the combinations resulting in the greatest differences between E_{l2} levels (−2) and (+2) reduced the number of choices to 4 (see Table 4).

Table 4. The last four feasible combinations of the operational variables values and the selected combination (in grey).

Variable	Level	Combination 1	Combination 2	Combination 3	Combination 4
	−2	3.50	3.50	3.50	4.00
	+2	6.50	6.50	6.50	7.00
V_{u1} (L)	−1	4.25	4.25	4.25	4.75
	+1	5.75	5.75	5.75	6.25
	0	5.00	5.00	5.00	5.50
	−2	1.00	1.00	1.00	1.00
	+2	5.00	5.00	5.00	5.00
E_{u1} [% (v/v)]	−1	2.00	2.00	2.00	2.00
	+1	4.00	4.00	4.00	4.00
	0	3.00	3.00	3.00	3.00
	−2	3.00	3.00	3.00	3.00
	+2	7.00	7.00	7.00	7.00
E_{l1} [% (v/v)]	−1	4.00	4.00	4.00	4.00
	+1	6.00	6.00	6.00	6.00
	0	5.00	5.00	5.00	5.00
	−2	1.00	1.50	1.00	1.00
	+2	5.00	5.50	6.00	5.00
E_{l2} [% (v/v)]	−1	2.00	2.50	2.25	2.00
	+1	4.00	4.50	4.75	4.00
	0	3.00	3.50	3.50	3.00

On the other hand, too low an E_{l2} value would slow down the process exceedingly at the loading stage. This fact allows the combinations involving the lowest E_{l2} value at level (−2) (viz., 1, 3 and 4 in Table 4) to be discarded, thus leaving a single one which fulfils all constraints imposed (viz., combination 2 in Table 4). Tables 5–7 show the set of experiments corresponding to such a combination as established according to the 2^{6-2} experimental design previously described by using an identical range for both working temperatures (T_1 and T_2): 26–34 °C.

As recommended when using experimental design methodology, only the experiments in Tables 5 and 6 would initially be performed (in random order). If the results were not conclusive enough, then the experiments in Table 7 would also be needed.

Table 5. Experiments performed according to the proposed 2^{6-2} fractional factorial design.

Experiment No.	V_{u1} (L)	E_{u1} [% (v/v)]	E_{l1} [% (v/v)]	E_{l2} [% (v/v)]	T_1 (°C)	T_2 (°C)
1	4.25	2.00	4.00	2.50	28	28
2	5.75	2.00	4.00	2.50	32	32
3	4.25	4.00	4.00	2.50	32	32
4	5.75	4.00	4.00	2.50	28	28
5	4.25	2.00	6.00	2.50	32	28
6	5.75	2.00	6.00	2.50	28	32
7	4.25	4.00	6.00	2.50	28	32
8	5.75	4.00	6.00	2.50	32	28
9	4.25	2.00	4.00	4.50	28	32
10	5.75	2.00	4.00	4.50	32	28
11	4.25	4.00	4.00	4.50	32	28
12	5.75	4.00	4.00	4.50	28	32
13	4.25	2.00	6.00	4.50	32	32
14	5.75	2.00	6.00	4.50	28	28
15	4.25	4.00	6.00	4.50	28	28
16	5.75	4.00	6.00	4.50	32	32

Table 6. Experiments for the central points of the proposed 2^{6-2} fractional factorial design.

Experiment No.	V_{u1} (L)	E_{u1} [% (v/v)]	E_{l1} [% (v/v)]	E_{l2} [% (v/v)]	T_1 (°C)	T_2 (°C)
17	5.00	3.00	5.00	3.50	30	30
18	5.00	3.00	5.00	3.50	30	30

Table 7. Extended experiments for the proposed 2^{6-2} fractional factorial design.

Experiment No.	V_{u1} (L)	E_{u1} [% (v/v)]	E_{l1} [% (v/v)]	E_{l2} [% (v/v)]	T_1 (°C)	T_2 (°C)
19	6.50	3.00	5.00	3.50	30	30
20	3.50	3.00	5.00	3.50	30	30
21	5.00	5.00	5.00	3.50	30	30
22	5.00	1.00	5.00	3.50	30	30
23	5.00	3.00	7.00	3.50	30	30
24	5.00	3.00	3.00	3.50	30	30
25	5.00	3.00	5.00	5.50	30	30
26	5.00	3.00	5.00	1.50	30	30
27	5.00	3.00	5.00	3.50	34	30
28	5.00	3.00	5.00	3.50	26	30
29	5.00	3.00	5.00	3.50	30	34
30	5.00	3.00	5.00	3.50	30	26

3.2. Experimental Results

Conducting the 18 experiments of Tables 5 and 6 (compiled in Table S2.1 in file "S2.docx" in Supplementary Materials) required changing the operating conditions between experiments, so cycles had to be repeated until steady conditions were reached with each new set of operation variables; after that, the experiments were replicated to obtain statistically valid results. Table 8 shows the total number of cycles and that of useful cycles for each experiment. As can be seen, the minimum number of cycles used to calculate the target variables was 5—some experiments required nearly 20, however. The fact that the total number of cycles differed markedly among experiments was a result of being performed in random order. Therefore, depending on the differences in the operating conditions between consecutive experiments, either a lower or a greater number of cycles may be required for the system to become stable again as a result.

Table 8. Experimental cycles.

Experiment No.	Number of Cycles	
	Total	Useful
1	25	5
2	19	8
3	15	10
4	7	6
5	19	7
6	16	7
7	14	9
8	9	5
9	17	13
10	13	5
11	15	13
12	15	10
13	40	12
14	15	8
15	25	8
16	20	8
17	22	9
18	26	19

By way of example, Figure 3 shows the results in terms of ethanol, acidity, volume, and total and viable cells in the first reactor in Experiment 1. Each result was the mean of 5 replicate cycles with its standard deviation (see Table 8). Similarly, Figure 4 shows the ethanol, acidity, and volume results for the second reactor as the means for 5 replicated runs and their standard deviations. The previous results were used to obtain the values of Table 9. The procedure used to calculate the values of non-measurable variables (e.g., mean rate of acetic acid formation, total acetic acid production and mean volume) is described in file "S1.docx" in Supplementary Materials. Figure S1.1 in such file shows the typical time course of ethanol concentration, acidity and volume in the first bioreactor and Figures S1.2–S1.3 show the time course of such variables in the second bioreactor when a fast loading or pre-depletion step exists, respectively. Also, the results obtained in the 18 experiments of Tables 5 and 6 are compiled in file "S2.docx" in Supplementary Materials (Figures S2.1 to S2.36 and Tables S2.2 to S2.19). On the other hand, Tables 10 and 11 show the experimental values of the key variables intended to be modelled using polynomial models relating them to the operational variables.

3.3. Obtained Polynomial Models

Below are stated the polynomial models (or response surfaces) corresponding to the target variables. By way of example, the specific procedure used to estimate the mean overall rate of acetic acid formation in the proposed two-bioreactor system, $(r_A)_{global\ est}$, is detailed below, and that for each of the other variables in file "S3.docx" in the Supplementary Materials.

3.3.1. Mean Overall Rate of Acetic Acid Formation

ANOVA exposed the presence of significant differences at a 99.9% confidence level in the mean overall rate of acetic acid formation, $(r_A)_{global}$, among experiments. In the absence of such differences, $(r_A)_{global}$ would be independent of the operational variables due to experimental error. Therefore, the 18 experiments of Tables 5 and 6 were representative enough of the operating conditions, so the other 12 potentially required (experiments in Table 7) were unnecessary. As shown below, the ANOVA led to identical conclusions for all target variables. Accordingly, the experimental $(r_A)_{global}$ data in Table 10 were fitted to a second-order polynomial by Forward Stepwise Regression, as described in Section 2.6.1, using the software SigmaStat 2.0 [49]. Although 27 terms (viz., the 6 individual operational variables and their 21 possible interactions) were initially considered, the maximum possible number was that of the experiments performed: 18.

Each fitted polynomial was checked for significant differences in the regressions between each step and the previous one by the effect of the addition or removal of terms. ANOVA revealed the presence of significant differences at a 95% confidence level in all cases and, therefore, the results are worth no additional detailed description here.

The fitting steps using the Forward Stepwise Regression method (Section 2.6.1) are detailed in Section S3.1 in file "S3.docx" in Supplementary Materials (Tables S3.1 to S3.14 show intermediate results). The final model was that of Equation (4) and allowed the mean overall rate of acetic acid formation to be estimated with an error of 0.01 g acetic acid·$(100 \text{ mL·h})^{-1}$. Such a model, however, only held over the operating ranges shown in Tables 5 and 6 for each variable.

$$(r_A)_{global\ est} = \begin{aligned} &-0.51 + 0.43 \cdot E_{u1} - 0.0654 \cdot E_{u1}^2 \\ &-0.00456 \cdot E_{l2} \cdot V_{u1} - 0.00468 \cdot E_{u1} \cdot E_{l1} \quad, \\ &+0.000672 \cdot T_1 \cdot E_{l1} + 0.000839 \cdot E_{l2} \cdot T_1 \end{aligned} \tag{4}$$

Based on statistical significance, not all operational variables in Equation (4) had a direct influence on $(r_A)_{global}$ (in fact, only E_{u1} had a direct impact) and this variable was independent of T_2. Also, only some of the interaction terms influenced $(r_A)_{global}$.

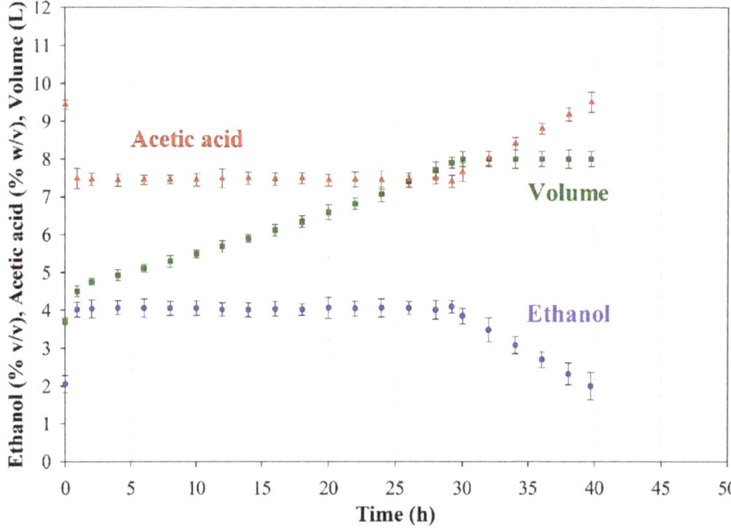

Figure 3. Time course of the mean ethanol concentration, acidity and volume in the first reactor in Experiment 1, and corresponding standard deviations.

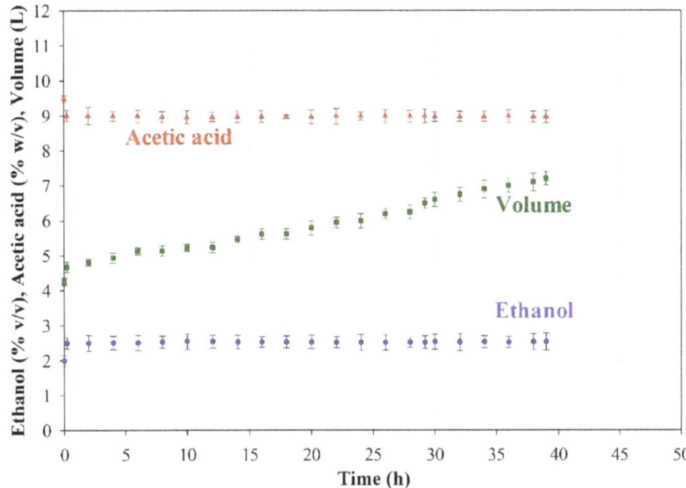

Figure 4. The time course of the mean ethanol concentration, acidity, and volume in the second reactor in Experiment 1, and corresponding standard deviations.

Table 9. The results obtained in Experiment 1.

Variable	Value
Total cycle duration (h)	39.7 ± 0.8
Duration of the fast loading stage in the first bioreactor (h)	0.9 ± 0.1
Duration of the slow loading stage in the first bioreactor (h)	28.3 ± 0.6
Duration of the depletion stage in the first bioreactor (h)	10.5 ± 1.0
Duration of the fast loading stage in the second bioreactor (h)	0.2 ± 0.1
Duration of the slow loading stage in the second bioreactor (h)	39.4 ± 1.2
Time fraction of the fast loading stage in the first bioreactor over the cycle duration	0.023 ± 0.003
Time fraction of the slow loading stage in the first bioreactor over the cycle duration	0.713 ± 0.021
Time fraction of the depletion stage in the first bioreactor over the cycle duration	0.264 ± 0.026
Time fraction of the fast loading stage in the second bioreactor over the cycle duration	0.006 ± 0.001
Time fraction of the slow loading stage in the second bioreactor over the cycle duration	0.994 ± 0.043
Mean volume in the fast loading stage in the first bioreactor (L)	4.21 ± 0.05
Mean volume in the slow loading stage in the first bioreactor (L)	6.35 ± 0.05
Mean volume in the depletion stage in the first bioreactor (L)	8.02 ± 0.05
Mean volume in the first bioreactor (L)	6.74 ± 0.25
Mean volume in the fast loading stage in the second bioreactor (L)	4.38 ± 0.05
Mean volume in the slow loading stage in the second bioreactor (L)	5.86 ± 0.05
Mean volume in the second bioreactor (L)	5.85 ± 0.25
Mean overall volume during a cycle in the two bioreactors as a whole (L)	12.59 ± 0.35
Mean volume of fermentation medium unloaded from the second reactor (L)	7.2 ± 0.05
Final ethanol concentration at the time the second bioreactor was unloaded (% *v/v*)	2.3 ± 0.2
Mean ethanol concentration in the first bioreactor (% *v/v*)	3.7 ± 0.1
Mean ethanol concentration in the second bioreactor (% *v/v*)	3.9 ± 0.1
Final acetic acid concentration at the time the first bioreactor was unloaded (% *w/v*)	9.4 ± 0.2
Final acetic acid concentration at the time the second bioreactor was unloaded (% *w/v*)	9.2 ± 0.2
Mean acetic acid concentration in the first bioreactor (% *w/v*)	7.8 ± 0.1
Mean acetic acid concentration in the second bioreactor (% *w/v*)	7.6 ± 0.1
Mean rate of acetic acid formation in the first bioreactor (% $w/v{\cdot}h^{-1}$)	0.15 ± 0.01
Mean rate of acetic acid formation in second bioreactor (% $w/v{\cdot}h^{-1}$)	0.11 ± 0.01
Mean overall rate of acetic acid formation in the two bioreactors as a whole (% $w/v{\cdot}h^{-1}$)	0.13 ± 0.01
Total acetic acid production in the first bioreactor (g acetic acid$\cdot h^{-1}$)	10.1 ± 0.3
Total acetic acid production in the second bioreactor (g acetic acid$\cdot h^{-1}$)	6.6 ± 0.5
Total acetic acid production in the two bioreactors as a whole (g acetic acid$\cdot h^{-1}$)	16.7 ± 0.5

Table 10. Experimental values of selected variables: $(r_A)_{global}$ is the mean overall rate of acetic acid formation in the two-bioreactor system, P_m the total acetic acid production in the two-bioreactor system, E_{u2} the ethanol concentration at the time the second bioreactor was unloaded, V_{u2} the volume unloaded from the second reactor, t_{cycle} the duration of a cycle, and V_m the mean overall volume in the two-bioreactor system during a cycle.

Exp. No.	$(r_A)_{global}$, g Acetic Acid $(100\ mL \cdot h)^{-1}$	P_m, g Acetic Acid$\cdot h^{-1}$	E_{u2}, %(v/v)	V_{u2}, L	t_{cycle}, h	V_m, L
1	0.13 ± 0.01	16.7 ± 0.5	2.3 ± 0.2	7.20 ± 0.05	39.7 ± 0.8	12.59 ± 0.35
2	0.13 ± 0.01	17.2 ± 0.5	1.2 ± 0.2	8.00 ± 0.05	48 ± 0.9	12.94 ± 0.29
3	0.19 ± 0.01	19.5 ± 0.6	2.3 ± 0.2	4.40 ± 0.05	20.8 ± 0.5	10.19 ± 0.36
4	0.16 ± 0.01	19.7 ± 0.6	0.0 ± 0.2	8.00 ± 0.05	46.8 ± 1.2	12.15 ± 0.31
5	0.18 ± 0.01	24.6 ± 0.8	2.5 ± 0.2	7.80 ± 0.05	28.5 ± 0.7	13.74 ± 0.36
6	0.14 ± 0.01	20.8 ± 0.5	0.0 ± 0.2	8.00 ± 0.05	44.3 ± 0.6	14.58 ± 0.27
7	0.18 ± 0.01	20.9 ± 0.8	2.5 ± 0.2	5.30 ± 0.05	22.8 ± 0.7	11.65 ± 0.37
8	0.18 ± 0.01	23.9 ± 0.5	0.0 ± 0.2	8.00 ± 0.05	38.5 ± 0.5	13.38 ± 0.32
9	0.14 ± 0.01	20.1 ± 0.6	1.5 ± 0.2	7.98 ± 0.05	39.7 ± 0.9	14.02 ± 0.33
10	0.14 ± 0.01	19.6 ± 0.4	0.0 ± 0.2	8.00 ± 0.05	47.3 ± 0.6	13.96 ± 0.23
11	0.21 ± 0.01	24.4 ± 1.0	4.5 ± 0.2	7.00 ± 0.05	20.1 ± 0.6	11.65 ± 0.49
12	0.16 ± 0.01	20.1 ± 0.6	0.0 ± 0.2	8.00 ± 0.05	45.7 ± 1.2	12.98 ± 0.37
13	0.19 ± 0.01	28.6 ± 0.9	0.9 ± 0.2	8.00 ± 0.05	29.6 ± 0.8	14.78 ± 0.46
14	0.13 ± 0.01	19.7 ± 0.6	0.0 ± 0.2	8.00 ± 0.05	46.8 ± 1.1	14.88 ± 0.39
15	0.18 ± 0.01	22.1 ± 0.9	4.5 ± 0.2	6.40 ± 0.05	20.3 ± 0.6	12.43 ± 0.41
16	0.17 ± 0.01	24.0 ± 0.6	0.0 ± 0.2	8.00 ± 0.05	38.3 ± 0.6	13.74 ± 0.30
17	0.23 ± 0.01	31.4 ± 1.0	0.4 ± 0.2	8.00 ± 0.05	28.3 ± 0.7	13.93 ± 0.35
18	0.22 ± 0.01	30.8 ± 1.0	0.5 ± 0.2	8.00 ± 0.05	28.6 ± 0.8	13.96 ± 0.42

Table 11. Experimental values of selected variables. $EtOH_{m1}$ is the mean ethanol concentration in the first bioreactor during a cycle, $EtOH_{m2}$ the mean ethanol concentration in the second bioreactor during a cycle, HAc_{m1} the mean acetic acid concentration in the first bioreactor during a cycle, and HAc_{m2} the mean acetic acid concentration in the second bioreactor during a cycle.

Exp. No.	$EtOH_{m1}$, % (v/v)	$EtOH_{m2}$, % (v/v)	HAc_{m1}, % (w/v)	HAc_{m2}, % (w/v)
1	3.7 ± 0.1	3.9 ± 0.1	7.8 ± 0.1	7.6 ± 0.1
2	3.8 ± 0.1	2.3 ± 0.1	7.7 ± 0.1	9.2 ± 0.1
3	4.0 ± 0.1	2.9 ± 0.2	7.5 ± 0.1	8.6 ± 0.2
4	4.0 ± 0.1	2.2 ± 0.2	7.5 ± 0.1	9.3 ± 0.2
5	4.4 ± 0.4	2.5 ± 0.1	7.1 ± 0.4	9.0 ± 0.1
6	4.7 ± 0.3	1.9 ± 0.2	6.8 ± 0.3	9.6 ± 0.2
7	5.5 ± 0.2	2.7 ± 0.1	6.0 ± 0.2	8.8 ± 0.1
8	5.6 ± 0.2	2.1 ± 0.2	5.9 ± 0.2	9.4 ± 0.2
9	3.6 ± 0.1	2.3 ± 0.1	7.9 ± 0.1	9.2 ± 0.1
10	3.7 ± 0.1	2.6 ± 0.3	7.8 ± 0.1	8.9 ± 0.3
11	4.0 ± 0.1	4.5 ± 0.1	7.5 ± 0.1	7.0 ± 0.1
12	4.0 ± 0.1	2.9 ± 0.4	7.5 ± 0.1	8.6 ± 0.4
13	4.5 ± 0.4	3.7 ± 0.3	7.0 ± 0.4	7.8 ± 0.3
14	4.8 ± 0.3	3.0 ± 0.4	6.7 ± 0.3	8.5 ± 0.4
15	5.5 ± 0.2	4.5 ± 0.2	6.0 ± 0.2	7.0 ± 0.2
16	5.7 ± 0.1	3.2 ± 0.4	5.8 ± 0.1	8.2 ± 0.4
17	4.7 ± 0.2	2.9 ± 0.3	6.8 ± 0.2	8.6 ± 0.3
18	4.7 ± 0.2	2.9 ± 0.2	6.8 ± 0.2	8.6 ± 0.2

A comparison of the $(r_A)_{global}$ predictions with the experimental values (Figure 5, which is the same as Figure S3.1 in in file "S3.docx" in Supplementary Materials) reveals that the two coincided exactly in 10 of the 18 experiments and differed only very slightly in the other 8. Also, all predictions fell within the 95% prediction interval.

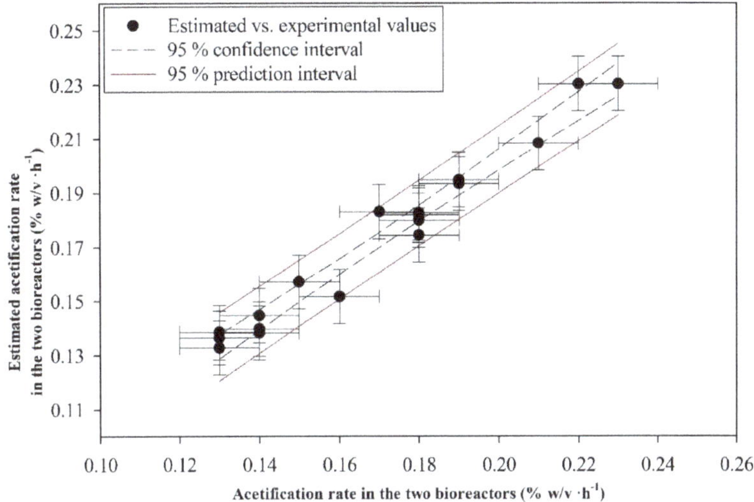

Figure 5. Plot of $(r_A)_{global\ est}$ against $(r_A)_{global}$, and curves for the 95% confidence and prediction intervals.

As can be seen from Figure 6 (same as Figure S3.2 in file "S3.docx" in Supplementary Materials), the residuals (viz., the differences between experimental and predicted values) had a zero or near-zero mean in most experiments and were normally distributed in all.

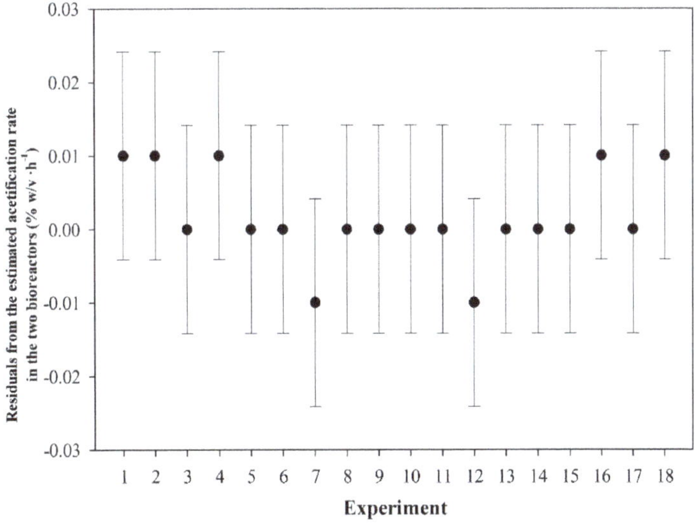

Figure 6. Residuals of the fitting of $(r_A)_{global\ est}$ for each experiment.

The process used with the other estimated variables is described in file "S3.docx" in Supplementary Materials. Their polynomial models are discussed below.

3.3.2. Total Acetic Acid Production

This variable, P_m, was also estimated from the experimental data of Table 10. ANOVA revealed statistically significant differences at a 99.9% confidence level between experiments and, hence,

P_m was dependent on the operational variables. Therefore, the experimental data of P_m were fitted to a second-order polynomial by using Forward Stepwise Regression to construct the model of Equation (5), which estimated this variable with an error of 0.7 g acetic acid·h^{-1}. The fitting steps are detailed in Section S3.2 in file "S3.docx" in Supplementary Materials (Tables S3.15 to S3.28 show intermediate results)

$$
\begin{aligned}
P_{m\,est} = \quad & -243.705 + 18.324 \cdot V_{u1} + 72.736 \cdot E_{l1} + 21.525 \cdot E_{u1} \\
& -9.708 \cdot E_{l1}^2 - 1.102 \cdot E_{u1} \cdot V_{u1} - 0.534 \cdot T_1 \cdot V_{u1} \\
& +0.742 \cdot T_1 \cdot E_{l1} - 0.416 \cdot E_{l2} \cdot E_{l1} + 0.175 \cdot T_2 \cdot E_{l1} \\
& -0.12 \cdot T_1 \cdot E_{u1} - 0.399 \cdot T_2 \cdot E_{u1} + 0.101 \cdot T_2 \cdot E_{l2}
\end{aligned}
\tag{5}
$$

As can be seen, $P_{m\,est}$ depends directly on V_{u1}, E_{u1} and E_{l1}, as well as on various interaction terms. A comparison of $P_{m\,est}$ and its experimental counterpart (Figure S3.3 in file "S3.docx" in Supplementary Materials) revealed the presence of scarcely significant differences. Also, the residuals of the fitting had a near-zero mean in most experiments and were normally distributed in all (see Figure S3.4 in file "S3.docx" in Supplementary Materials).

3.3.3. Final Ethanol Concentration at the Time the Second Reactor Was Unloaded

As with the previous two variables, the ANOVA on E_{u2} (Table 10) revealed the presence of statistically significant differences at a 99.9% confidence level between the predicted and experimental results. As before, the experimental results were fitted to a second-order polynomial by using Forward Stepwise Regression. The fitting steps are detailed in Section S3.3 in file "S3.docx" in Supplementary Materials (Tables S3.29 to S3.41 show intermediate results). The resulting model (Equation (6)) predicted the ethanol concentration with an error of 0.3% (v/v).

$$
\begin{aligned}
E_{u2\,est} = \quad & 14.935 - 5.371 \cdot E_{u1} + 0.988 \cdot E_{u1}^2 - 0.0592 \cdot E_{l1} \cdot V_{u1} \\
& -0.456 \cdot E_{u1} \cdot V_{u1} + 0.0678 \cdot E_{u1} \cdot E_{l1} \\
& +0.494 \cdot E_{l2} \cdot E_{u1} - 0.049 \cdot T_2 \cdot E_{l2}
\end{aligned}
\tag{6}
$$

As can be seen, the variable $E_{u2\,est}$ is directly dependent on E_{u1} and on some interaction terms; however, it is independent of T_1.

As with the previous variables, a comparison of $E_{u2\,est}$ and its experimental counterpart (Figure S3.5 in file "S3.docx" in Supplementary Materials) confirmed that the predictions of the model were quite accurate; also, they fell within the 95% prediction interval except in one case.

Finally, as can be seen from Figure S3.6 in file "S3.docx" in Supplementary Materials, the residuals had a near-zero mean in most experiments and were normally distributed in all.

3.3.4. Volume of Fermentation Medium Unloaded from the Second Reactor

The ANOVA of the volume unloaded from the second reactor, V_{u2}, revealed the presence of statistically significant differences between experiment means at a 99.9% confidence level (see Table 10).

Like the previous variables, V_{u2} was fitted by Forward Stepwise Regression. The fitting steps are detailed in Section S3.4 in file "S3.docx" in Supplementary Materials (Tables S3.42 to S3.53 show intermediate results). The resulting model (Equation (7)) predicted it with an error of 0.22 L.

$$
\begin{aligned}
V_{u2\,est} = \quad & 4.921 + 1.399 \cdot E_{l2} - 0.59 \cdot E_{u1}^2 \\
& +0.665 \cdot E_{u1} \cdot V_{u1} - 0.324 \cdot E_{l2} \cdot V_{u1} \\
& +0.172 \cdot E_{l2} \cdot E_{u1} - 0.0275 \cdot T_2 \cdot E_{u1}
\end{aligned}
\tag{7}
$$

As can be seen, $V_{u2\,est}$ is independent of E_{l1} and T_1. A comparison of $V_{u2\,est}$ and its experimental counterpart, V_{u2}, confirmed the goodness of the predictions (see Figure S3.7 in file "S3.docx" in Supplementary Materials). In fact, all fell within the 95% prediction interval. Also, the residuals had a zero or near-zero mean in most experiments and were normally distributed in all (see Figure S3.8 in file "S3.docx" in Supplementary Materials).

3.3.5. Total Cycle Duration

As in the previous cases, the ANOVA on t_{cycle} revealed the presence of statistically significant differences at a 99.9% confidence level (see Table 10). Applying Forward Stepwise Regression to the experimental data provided the model of Equation (8), which predicted t_{cycle} with an error of 2.5 h. The fitting steps are detailed in Section S3.5 in file "S3.docx" in Supplementary Materials (Tables S3.54 to S3.60 show intermediate results)

$$t_{cycle\ est} = \begin{array}{l} 518.591 - 156.652 \cdot V_{u1} \\ -6.474 \cdot T_1 + 13.556 \cdot V_{u1}^2 \\ +1.076 \cdot T_1 \cdot V_{u1} - 0.864 \cdot E_{u1} \cdot E_{l1} \end{array} , \tag{8}$$

As can be seen, $t_{cycle\ est}$ is influenced by the variables V_{u1}, E_{l1}, E_{u1} and T_1, which is logical since the overall behaviour of the first bioreactor dictates when the second is to be unloaded. A comparison of $t_{cycle\ est}$ and its experimental counterpart, t_{cycle}, revealed that all predictions fell within the 95% interval (see Figure S3.9 in file "S3.docx" in Supplementary Materials). Finally, as with the previous variables, the residuals had a near-zero mean in most cases and were normally distributed in all (see Figure S3.10 in file "S3.docx" in Supplementary Materials).

3.3.6. Mean Overall Volume in the Two-Bioreactor System

The ANOVA on the experimental data of V_m (Table 10) revealed statistically significant differences between experiment means at a 99.9% confidence level. The multiple regression model obtained (Equation (9)) estimated the mean overall volume in each cycle with an error of 0.33 L. The fitting steps are detailed in Section S3.6 in file "S3.docx" in Supplementary Materials (Tables S3.61 to S3.74 show intermediate results). Also, as can be seen from the equation, $V_{m\ est}$ is independent of T_1.

$$V_{m\ est} = \begin{array}{l} 4.306 + 4.739 \cdot E_{u1} - 0.841 \cdot E_{u1}^2 \\ +0.336 \cdot E_{u1} \cdot V_{u1} - 0.0127 \cdot T_2 \cdot V_{u1} \\ -0.125 \cdot E_{l2} \cdot E_{l1} + 0.0326 \cdot T_2 \cdot E_{l1} \\ -0.0735 \cdot T_2 \cdot E_{u1} + 0.0358 \cdot T_2 \cdot E_{l2} \end{array} , \tag{9}$$

As shown by a plot of $V_{m\ est}$ against the experimental data V_m (Figure S3.11 in file "S3.docx" in Supplementary Materials), all predictions fell within the 95% interval, so the degree of fitting was acceptable. This was further confirmed by the residuals for each experiment (Figure S3.12 in file "S3.docx" in Supplementary Materials), which had a near-zero mean in virtually all experiments and were normally distributed.

3.3.7. Mean Ethanol Concentration in the First Bioreactor

The ANOVA of $EtOH_{m1}$ (Table 11) also exposed statistically significant differences between experiment means at a 99.9% confidence level. The experimental data were fitted by Forward Stepwise Regression to construct Equation (10), which reproduced them with an estimation error of 0.2% (v/v). The fitting steps are detailed in Section S3.7 in file "S3.docx" in Supplementary Materials (Tables S3.75 to S3.81 show intermediate results)

$$EtOH_{m1\ est} = \begin{array}{l} 2.312 - 0.0932 \cdot E_{u1}^2 \\ +0.0191 \cdot E_{l1} \cdot V_{u1} + 0.175 \cdot E_{u1} \cdot E_{l1} \end{array} , \tag{10}$$

As expected, $EtOH_{m1\ est}$ is independent of T_1, E_{l2} and T_2. Comparing $EtOH_{m1\ est}$ and its experimental counterpart $EtOH_{m1}$ (Figure S3.13 in in file "S3.docx" Supplementary Materials) revealed that all predictions of the model fell within the 95% interval. Therefore, the fitting was good, as further confirmed by the residuals (Figure S3.14 in file "S3.docx" in Supplementary Materials), which had a near-zero mean in most experiments and were normally distributed.

3.3.8. Mean Ethanol Concentration in the Second Bioreactor

Based on the results of the ANOVA on $EtOH_{m2}$ (Table 11), there were statistically significant differences between experiment means at a 99.9% confidence level. Fitting the experimental results by Forward Stepwise Regression provided the model of Equation (11), which estimated $EtOH_{m2}$ with an error of 0.4% (*v/v*). The fitting steps are detailed in Section S3.8 in file "S3.docx" in Supplementary Materials (Tables S3.82 to S3.88 show intermediate results)

$$EtOH_{m2\ est} = \begin{array}{l} 4.327 - 0.179 \cdot E_{u1} \cdot V_{u1} \\ +0.306 \cdot E_{l2} \cdot E_{u1} - 0.0182 \cdot T_2 \cdot E_{l2} \end{array} \text{,} \tag{11}$$

As can be seen, $EtOH_{m2\ est}$ is independent of E_{l1} and T_1. A plot of $EtOH_{m2\ est}$ against experimental data $EtOH_{m2}$ (Figure S3.15 in file "S3.docx" in Supplementary Materials) revealed that the former all fell within the 95% prediction interval. Also, the residuals (Figure S3.16 in file "S3.docx" in Supplementary Materials) were all normally distributed and had a near-zero mean in most cases.

3.3.9. Mean Acetic Acid Concentration in the First Bioreactor

Based on the results of the ANOVA on HAc_{m1} (Table 11), there were statistically significant differences between experiments at a 99.9% confidence level. Fitting the data provided the model represented by Equation (12), which predicted the mean acetic acid concentration in the first bioreactor with an error of 0.2% (*w/v*). The fitting steps are detailed in Section S3.9 in file "S3.docx" in Supplementary Materials (Tables S3.89 to S3.95 show intermediate results)

$$HAc_{m1\ est} = \begin{array}{l} 9.188 + 0.0932 \cdot E_{u1}^2 \\ -0.0191 \cdot E_{l1} \cdot V_{u1} - 0.175 \cdot E_{u1} \cdot E_{l1} \end{array} \text{,} \tag{12}$$

As can be seen, $HAc_{m1\ est}$ depends on the same variables as $EtOH_{m1\ est}$ and is also independent of T_1, E_{l2} and T_2—a logical result, since the two variables are mutually related.

A comparison of $HAc_{m1\ est}$ and its experimental counterpart HAc_{m1} (Figure S3.17 in file "S3.docx" in Supplementary Materials) revealed that the predictions of the model all fell within the 95% interval. Also, as can be seen from Figure S3.18 in file "S3.docx" in the Supplementary Materials, the residuals had a near-zero mean in virtually all cases and were normally distributed.

3.3.10. Mean Acetic Acid Concentration in the Second Bioreactor

As with the previous variables, the ANOVA on HAc_{m2} (Table 11) exposed statistically significant differences at a 99.9% confidence level in mean acetic acid concentration between experiments. Equation (13) represents the linear model obtained by fitting the experimental data with the Forward Stepwise Regression method. The fitting steps are detailed in Section S3.10 in file "S3.docx" in Supplementary Materials (Tables S3.96 to S3.102 show intermediate results). The model predicted HAc_{m2} with an error of 0.4% (*w/v*).

$$HAc_{m2\ est} = \begin{array}{l} 7.227 + 0.177 \cdot E_{u1} \cdot V_{u1} \\ -0.305 \cdot E_{l2} \cdot E_{u1} + 0.0178 \cdot T_2 \cdot E_{l2} \end{array} \text{,} \tag{13}$$

Similarly to $HAc_{m1\ est}$ and $EtOH_{m1\ est}$, $HAc_{m2\ est}$ is dependent on the same operational variables as HAc_{m2}. A comparison of $HAc_{m2\ est}$ predictions and experimental data (HAc_{m2}, Figure S3.19 in file "S3.docx" in Supplementary Materials) revealed that the former invariably fell within the 95% interval. Also, the residuals (Figure S3.20 in file "S3.docx" in Supplementary Materials) had a near-zero mean in most cases and were normally distributed in all. Therefore, the model can be deemed acceptably accurate.

3.4. Discussion about the Obtained Polynomial Models

Although a black-box model does not allow one to ascertain why some operational variables are influential whereas others are not, it could be interesting to identify the most influential variables and their interaction terms. The influence (statistical significance) of each term in a polynomial equation can be assessed through statistic F, which was used here to decide whether a term was to be included or excluded. By way of example, Table S3.14 in file "S3.docx" in Supplementary Materials reveals that the highest F values were those for E_{u1} and E_{u1}^2. Therefore, the variable $(r_A)_{global\ est}$ was especially sensitive to the ethanol concentration at the time the first bioreactor was unloaded—it was directly influenced by E_{u1} and by its quadratic term.

As a rule, the operational variables associated to the first bioreactor were more markedly influential on most of the dependent variables than were those pertaining to the second. This is unsurprising if one considers that the first reactor not only contributed to the total acetic acid production but also supplied the second with the microorganisms which must work under more extreme conditions in the second bioreactor, since one of the main goals was to deplete ethanol in the medium. Therefore, the conditions prevailing in the first reactor should allow a high concentration of very active acetic acid bacteria to be maintained. As stated in the introduction, such conditions are obtained by keeping the ethanol and acetic acid concentration at not too high levels. It is thus unsurprising that the polynomials used to estimate $(r_A)_{global\ est}$ and $P_{m\ est}$ were so strongly dependent on E_{u1} and E_{l1} as the latter two variables are directly related to acidity in the reaction medium. V_{u1} is also highly influential; in fact, the greater the volume unloaded into the second bioreactor is, the more marked will be the potential changes in ethanol and acidity levels in the first as a result of the need for a greater volume of fresh medium for replenishment. As expected, the interaction term $E_{u1} \cdot E_{l1}$ in the polynomial for $EtOH_{m1\ est}$ is especially important; in fact, changes in E_{u1} and E_{l1} must have a strong impact on the mean ethanol concentration in each transformation cycle in the first bioreactor.

One other interesting inference is that the polynomials for $EtOH_{m1\ est}$ and $HAc_{m1\ est}$ are complementary; in fact, they only differ in the independent term and in the signs of the others. The sum of the independent term coincides with the overall content of the medium [% (v/v) ethanol + % (w/v) acetic acid]. Since the total concentration remains constant, in the absence of volatile losses by sweeping—which was the case with our experiments—this result is unsurprising and provides support for the correlation procedure used to develop the equations. Similar reasoning can be applied to $EtOH_{m2\ est}$ and $HAc_{m2\ est}$.

The variable $E_{d2\ est}$ is of special interest as it is a measure of ethanol depletion in each biotransformation cycle. One aim of the acetification process may be not to operate at the highest possible rate but rather to deplete or nearly deplete the substrate in each cycle—in which case $E_{d2\ est}$ will be zero or near-zero. Again, the variables E_{u1}, E_{l1} and V_{u1} will be especially influential—not directly, but through their interaction terms—, but so will T_2 and E_{l2}.

Once the previously described models have been obtained, the operating conditions can only be optimized, for specific purposes, through a well-designed optimization process using several objective functions. Hence, additional work would be necessary in this regard.

4. Conclusions

Despite the broad available experience and technical knowledge available on the biotransformation of ethanol into acetic acid in the vinegar production process, a number of essential questions remain unanswered. Such is the case, for example, with the nature of the microbiota that effects the process and with its complex metabolic interactions. In practice, the process continues to require more or less extensive modelling in order to relate operational variables to specific objective functions.

Black-box models based on generalized polynomials have proved especially suitable for representing the behaviour of acetification systems in the form of response surfaces.

In this work, an effective experimental design based on useful data for the unequivocal calculation of the coefficients of the polynomial equations has been developed. Once the main operational variables have been identified, their admissible ranges have been established. That information has been used into an experimental design in order to identify the combinations of values of the operational variables that would allow the number of experiments maximizing the predictive ability of a model to be minimized. Another aim carried out in this work was to model key variables related to the acetification process performed with two serial bioreactors working in the semi-continuous mode by using quadratic polynomial equations.

The operational variables considered were the ethanol concentration at the time of unloading (E_{u1}), unloaded volume (V_{u1}), ethanol concentration during loading (E_{l1}) and operating temperature (T_1) in the first reactor, and the ethanol concentration during loading (E_{l2}) and operating temperature (T_2) in the second. As it has been shown, the variation ranges for these variables are subject not only to physico–chemical constraints, but also to others arising from the fact that the two reactors operate in a serial mode and from the limited number of combinations allowed for by the experimental design.

As a result, a fractional factorial design with 30 experiments (see Tables 5–7) considering the previous constraints has been obtained, with it being necessary to perform only 18 of them (see Tables 5 and 6).

With the gathered experimental data, second-order polynomials for several target variables of interest in terms of the considered operational variables involved in the industrial production process were developed. Specifically, models for the mean overall rate of acetic acid formation in the two-bioreactor system $[(r_A)_{global}]$, total acetic acid production in the system (P_m), ethanol concentration at the time the second bioreactor was unloaded (E_{u2}), the volume unloaded from the second bioreactor (V_{u2}), cycle duration (t_{cycle}), the mean overall volume in the two bioreactors during a cycle (V_m), the mean ethanol concentration in the first bioreactor during a cycle $(EtOH_{m1})$, the mean ethanol concentration in the second bioreactor during a cycle $(EtOH_{m2})$, the mean acetic acid concentration in the first reactor during a cycle (HAc_{m1}) and the mean acetic acid concentration in the second reactor during a cycle (HAc_{m2}) were obtained. The experimental results and their estimates were correlated via Forward Stepwise Regression, which allowed models with high predictive ability and minimal errors to be obtained. The resulting goodness of fit allowed a set of polynomials to be established that accurately reproduced the experimental results with only 18 experiments rather than the 30 needed in theory. Such models should allow us to carry out further optimization studies.

With all models, the variables associated to the first bioreactor were the more influential on the process. This was particularly so with E_{u1}, E_{l1} and V_{u1}, and can be ascribed not only to the fact that these variables depend on the fermentation conditions in the first bioreactor—and hence contribute to the total production of acetic acid—but also to the operating conditions in the second—usually more extreme—being influenced by those under which the first is operated.

Supplementary Materials: The following are available online at http://www.mdpi.com/2076-3417/10/24/9064/s1. "Get_feasible_combinations.m": MATLAB script for systematic analysis of feasible combinations of operational variables. File "Results.xlsx": Excel file with successively obtained feasible combinations of operational variables. File "S1.pdf": Description of the procedure used to determine the non-measurable variables of the process. File "S2.pdf": Description of the experimental results. File "S3.pdf": Detailed description of the procedure used to obtain the polynomial models of all analysed variables.

Author Contributions: Conceptualization, I.G.-G.; methodology, C.M.Á.-C., I.M.S.-D. and I.G.-G.; formal analysis, C.M.Á.-C., I.M.S.-D., J.E.J.-H. and I.G.-G.; Data curation, C.M.Á.-C. and I.M.S.-D.; writing—original draft preparation, J.E.J.-H. and I.G.-G.; writing—review and editing, J.E.J.-H., I.G.-G and I.M.S.-D.; funding acquisition, I.G.-G. All authors have read and agreed to the published version of the manuscript.

Funding: This research was funded by "XXIII Programa Propio de Fomento de la Investigación 2018" (MOD 4.2. SINERGIAS, Ref XXIII. PP Mod 4.2) from University of Córdoba (Spain) and by "Programa PAIDI" from Junta de Andalucía (RNM-271).

Conflicts of Interest: The authors declare no conflict of interest.

References

1. Valero, E.; Berlanga, T.M.; Roldán, P.M.; Jiménez, C.; García-García, I.; Mauricio, J.C. Free amino acids and volatile compounds of vinegars obtained from different types of substrate. *J. Sci. Food Agric.* **2005**, *85*, 603–608. [CrossRef]

2. García-García, I. *Second Symposium on R + D + I for Vinegar Production*, 1st ed.; Publication Services, University of Córdoba: Córdoba, Spain, 2006; pp. 1–296.

3. Solieri, L.; Giudici, P. *Vinegars of the World*; Springer: Milan, Italy, 2009; pp. 1–300.

4. Bekatorou, A. *Advances in Vinegar Production*, 1st ed.; CRC Press: Boca Raton, FL, USA, 2019; pp. 1–525. [CrossRef]

5. Emde, F. Improvements for an optimized process strategy in vinegar fermentation. In *Book of Abstracts, International Symposium of Vinegars and Acetic Acid Bacteria, Reggio Emilia, Italy, 8–12 May 2005*; Giudici, P., Lisa, S., de Vero, L., Eds.; University of Modena and Reggio Emilia: Modena, Italy, 2005.

6. García-García, I.; Cantero-Moreno, D.; Jiménez-Ot, C.; Baena-Ruano, S.; Jiménez-Hornero, J.; Santos-Duenas, I.; Bonilla-Venceslada, J.L.; Barja, F. Estimating the mean acetification rate via on-line monitored changes in ethanol during a semi-continuous vinegar production cycle. *J. Food Eng.* **2007**, *80*, 460–464. [CrossRef]

7. García-García, I.; Jiménez-Hornero, J.E.; Santos-Dueñas, I.M.; González-Granados, Z.; Cañete-Rodríguez, A.M. Modelling and optimization of acetic acid fermentation (Chapter 15). In *Advances in Vinegar Production*; Bekatorou, A., Ed.; CRC Press (Taylor & Francis Group): Boca Raton, FL, USA, 2019; pp. 299–325. [CrossRef]

8. Jiménez-Hornero, J.E.; Santos-Dueñas, I.M.; García-García, I. Optimization of biotechnological processes. The acetic acid fermentation. Part I: The proposed model. *Biochem. Eng. J.* **2009**, *45*, 1–6. [CrossRef]

9. Jiménez-Hornero, J.E.; Santos-Dueñas, I.M.; Garcia-Garcia, I. Optimization of biotechnological processes. The acetic acid fermentation. Part II: Practical identifiability analysis and parameter estimation. *Biochem. Eng. J.* **2009**, *45*, 7–21. [CrossRef]

10. Jiménez-Hornero, J.E.; Santos-Dueñas, I.M.; Garcia-Garcia, I. Optimization of biotechnological processes. The acetic acid fermentation. Part III: Dynamic optimization. *Biochem. Eng. J.* **2009**, *45*, 22–29. [CrossRef]

11. García-García, I.; Santos-Dueñas, I.M.; Jiménez-Ot, C.; Jiménez-Hornero, J.E.; Bonilla-Venceslada, J.L. Vinegar engineering. In *Vinegars of the World*; Solieri, L., Giudici, P., Eds.; Springer: Milano, Italy, 2009; Chapter 9; pp. 97–120. [CrossRef]

12. JJiménez-Ot, C.; Santos-Dueñas, I.M.; Jiménez-Hornero, J.; Baena-Ruano, S.; Martín-Santos, M.A.; Bonilla-Venceslada, J.L.; García-García, I. Influencia de la graduación total de un vino Montilla-Moriles sobre la velocidad de acetificación en el proceso de elaboración de vinagre. In *Proceedings of the XVI Congreso Nacional de Microbiología de los Alimentos, Córdoba, Spain, 14–17 September 2008*; Fernández-Salguero, J., García-Jimeno, R., Medina-Canalejo, L., Cabezas Redondo, L., Eds.; Publication Services of Diputación de Córdoba: Córdoba, Spain, 2008; pp. 255–256.

13. Baena-Ruano, S.; Jiménez-Ot, C.; Santos-Dueñas, I.M.; Cantero-Moreno, D.; Barja, F.; García-García, I. Rapid method for total, viable and non-viable acetic acid bacteria determination during acetification process. *Process Biochem.* **2006**, *41*, 1160–1164. [CrossRef]

14. Baena-Ruano, S.; Jiménez-Ot, C.; Jiménez-Hornero, J.; Santos-Dueñas, I.M.; Bonilla-Venceslada, J.L.; Cantero-Moreno, D.; García-García, I. Optimización de la producción de vinagre de vino. Influencia del volumen de descarga. In *Proceedings of the Second Symposium on Research + Development + Innovation for Vinegars Production, Córdoba, Spain, 17–20 April 2006*; García-García, I., Ed.; Publication Services, University of Córdoba: Córdoba, Spain, 2006; pp. 180–183.

15. Baena-Ruano, S.; Jiménez-Ot, C.; Santos-Dueñas, I.M.; Jiménez-Hornero, J.E.; Bonilla-Venceslada, J.L.; Álvarez-Cáliz, C.; García-García, I. Influence of the final ethanol concentration on the acetification and production rate in the wine vinegar process. *J. Chem. Technol. Biotechnol.* **2010**, *85*, 908–912. [CrossRef]

16. Álvarez-Cáliz, C.; Santos-Dueñas, I.M.; Cañete-Rodríguez, A.M.; García-Martínez, T.; Mauricio, J.C.; García-García, I. Free amino acids, urea and ammonium ion contents for submerged wine vinegar production: Influence of loading rate and air-flow rate. *Acetic Acid Bacteria* **2012**, *1*, 1–6. [CrossRef]

17. Santos-Dueñas, I.M. Modelización Polinominal y Optimización de la Acetificación de Vino. Ph.D. Thesis, Universidad de Córdoba, Córdoba, Spain, 2009.

18. Baena-Ruano, S.; Santos-Dueñas, I.M.; Mauricio, J.C.; García-García, I. Relationship between changes in the total concentration of acetic acid bacteria and major volatile compounds during the acetic acid fermentation of white wine. *J. Sci. Food Agric.* **2010**, *90*, 2675–2681. [CrossRef]

19. Jiménez-Hornero, J.E.; Santos-Dueñas, I.M.; García-García, I. Modelling acetification with artificial neural networks and comparison with alternative procedures. *Processes* **2020**, *8*, 749. [CrossRef]

20. García-García, I.; Gullo, M. Acetic acid bacteria: Features and impact in bio-applications. *Acetic Acid Bact.* **2013**, *1*, 1. [CrossRef]

21. Gullo, M.; Verzelloni, E.; Canonico, M. Aerobic submerged fermentation by acetic acid bacteria for vinegar production: Process and biotechnological aspects. *Process Biochem.* **2014**, *49*, 1571–1579. [CrossRef]

22. Mamlouk, D.; Gullo, M. Acetic Acid Bacteria: Physiology and carbon sources oxidation. *Indian J. Microbiol.* **2013**, *53*, 377–384. [CrossRef]

23. Matsushita, K.; Toyama, H.; Tonouchi, N.; Okamoto-Kainuma, A. *Acetic Acid Bacteria: Ecology and Physiology*; Springer: Osaka, Japan, 2016; pp. 1–350. [CrossRef]

24. Agger, T.; Nielsen, J. Mathematical Modelling of Microbial Processes-Motivation and Means. In *Engineering and Manufacturing for Biotechnology: Focus on Biotechnology*; Hofman, M., Thonart, P., Eds.; Springer: Dordrecht, The Netherlands, 2001; Volume 4, pp. 61–75. [CrossRef]

25. Emde, F. State of the art technologies in submersible vinegar production. In *Proceedings of the Second Symposium on Research + Development + Innovation for Vinegars Production, Córdoba, Spain, 17–20 April 2006*; García-García, I., Ed.; Publication Services, University of Córdoba: Córdoba, Spain, 2006; pp. 101–109.

26. Sellmer, S. New strategies in process control for the production of wine vinegar. In *Proceedings of the Second Symposium on Research + Development + Innovation for Vinegars Production, Córdoba, Spain, 17–20 April 2006*; García-García, I., Ed.; Publication Services, University of Córdoba: Córdoba, Spain, 2006; pp. 127–132.

27. González-Sáiz, J.M.; Pizarro, C.; Garrido-Vidal, D. Evaluation of kinetic models for industrial acetic fermentation: Proposal of a new model optimized by genetic algorithms. *Biotechnol. Prog.* **2003**, *19*, 599–611. [CrossRef]

28. Garrido-Vidal, D.; Pizarro, C.; Gonzalez-Saiz, J.M. Study of process variables in industrial acetic fermentation by a continuous pilot fermentor and response surfaces. *Biotechnol. Prog.* **2003**, *19*, 1468–1479. [CrossRef]

29. Jiménez-Hornero, J.E.; Santos-Dueñas, I.M.; Garcia-Garcia, I. Structural identifiability of a model for the acetic acid fermentation process. *Math. Biosci.* **2008**, *216*, 154–162. [CrossRef]

30. Santos-Dueñas, I.M.; Jimenez-Hornero, J.E.; Cañete-Rodríguez, A.M.; García-García, I. Modeling and optimization of acetic acid fermentation: A polynomial-based approach. *Biochem. Eng. J.* **2015**, *99*, 35–43. [CrossRef]

31. Román-Camacho, J.J.; Santos-Dueñas, I.M.; García-García, I.; Moreno-García, J.; García-Martínez, T.; Mauricio, J.C. Metaproteomics of microbiota involved in submerged culture production of alcohol wine vinegar: A first approach. *Int. J. Food Microbiol.* **2020**, *333*, 108797. [CrossRef]

32. Nguyen, N.K.; Borkowski, J.J. New 3-level response surface designs constructed from incomplete block designs. *J. Stat. Plan Inference* **2008**, *138*, 294–305. [CrossRef]

33. Abilov, A.G.; Aliev, V.S.; Rustamov, M.I.; Aliev, N.M.; Lutfaliev, K.A. Problems of control and chemical engineering experiment. In Proceedings of the IFAC 6th Triennal World Congress, Boston, MA, USA, 24–30 August 1975; Volume 45, pp. 1–7.

34. Box, G.E.P.; Hunter, J.S.; Hunter, W.G. *Statistics for Experimenters: Design, Innovation and Discovery*, 2nd ed.; John Wiley & Sons, Inc.: Hoboken, NJ, USA, 2008; pp. 1–672.

35. Huerta-Ochoa, S.; Castillo-Araiza, C.O.; Roman-Guerrero, A.; Prado-Barragan, A. Whole-cell bioconversion of citrus flavonoids to enhance their biological properties. *Stud. Nat. Prod. Chem.* **2019**, *61*, 335–367. [CrossRef]

36. Ramis Ramos, G.; García Álvarez-Coque, M.C. *Quimiometría*; Sintesis: Madrid, Spain, 2001; p. 240.

37. Álvarez-Cáliz, C. Modelización Polinominal y Optimización Empleando dos Fermentadores en Serie Para la Producción de Vinagre de Vino. Ph.D. Thesis, Universidad de Córdoba, Córdoba, Spain, 2016.

38. Blasco, X.; Herrero, J.M.; Sanchos, J.; Martínez, M. A new graphical visualization of n-dimensional Pareto front for decision-making in multiobjetive optimization. *Inf. Sci.* **2008**, *178*, 3908–3924. [CrossRef]

39. Grierson, D.E. Pareto multi-criteria decision making. *Adv. Eng. Inform.* **2008**, *22*, 371–384. [CrossRef]

40. Pope, P.T.; Webster, J.T. The use of an F-statistic in stepwise regression procedures. *Technometrics* **1972**, *14*, 327–340. [CrossRef]

Appl. Sci. **2020**, *10*, 9064

41. Wilkinson, L.; Dallal, G.E. Tests of significance in Forward Selection Regression with an F-to-enter stopping rule. *Technometrics* **1981**, *23*, 377–380. [CrossRef]
42. Pinsker, I.S.; Kipnis, V.; Grechanovsky, E. The use of F-statistics in the Forward Selection Regression Algorithm. In *Proceedings of the Statistical Computing Section, Washington, DC, USA, 1985*; American Statistical Association: Washington, DC, USA, 1985.
43. Derksen, S.; Keselman, H.J. Backward, forward and stepwise automated subset selection algorithms: Frequency of obtaining authentic and noise variables. *Br. J. Math. Stat. Psychol.* **1992**, *45*, 265–282. [CrossRef]
44. Miller, N.; Miller, C. *Estadística y Quimiometria Para Quimica Analitica*, 4th ed.; Pearson Educación SA: Madrid, Spain, 2002; p. 296.
45. Jiménez-Hornero, J.E. Contribuciones al Modelado y Optimización del Proceso de Fermentación Acética. Ph.D. Thesis, Universidad Nacional de Educación a Distancia, Madrid, Spain, 2007.
46. De Ley, J.; Gosselé, F.; Swings, J. Genus I Acetobacter. In *Bergery's Manual of Systematic Bacteriology*; Williams & Wilkens: Baltimore, MD, USA, 1984; pp. 267–279.
47. Fregapane, G.; Rubio-Fernandez, H.; Salvador, M. Influence of fermentation temperature on semi-continuous acetification for wine vinegar production. *Eur. Food Res. Technol.* **2001**, *213*, 62–66. [CrossRef]
48. *MATLAB Version 9.4*; Mathworks Inc.: Natick, MA, USA, 2018.
49. *SigmaStat*; Systat Software Inc.: San Jose, CA, USA, 2018.

Article

Optimization of the Acetification Stage in the Production of Wine Vinegar by Use of Two Serial Bioreactors

Carmen M. Álvarez-Cáliz [1], Inés María Santos-Dueñas [1,*], Jorge E. Jiménez-Hornero [2] and Isidoro García-García [1]

[1] Department of Chemical Engineering, University of Cordoba, Campus de Rabanales, 14071 Cordoba, Spain; q42alcac@yahoo.es (C.M.Á.-C.); isidoro.garcia@uco.es (I.G.-G.)

[2] Department of Electrical Engineering and Automatic Control, University of Cordoba, Campus de Rabanales, 14071 Cordoba, Spain; jjimenez@uco.es

* Correspondence: ines.santos@uco.es; Tel.: +34-957-218-658

Abstract: In the scope of a broader study about wine acetification, previous works concluded that using a single bioreactor hindered simultaneously reaching high productivities with high substrate consumption and the use of two serially arranged bioreactors (TSAB) could achieve such goal. Then, the aim of this work is the optimization, using Karush–Kuhn–Tucker (KKT) conditions, of this TSAB using polynomial models previously obtained. The ranges for the operational variables leading to either maximum and minimum mean rate of acetification of $0.11 \leq (r_A)_{global} \leq 0.27$ g acetic acid$\cdot$(100 mL$\cdot$h)$^{-1}$ and acetic acid production of $14.7 \leq P_m \leq 36.6$ g acetic acid$\cdot$h^{-1} were identified; the results show that simultaneously maximizing $(r_A)_{global}$ and P_m is not possible so, depending on the specific objective, different operational ranges must be used. Additionally, it is possible to reach a productivity close to the maximum one ($34.6 \leq P_m \leq 35.5$ g acetic acid$\cdot$h^{-1}) with an almost complete substrate use [$0.2\% \leq E_{u2} \leq 1.5\%$ (v/v)]. Finally, comparing the performance of the bioreactors operating in series and in parallel revealed that the former choice resulted in greater production.

Keywords: vinegar; wine; acetification; bioreactor systems; bioprocesses; optimization

Citation: Álvarez-Cáliz, C.M.; Santos-Dueñas, I.M.; Jiménez-Hornero, J.E.; García-García, I. Optimization of the Acetification Stage in the Production of Wine Vinegar by Use of Two Serial Bioreactors. *Appl. Sci.* **2021**, *11*, 1217. https://doi.org/10.3390/app11031217

Academic Editor: Alessandro Leone
Received: 31 December 2020
Accepted: 23 January 2021
Published: 28 January 2021

Publisher's Note: MDPI stays neutral with regard to jurisdictional claims in published maps and institutional affiliations.

1. Introduction

Bioprocesses, in general, and acetification as an example, are especially complex owing to the large number of variables involved and their mutual interactions. When a bioprocess is microbially driven—which is the case with acetification—most of the complexity arises from the biological activity of a cell population that is usually highly diverse and capable of adjusting to a variety of environments—and hence of behaving in rather different manners. The problem is even greater when the microbial population comprises not a single species but rather a mixture of species or even genera. Although omics techniques have considerably increased the available knowledge on industrial bioprocesses, there is still a long way to go before they can be elucidated at the molecular scale, so their modelling is an invaluable tool to design them as efficiently as possible.

Because such systems are normally industrially or economically significant, developed models are frequently used to identify the operating conditions that will optimize the output in terms of specific performance indicators or variables. The scientific literature abounds with very recent references to modelled bioprocesses: the xanthan biosynthesis by *Xanthomonas campestris* ATCC 13951 using winery wastewater [1], the uricase production by *Aspergillus welwitschiae* strain 1–4 [2], the use of polyethylene terephthalate (PET) for the cultivation of *Pseudomonas* sp. GO16 [3], the extracellular protease production by a native *Bacillus aryabhattai* Ab15-ES [4] and many other examples could be found. Usually, however, the large number of variables involved in a process, and their mutual interactions, require careful analysis and specific optimization methods with a view to establishing their optimum values for the intended purpose.

The acetification process, which involves a bio-oxidative transformation effected by a mixture of acetic acid bacteria (AAB) [5], leads to the biological conversion of ethanol into acetic acid. Industrial acetification bioreactors usually operate in an automated repeated semi-continuous mode, so that once the fermenters are fully loaded, the ethanol concentration is allowed to decrease to a preset level and then a certain fraction of the reactor volume is unloaded, actuating the remainder as an inoculum for the next conversion cycle [6–11]. After the bioreactor is unloaded, it is slowly replenished with fresh alcoholic substrate to start a new ethanol cycle. For any specific substrate, typical operational variables are the ethanol concentration at the time the reactor is unloaded, the percentage of unloaded volume and the loading rate [12–17], which influence the mean ethanol and acetic acid concentrations in the culture medium leading to more or less stressing environmental conditions for the acetic acid bacteria (AAB) and so affecting to the rate and efficiency of the process. In practice, it is necessary to optimize the outcome of the acetification process under operationally restricted conditions, for instance, the end product must have a very little ethanol concentration but if ethanol in the culture medium is almost depleted, the high acidity and lack of substrate would affect very negatively the action of AAB in next cycles. In this regard, the use of two identical serially arranged bioreactors working in a repeated semi-continuous mode, see Figure 1, could be a proper alternative to overcome the problem; this possibility is also suggested from several reported models for the acetification process [7–11,18–27]. In many vinegar plants the main reactors (bioreactor B1 in Figure 1) are partially unloaded in additional depletion fermenters (bioreactor B2 in Figure 1) where the ethanol concentration could be exhausted. This set up, in practice, results in six operational variables the values of which must be found: E_{u1}, ethanol concentrations at the time the first bioreactor is unloaded; E_{l1} and E_{l2} the ethanol concentrations during loading of the first and second bioreactor, respectively; V_{u1} volume of medium unloaded from the first bioreactor and T_1 and T_2 the constant working temperatures of the first and second bioreactor, respectively.

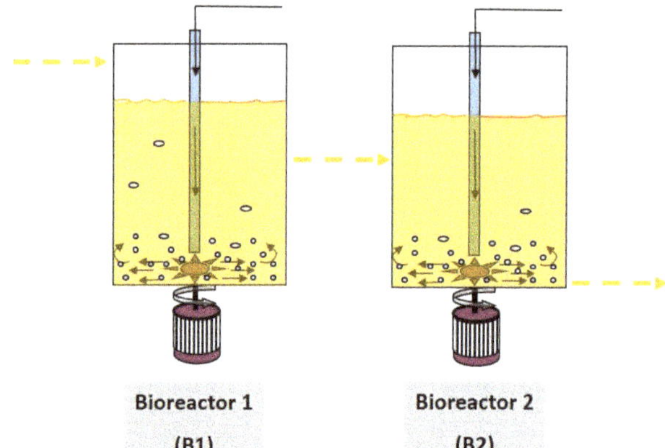

Figure 1. Two identical serially arranged bioreactors working in a repeated semi-continuous mode for the production of wine vinegar.

Among the different approaches used for modelling this process, the black-box models based on second order generalized polynomials [7,11,24–27] showed to be the best alternatives because of their simplicity and accurate predictions.

Because of the previous comments and with the aim of optimizing the process, a first work was carried out for modelling the behavior of two serially arranged bioreactors working in a repeated semi-continuous mode [27]. Now, in this work, the obtained black box models are being used to find the operating conditions leading to the maximum and minimum values of specific industrially useful variables that were used as objective functions.

2. Materials and Methods

2.1. Raw Material and Microorganisms

As substrate, white wine from the Montilla–Moriles D.O. (Córdoba, Spain) containing (11.5 ± 0.5) % (*v/v*) ethanol and an initial acidity of (0.4 ± 0.1)% (*w/v*) as acetic acid was used. Imitating industrial procedures, the inoculum used was a natural mixed culture [5] maintained and stored in our laboratories from experiments in a long-term fully operational bench acetator working with either wine or a synthetic ethanol medium; the original inoculum was taken, approximately a couple of years before, from an industrial tank in full operation (UNICO Vinagres y Salsas, S.L.L., Córdoba, Spain). First, a stage for reactivating and for the adaptation of the inoculum, which includes several previous acetification cycles to achieve repetitive results, was performed [27]. Additional details about some methodological aspects for obtaining the experimental data used to build the models necessary for the optimization study carried out in this work, were described elsewhere [27]; in any case, as a summary, the only variable not determined automatically was acidity, which was measured by acid—base titration with an NaOH solution approximately 0.5 N that was previously standardized with potassium hydrogen phthalate. The volume and ethanol concentration were measured in a continuous manner by using an EJA 110 differential pressure probe from Yokogawa Electric Corp. (Tokyo, Japan) and an Alkosens probe equipped with an Acetomat transducer from Heinrich Frings, respectively.

2.2. Operating Mode

Operation of the two bioreactors working serially, see Figure 2, is summarized as follows: once the ethanol concentration in the first reactor decreased to a preset value E_{u1}, a volume of medium V_{u1} was unloaded into the second. Then, both bioreactors were loaded with fresh wine to an also preset level (E_{l1} and E_{l2}, respectively). Once the maximum volume of medium with which each bioreactor could be loaded (8 L) was reached, the acetification system entered an ethanol depletion stage. When the ethanol concentration again fell to E_{u1}, the second bioreactor was completely unloaded for replenishment with medium from the first. Depending on the particular operating conditions, the second bioreactor may not be loaded to full capacity and V_{u2} be less than 8 L as a result. In addition, each reactor can be operated at a different temperature (T_1 and T_2, respectively). Different values of each of the previous variables can therefore lead to minimum or maximum levels of the dependent variables.

One should bear in mind that the primary aim was to maintain the first bioreactor under non-stressing conditions for the culture to retain its activity; in this way, unloading a fraction of fermentation medium into the second bioreactor would supply it with healthy bacterial biota capable of operating under the substrate depletion conditions they would inevitably find in that reactor. By way of example, the profiles of some state variables of the system, namely, volume of medium, ethanol concentration and acetic acid concentration in each bioreactor are shown in Figure 2.

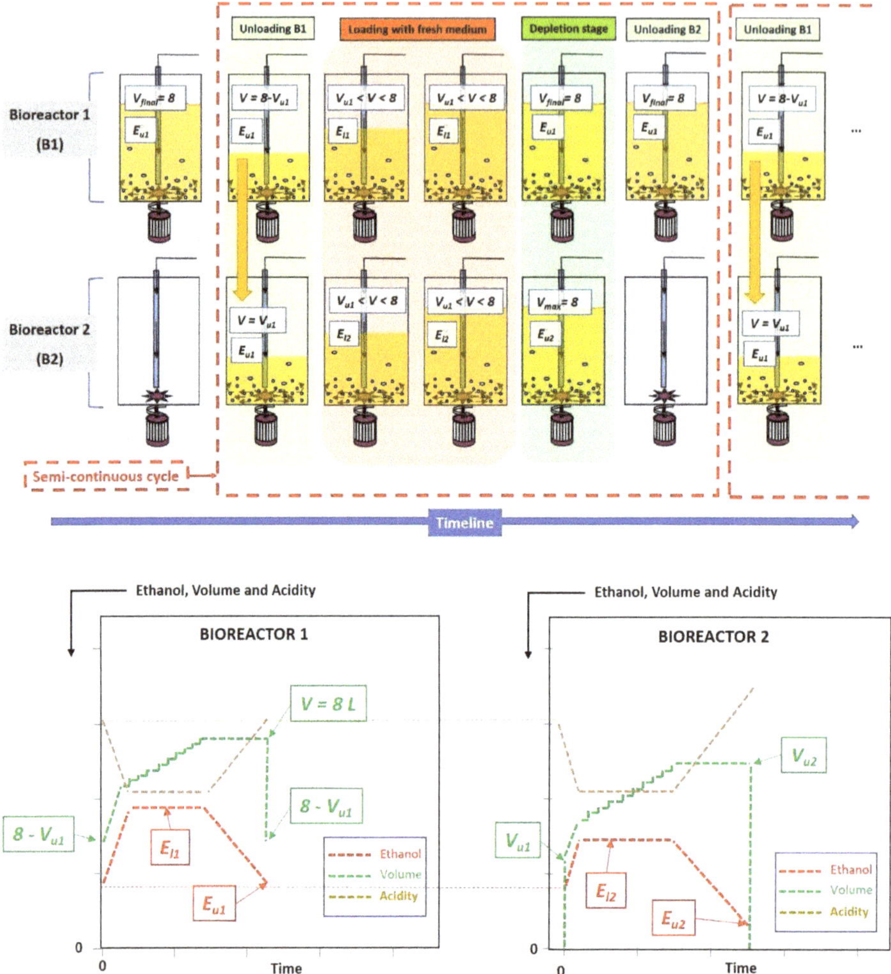

Figure 2. Scheme of two serial bioreactors operating in a repeated semi-continuous mode as well as example of profiles for the volume of medium, ethanol concentration and acetic acid concentration in each bioreactor. E_{u1} and E_{u2} are the ethanol concentrations at the time the first and second bioreactor, respectively, are unloaded; E_{l1} and E_{l2} the ethanol concentrations during loading of the first and second bioreactor, respectively; and V_{u1} and V_{u2} the volumes of medium unloaded from the first and second, respectively.

2.3. Optimization Method

The values of the operational variables providing the optimum polynomial functions were determined by solving the constrained non-linear optimization problem defined by Equation (1):

$$\begin{aligned} & Max\ f(x) \\ & \text{s.t.}\ \ h(x) = 0\ , \\ & \qquad g(x) \leq 0 \end{aligned} \tag{1}$$

where $f(x)$ denotes the objective function, x the vector of operational (decision) variables, $h(x)$ the set of equality constraints and $g(x)$ that of inequality constraints.

If these functions are assumed to be differentiable, then optimal points fulfilling the Karush–Kuhn–Tucker (KKT) conditions can be determined [28,29] from the Lagrange function shown in Equation (2) for a maximization problem (for a minimization problem, the Lagrange function is the same as Equation (2) simply by changing the sign of $f(x)$).

$$\mathcal{L}(x,\lambda,\mu) = f(x) - \sum_j \lambda_j h_j(x) - \sum_i \mu_i g_i(x), \tag{2}$$

where λ_j and μ_i are the KKT multipliers.

Equation (3) sets the first-order KKT conditions needed for optimality:

$$\begin{aligned}
\nabla_x \mathcal{L}(x,\lambda,\mu) &= 0 \\
h_j(x) &= 0 \\
g_i(x) &\leq 0 \\
\mu_i g_i(x) &= 0 \\
\mu_i &\geq 0
\end{aligned} \tag{3}$$

where ∇_x is the gradient operator with respect to x.

$\nabla_x g_i$ and $\nabla_x h_j$ must be linearly independent of the active constraints at the critical points x^* obtained as solutions to Equation (3). The active constraints at x^* are those with associated non-zero KKT multipliers. In a maximization problem, if $f(x)$ is concave and all $g(x)$ functions are convex, then x^* will be the global solution to the optimization problem. Otherwise, it will be necessary to check whether the second order KKT conditions for x^* are fulfilled in order to confirm that they provide a local maximum on $f(x)$. A function is concave if its Hessian matrix is negative semi-definite (i.e., if all its eigenvalues are equal to or less than 0) and convex if it is positive semi-definite.

Sufficient second-order KKT conditions can be verified by examining the sign of the last $n - m$ leading principal minors of the bordered Hessian B (Equation (4)) evaluated on a critical point.

$$B(x,\lambda,\mu) = \begin{pmatrix} \nabla_{xx}^2 \mathcal{L}(x,\lambda,\mu) & \nabla_x g(x)^T \\ \nabla_x g(x) & 0_{m \times m} \end{pmatrix}_{(n+m) \times (n+m)}, \tag{4}$$

where n is the number of operational variables, m that of active constraints at the critical point, $\nabla_{xx}^2 \mathcal{L}(x,\lambda,\mu)$ the Hessian matrix of $\mathcal{L}(x,\lambda,\mu)$ and $g(x)$ the set of active constraints.

The kth leading principal minor of $B(x,\lambda,\mu)$ equals $\left|B(x,\lambda,\mu)_{k \times k}\right|$ [i.e., the determinant of the $k \times k$ submatrix taken from the upper-left of $B(x,\lambda,\mu)$]. Therefore, the last leading principal minor of $B(x,\lambda,\mu)$ equals $\left|B(x,\lambda,\mu)_{(n+m) \times (n+m)}\right|$ [i.e., the determinant of the whole matrix B], the penultimate one being $\left|B(x,\lambda,\mu)_{(n+m-1) \times (n+m-1)}\right|$ and so on. For a critical point to be a local maximum, $sign\left(B(x,\lambda,\mu)_{(n+m) \times (n+m)}\right)$ must be equal to $(-1)^n$ and the previous $n - m - 1$ leading principal minors must have an alternating sign.

3. Results and Discussion

3.1. Optimization of the Mean Rate of Acetic Acid Formation in the Two-Bioreactor System

One potentially useful variable for assessing the "health" of a bioprocess and its productivity is the rate of biotransformation. Because the bioreactors operated in a non-steady state here, the target variable was the mean reaction rate, which can be estimated from the rate of ethanol consumption or that of acetic acid formation [6]. When volatile losses are negligible, as it is the case [27], both rates are identical. We focused on the mean rate of acetic acid formation, $(r_A)_{global}$, because it is easier to estimate.

In [27], the polynomial model for the mean rate of acetic acid formation in the two-bioreactor system was obtained (Equation (5)). The model allowed $(r_A)_{global\ est}$ to be estimated with an error of 0.01 g acetic acid·$(100\ mL\cdot h)^{-1}$.

$$(r_A)_{global\ est} = \begin{aligned} &-0.51 + 0.43\cdot E_{u1} - 0.0654\cdot E_{u1}^2 \\ &-0.00456\cdot E_{l2}\cdot V_{u1} - 0.00468\cdot E_{u1}\cdot E_{l1} \ , \\ &+0.000672\cdot T_1\cdot E_{l1} + 0.000839\cdot E_{l2}\cdot T_1 \end{aligned} \tag{5}$$

As Equation (5) shows, not all the operational variables have a direct influence on $(r_A)_{global\ est}$; additionally, as explained in [27], particularly in its Supplementary Material ("S3.docx" file), not all the terms in Equation (5) have the same influence on $(r_A)_{global\ est}$; the higher the value of the statistic F (see Table 1), the more significant the term is for $(r_A)_{global\ est}$; it can be seen from Table 1 that the most influential term is that of E_{u1}.

Table 1. Significance of the terms of Equation (5). (Adapted from Table S3.14 in [27], Supplementary Material ("S3.docx" file)).

Terms	F	Coefficient
Constant		−0.51
E_{u1}^2	342.526	−0.0654
E_{u1}	392.086	0.43
$T_1 E_{l1}$	56.948	0.000672
$E_{u1} E_{l1}$	28.095	−0.00468
$E_{l2} T_1$	119.001	0.000839
$E_{l2} V_{u1}$	122.922	−0.00456

This model is of especial practical interest as it allows the operating conditions leading to maximum—desirable—or minimum levels of the industrially useful variables to be identified. We used the procedure described in Section 2.3 to develop a MATLAB script [30] that allowed the maximum or minimum value of $(r_A)_{global\ est}$ to be calculated (see file "Optimal_rA.m" in Supplementary Materials).

The optimization problem addressed to maximize $(r_A)_{global\ est}$ is represented by Equations (6) and (7). The latter sets the constraints imposed by the experimental conditions on the operational variables.

$$\max\ (r_A)_{global\ est} \ , \tag{6}$$
$$\text{s.t.}$$

$$\begin{aligned} 4.25 &\le V_{u1} \le 5.75 \\ 4 &\le E_{l1} \le 6 \\ 2 &\le E_{u1} \le 4 \ , \\ 28 &\le T_1 \le 32 \\ 2.5 &\le E_{l2} \le 4.5 \end{aligned} \tag{7}$$

where V_{u1} is the volume of medium unloaded from the first bioreactor into the second, E_{l1} the ethanol concentration during loading of the first bioreactor, E_{u1} the ethanol concentration at the time the first bioreactor is unloaded, T_1 the temperature in the first bioreactor and E_{l2} the ethanol concentration in the second bioreactor during loading.

Solving the previous optimization problem provided the operating conditions of Table 2.

Table 2. Values of the operational variables needed to maximize $(r_A)_{global\ est}$.

V_{u1}, L	E_{l1}, % (v/v)	E_{u1}, % (v/v)	T_1, °C	E_{l2}, % (v/v)	T_2, °C	$(r_A)_{global\ est}$, g Acetic Acid·$(100\ mL\cdot h)^{-1}$
4.25	6.0	3.07	32.0	4.5	−	0.27

Because of the number of involved operational variables, it is not straightforward to graphically show the optimum obtained with the conditions of Table 2. Nevertheless, considering the previous comments on the significance of the terms in Equation (5), when plotting the response surfaces of $(r_A)_{global\ est}$ varying just a pair of the operational variables at once, it is possible to show that significant changes can only be found when the variable E_{u1} is included. By way of example, response surfaces varying E_{u1} and E_{l2} on one side and E_{l1} and E_{l2} on the other are shown in Figure 3a,b, respectively. Values from Table 2 are used for the remaining variables.

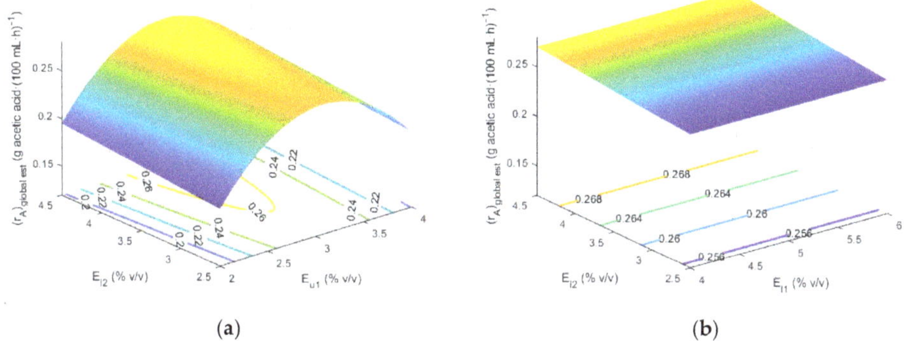

Figure 3. Response surfaces of $(r_A)_{global\ est}$ varying E_{u1} and E_{l2} (**a**) as well as E_{l1} and E_{l2} (**b**), using values from Table 2 for the remaining operational variables.

Since $(r_A)_{global\ est}$ could be subject to an error of up to 0.01 g acetic acid$\cdot$(100 mL$\cdot$h)$^{-1}$, we checked whether other combinations of values of the operational variables falling within the ranges imposed by their errors would lead to maximum values over the range $0.26 \leq (r_A)_{global\ est} \leq 0.27$ g acetic acid$\cdot$(100 mL$\cdot$h)$^{-1}$. A systematic analysis provided the ranges shown in Table 3. With a 0.2 interval for each variable, such ranges led to a total of 162 combinations of which only 109 (see Table S1 in file "S1.docx" in Supplementary Materials) fell within the previous range of $(r_A)_{global\ est}$. However, using all 162 combinations ensured that this variable would be very close to its maximum value: 0.25–0.27 g acetic acid$\cdot$(100 mL$\cdot$h)$^{-1}$. Therefore, any of the 162 combinations would, in theory, lead to the highest possible mean rate of acetic acid formation in the two-bioreactor system—and hence to values of each operational variable falling in the ranges of Table 3.

Table 3. Upper and lower end of the ranges over which the operational variables maximized $(r_A)_{global\ est}$.

End	V_{u1}, L	E_{l1}, % (v/v)	E_{u1}, % (v/v)	T_1, °C	E_{l2}, % (v/v)	T_1, °C
Lower	4.25	5.6	2.8	31.6	4.1	
$\updownarrow$	$\updownarrow$	$\updownarrow$	$\updownarrow$	$\updownarrow$	$\updownarrow$	–
Upper	4.50	6.0	3.2	32.0	4.5	

Conversely, minimizing $(r_A)_{global\ est}$ led to the results of Table 4 (see file "Optimal_rA.m" in Supplementary Materials). The problem was like that stated in Equations (6) and (7) except that the aim was to minimize $(r_A)_{global\ est}$ rather than maximize it.

Table 4. Values of the operational variables needed to minimize $(r_A)_{global\ est}$.

V_{u1}, L	El_1, % (v/v)	E_{u1}, % (v/v)	T_1, °C	E_{l1}, % (v/v)	T_2, °C	$(r_A)_{global\ est}$, g Acetic Acid·(100 mL·h)$^{-1}$
5.75	4.0	2.0	28	4.5	–	0.11

As in the previous case, the error made in modelling $(r_A)_{global\ est}$ was used to examine the combinations of values of the operational variables, with provision for their errors (see Table 5), leading to minimum values over the range $0.11 \leq (r_A)_{global\ est} \leq 0.12$ g acetic acid·(100 mL·h)$^{-1}$. The ranges of Table 5 for the operational variables, with 0.2 unit intervals, led to a total of 550 combinations only 29 of which provided an $(r_A)_{global\ est}$ value falling in the previous range of minimum values (see Table S2 in file "S1.docx" in Supplementary Materials). However, any of the 550 combinations ensured that $(r_A)_{global\ est}$ would fall in the range 0.116–0.135 g acetic acid·(100 mL·h)$^{-1}$ and thus be very close to the minimum of Table 4. Therefore, any combination would provide the lowest mean rate of acetic acid formation in the two-bioreactor system and lead to values of the operational variables falling in the ranges of Table 5.

Table 5. Upper and lower end of the ranges over which the operational variables minimized $(r_A)_{global\ est}$.

End	V_{u1}, L	E_{l1}, % (v/v)	E_{u1}, % (v/v)	T_1, °C	E_{l1}, % (v/v)	T_2, °C
Lower	5.45	4.0		28.0	2.5	
↕	↕	↕	2.0	↕	↕	–
Upper	5.75	4.8		28.8	4.5	

Applying the equations obtained in [27] to the other variables (Equations (8)–(16)) provided the estimated values of Table 6 for $P_{m\ est}$, $E_{u2\ est}$, $V_{u2\ est}$, $t_{cycle\ est}$, $V_{m\ est}$, $EtOH_{m1\ est}$, $EtOH_{m2\ est}$, $HAc_{m1\ est}$ and $HAc_{m2\ est}$ under the operating conditions maximizing or minimizing $(r_A)_{global\ est}$. The ranges of some variables reflect differences due to the influence of T_2 —like $(r_A)_{global\ est}$, other variables were independent of the temperature in the second reactor, however.

$$
\begin{aligned}
P_{m\ est} = \ &-243.705 + 18.324 \cdot V_{u1} + 72.736 \cdot E_{l1} + 21.525 \cdot E_{u1} \\
&-9.708 \cdot E_{l1}^2 - 1.102 \cdot E_{u1} \cdot V_{u1} - 0.534 \cdot T_1 \cdot V_{u1} \\
&+0.742 \cdot T_1 \cdot E_{l1} - 0.416 \cdot E_{l2} \cdot E_{l1} + 0.175 \cdot T_2 \cdot E_{l1} \\
&-0.12 \cdot T_1 \cdot E_{u1} - 0.399 \cdot T_2 \cdot E_{u1} + 0.101 \cdot T_2 \cdot E_{l2}
\end{aligned}
\tag{8}
$$

$$
\begin{aligned}
E_{u2\ est} = \ &14.935 - 5.371 \cdot E_{u1} + 0.988 \cdot E_{u1}^2 - 0.0592 \cdot E_{l1} \cdot V_{u1} \\
&-0.456 \cdot E_{u1} \cdot V_{u1} + 0.0678 \cdot E_{u1} \cdot E_{l1} \\
&+0.494 \cdot E_{l2} \cdot E_{u1} - 0.049 \cdot T_2 \cdot E_{l2}
\end{aligned}
\tag{9}
$$

$$
\begin{aligned}
V_{u2\ est} = \ &4.921 + 1.399 \cdot E_{l2} - 0.59 \cdot E_{u1}^2 \\
&+0.665 \cdot E_{u1} \cdot V_{u1} - 0.324 \cdot E_{l2} \cdot V_{u1} \\
&+0.172 \cdot E_{l2} \cdot E_{u1} - 0.0275 \cdot T_2 \cdot E_{u1}
\end{aligned}
\tag{10}
$$

$$
\begin{aligned}
t_{cycle\ est} = \ &518.591 - 156.652 \cdot V_{u1} \\
&-6.474 \cdot T_1 + 13.556 \cdot V_{u1}^2 \\
&+1.076 \cdot T_1 \cdot V_{u1} - 0.864 \cdot E_{u1} \cdot E_{l1}
\end{aligned}
\tag{11}
$$

$$
\begin{aligned}
V_{m\ est} = \ &4.306 + 4.739 \cdot E_{u1} - 0.841 \cdot E_{u1}^2 \\
&+0.336 \cdot E_{u1} \cdot V_{u1} - 0.0127 \cdot T_2 \cdot V_{u1} \\
&-0.125 \cdot E_{l2} \cdot E_{l1} + 0.0326 \cdot T_2 \cdot E_{l1} \\
&-0.0735 \cdot T_2 \cdot E_{u1} + 0.0358 \cdot T_2 \cdot E_{l2}
\end{aligned}
\tag{12}
$$

$$EtOH_{m1\,est} = \begin{array}{c} 2.312 - 0.0932 \cdot E_{u1}^2 \\ +0.0191 \cdot E_{l1} \cdot V_{u1} + 0.175 \cdot E_{u1} \cdot E_{l1} \end{array} \, , \tag{13}$$

$$EtOH_{m2\,est} = \begin{array}{c} 4.327 - 0.179 \cdot E_{u1} \cdot V_{u1} \\ +0.306 \cdot E_{l2} \cdot E_{u1} - 0.0182 \cdot T_2 \cdot E_{l2} \end{array} \, , \tag{14}$$

$$HAc_{m1\,est} = \begin{array}{c} 9.188 + 0.0932 \cdot E_{u1}^2 \\ -0.0191 \cdot E_{l1} \cdot V_{u1} - 0.175 \cdot E_{u1} \cdot E_{l1} \end{array} \, , \tag{15}$$

$$HAc_{m2\,est} = \begin{array}{c} 7.227 + 0.177 \cdot E_{u1} \cdot V_{u1} \\ -0.305 \cdot E_{l2} \cdot E_{u1} + 0.0178 \cdot T_2 \cdot E_{l2} \end{array} \, , \tag{16}$$

Table 6. Values of the state variables under the operating conditions maximizing and minimizing $(r_A)_{global\,est}$.

Variable	Conditions Minimizing $(r_A)_{global\,est}$ (0.11 g Acetic Acid$\cdot$(100 mL$\cdot$h)$^{-1}$)	Conditions Maximizing $(r_A)_{global\,est}$ (0.27 g Acetic Acid$\cdot$(100 mL$\cdot$h)$^{-1}$)
$P_{m\,est}$, g acetic acid$\cdot$h^{-1}	$20.6 \pm 0.7 \leftrightarrow 22.0 \pm 0.7$	$27.5 \pm 0.7 \leftrightarrow 28.6 \pm 0.7$
$E_{u2\,est}$, % (v/v)	$0 \pm 0.3 \leftrightarrow 0.4 \pm 0.3$	$1.4 \pm 0.3 \leftrightarrow 2.2 \pm 0.3$
$V_{u2\,est}$, L	$7.77 \pm 0.22 \leftrightarrow 7.99 \pm 0.22$	$7.63 \pm 0.22 \leftrightarrow 7.97 \pm 0.22$
$t_{cycle\,est}$, h	51.1 ± 2.5	20.8 ± 2.5
$V_{m\,est}$, L	$14.04 \pm 0.33 \leftrightarrow 14.32 \pm 0.33$	$14.06 \pm 0.33 \leftrightarrow 14.36 \pm 0.33$
$EtOH_{m1\,est}$, % (v/v)	3.8 ± 0.2	5.2 ± 0.2
$EtOH_{m2\,est}$, % (v/v)	$2.4 \pm 0.4 \leftrightarrow 2.7 \pm 0.4$	$3.6 \pm 0.4 \leftrightarrow 3.9 \pm 0.4$
$HAc_{m1\,est}$, % (w/v)	7.7 ± 0.2	6.3 ± 0.2
$HAc_{m2\,est}$, % (w/v)	$8.8 \pm 0.4 \leftrightarrow 9.1 \pm 0.4$	$7.5 \pm 0.4 \leftrightarrow 7.9 \pm 0.4$

Based on the previous results, the highest and lowest mean rate of acetic acid production were obtained by using extreme values of some operational variables. Thus, the minimum $(r_A)_{global\,est}$ value (Table 4) was obtained with the lowest temperature in the first bioreactor ($T_1 = 28\,°C$), and the smallest values in the ranges of E_{l1} and E_{u1} [viz., 4% and 2% (v/v), respectively]. These results suggest that the smaller the amount of substrate available to AAB and the higher the acidity to which the bacteria are exposed during a fermentation cycle (see $EtOH_{m1\,est}$ and $HAc_{m1\,est}$ values in Table 6) are, the less suitable will be the medium for their metabolic action —and hence for acetification. Also, because the volume of medium unloaded from the first bioreactor was quite substantial ($V_{u1} = 5.75$ L), the amount of inoculum remaining in it for the next cycle was considerably reduced, and so was $(r_A)_{global\,est}$ as a consequence. This result is consistent with those of modelling studies on a single bioreactor using both first-principles and black-box models [8–10,22,23,31,32].

On the other hand, if the aim is to maximize $(r_A)_{global\,est}$, the operating conditions should be essentially the opposite of those described in the previous paragraph. As can be seen from Table 6, such conditions resulted in increased mean substrate availability and decreased mean acidity in both bioreactors.

One other interesting conclusion is that $P_{m\,est}$ is rather different under the conditions maximizing and minimizing $(r_A)_{global\,est}$ (see Table 6). As expected, the greater $(r_A)_{global\,est}$ is, the higher will be acetic acid production. However, as shown in the following section, the maximum mean rate did not result in the highest possible $P_{m\,est}$ level.

3.2. Optimization of $P_{m\,est}$

The variable $(r_A)_{global\,est}$ possesses a high industrial interest. In practice, however, the aim of a vinegar producing plant is to obtain as much acetic acid per hour of operation (i.e., to maximize the overall production of acid, P_m). This variable can be estimated with an error of 0.7 g acetic acid$\cdot$h^{-1} from Equation (8). In this Equation, only three of the operational variables (E_{l1}, E_{u1} and V_{u1}) have a direct influence on $P_{m\,est}$, but as explained before in the case of Equation (5), not all the terms in Equation (8) have the same influence on $P_{m\,est}$ (see Table 7); the higher the value of the statistic F, the more significant the term is for $P_{m\,est}$; it can be seen from Table 7 that the most influential term is that of E_{l1}.

Table 7. Significance of the terms of Equation (8). (Adapted from Table S3.27 in [27], Supplementary Material ("S3.docx" file)).

Terms	F	Coefficient
Constant		-243.705
$T_1 E_{l1}$	92.918	0.742
E_{l1}^2	961.793	-9.708
E_{l1}	326.099	72.736
$T_2 E_{l2}$	34.06	0.101
$T_1 V_{u1}$	58.929	-0.534
V_{u1}	46.188	18.324
E_{u1}	63.689	21.525
$T_2 E_{u1}$	77.65	-0.399
$T_2 E_{l1}$	35.845	0.175
$E_{l2} E_{l1}$	16.55	-0.416
$E_{u1} V_{u1}$	19.078	-1.102
$T_1 E_{u1}$	5.284	-0.12

Below is described the procedure followed to identify the operating conditions maximizing (the industrial target) or minimizing (an undesirable outcome) $P_{m\ est}$.

The procedure of Section 2.3 to develop a MATLAB script to identify the optimum (maximum or minimum) $P_{m\ est}$ value was used (see file "Optimal_Prod.m" in Supplementary Materials). The conditions maximizing $P_{m\ est}$ can be identified by solving Equations (17) and (18), the latter containing the constraints imposed by the values of the experimental variables under the experimental conditions used.

$$\max\ P_{m\ est} \atop \text{s.t.} \tag{17}$$

$$\begin{aligned} 4.25 \leq V_{u1} \leq 5.75 \\ 4 \leq E_{l1} \leq 6 \\ 2 \leq E_{u1} \leq 4 \\ 28 \leq T_1 \leq 32 \\ 2.5 \leq E_{l2} \leq 4.5 \\ 28 \leq T_2 \leq 32 \end{aligned} \tag{18}$$

where T_2 is the temperature in the second bioreactor. Table 8 shows the value of each operational variable maximizing $P_{m\ est}$ and hence the solution to the optimization problem.

Table 8. Values of the operational variables needed to maximize $P_{m\ est}$.

V_{u1}, L	E_{l1}, % (v/v)	E_{u1}, % (v/v)	T_1, °C	E_{l2}, % (v/v)	T_2, °C	$P_{m\ est}$, g Acetic Acid·h^{-1}
4.25	5.1	4.0	32	4.5	28	36.6

In a similar way to the previous discussion on $(r_A)_{global\ est}$, plotting the response surfaces of $P_{m\ est}$ varying just a pair of the operational variables at once, it is possible to show that significant changes can only be found when the variable E_{l1} is included. By way of example, response surfaces varying E_{l1} and E_{u1} on one side and E_{u1} and E_{l2} on the other are shown in Figure 4a,b, respectively. Values from Table 8 are used for the remaining variables.

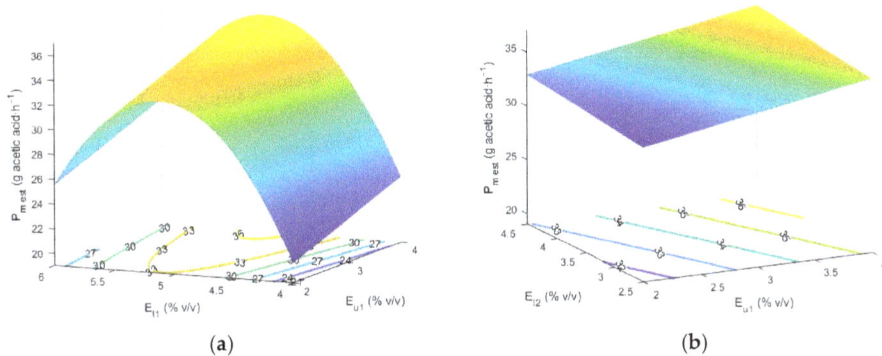

Figure 4. Response surfaces of $P_{m\ est}$ varying E_{l1} and E_{u1} (**a**) as well as E_{u1} and E_{l2} (**b**), using values from Table 8 for the remaining operational variables.

Similarly to $(r_A)_{global\ est}$, the errors in $P_{m\ est}$ and the operational variables were used to identify the combinations of variables leading to values over the range $35.9 \leq P_{m\ est} \leq 36.6$ g acetic acid$\cdot$h^{-1} (see Table 9). The ranges of Table 9, at 0.2 unit intervals, allowed a total of 4320 combinations to be identified of which 50 provided a $P_{m\ est}$ value within the previous range (see Table S3 in file "S1.docx" in Supplementary Materials). However, any of the 4320 combinations led to a very high $P_{m\ est}$ value (32.8–36.6 g acetic acid$\cdot$h^{-1}, which are close to that range). Therefore, any of the combinations of operational variables at values within the ranges of Table 9 would lead to a near-maximum $P_{m\ est}$ value.

Table 9. Upper and lower end of the ranges over which the operational variables maximized $P_{m\ est}$.

End	V_{u1}, L	E_{l1}, % (v/v)	E_{u1}, % (v/v)	T_1, °C	E_{l2}, % (v/v)	T_2, °C
Lower	4.25	4.9	3.7	31.3	3.5	28.0
$\updownarrow$	$\updownarrow$	$\updownarrow$	$\updownarrow$	$\updownarrow$	$\updownarrow$	$\updownarrow$
Upper	4.45	5.3	4.0	32.0	4.5	30.9

Minimizing $P_{m\ est}$ provided the results of Table 10 (see file "Optimal_Prod.m" in Supplementary Materials).

Table 10. Values of the operational variables needed to minimize $P_{m\ est}$.

V_{u1}, L	E_{l1}, % (v/v)	E_{u1}, % (v/v)	T_1, °C	E_{l2}, % (v/v)	T_2, °C	$P_{m\ est}$, g Acetic Acid$\cdot$h^{-1}
5.75	4.0	4.0	31.9	2.5	32	14.7

The combinations of values of the operational variables leading to minimum $P_{m\ est}$ values over the range $14.7 \leq P_{m\ est} \leq 15.4$ g acetic acid$\cdot$h^{-1} were identified similarly to those maximizing this dependent variable. Using the values in the ranges of Table 11 (0.2 unit intervals) provided 360 combinations of which 36 leading to $P_{m\ est}$ values within the previous range (see Table S4 in file "S1.docx" in Supplementary Materials). However, all combinations led to very low $P_{m\ est}$ values: 14.6–17.1 g acetic acid$\cdot$h^{-1}, which are very close to the minimum value.

Table 11. Upper and lower end of the ranges over which the operational variables minimized $P_{m\,est}$.

End	V_{u1}, L	E_{l1}, % (v/v)	E_{u1}, % (v/v)	T_1, °C	E_{l2}, % (v/v)	T_2, °C
Lower	5.55		3.6	31.2	2.5	31.3
$\updownarrow$	$\updownarrow$	4.0	$\updownarrow$	$\updownarrow$	$\updownarrow$	$\updownarrow$
Upper	5.75		4.0	32.0	2.9	32.0

Table 12 compares the values of the estimated state variables for the maximum and minimum $P_{m\,est}$ values.

Table 12. Values of the state variables under the operating conditions maximizing and minimizing $P_{m\,est}$.

Variable	Conditions Minimizing $P_{m\,est}$ (14.7 g Acetic Acid·h^{-1})	Conditions Maximizing $P_{m\,est}$ (36.6 g Acetic Acid·h^{-1})
$(r_A)_{global\,est}$, g acetic acid·(100 mL·h)$^{-1}$	0.18 ± 0.01	0.21 ± 0.01
$E_{u2\,est}$, % (v/v)	0.0 ± 0.3	4.3 ± 0.3
$V_{u2\,est}$, L	7.56 ± 0.22	6.71 ± 0.22
$t_{cycle\,est}$, h	43.0 ± 2.5	19.2 ± 2.5
$V_{m\,est}$, L	11.58 ± 0.33	12.07 ± 0.33
$EtOH_{m1\,est}$, % (v/v)	4.1 ± 0.2	4.8 ± 0.2
$EtOH_{m2\,est}$, % (v/v)	1.8 ± 0.4	4.5 ± 0.4
$HAc_{m1\,est}$, % (w/v)	7.4 ± 0.2	6.7 ± 0.2
$HAc_{m2\,est}$, % (w/v)	9.7 ± 0.4	7.0 ± 0.4

As can be seen, the conditions maximizing $P_{m\,est}$ were different from those maximizing $(r_A)_{global\,est}$. In fact, $P_{m\,est}$ and $(r_A)_{global\,est}$ are related by the mean volume of medium, $V_{m\,est}$, in Equation (19). Since the bioreactors worked in a semi-continuous mode, the mean volume was dependent on the particular operating conditions and those resulting in the maximum possible values of $P_{m\,est}$ and $(r_A)_{global\,est}$ need not coincide (see $V_{m\,est}$ values in Tables 6 and 12).

$$(r_A)_{global\,est} = \frac{P_{m\,est}}{V_{m\,est}}, \tag{19}$$

As stated above for $(r_A)_{global\,est}$, an increased mean availability of ethanol and a decreased mean acidity boosted acetic acid production (P_m). However, ethanol and acetic acid concentrations above certain levels have an inhibitory effect whereas too low levels can even boost bacterial activity. These effects have been extensively examined in modelling studies [8–10,22,23,31,32]. For example, ethanol concentrations above 6% (v/v) and acetic acid concentrations above 6% (w/v) have been found to have substantial inhibitory effects [8–10]. This knowledge, and the mean acidity and ethanol values of Tables 6 and 12, allow the $(r_A)_{global\,est}$ and $P_{m\,est}$ results to be more easily understood.

One other major inference is that, based on Table 12, maximizing $P_{m\,est}$ leaves a substantial amount of unavailable ethanol in the second bioreactor [E_{u2} = 4.3% (v/v)], which is, in general, industrially unacceptable. It is, therefore, interesting to identify the specific conditions that will maximize $P_{m\,est}$ while ensuring that the substrate is depleted to an acceptable degree. The procedure followed for this purpose is described in the following section.

3.3. Optimizing $P_{m\,est}$ While Ensuring Enough Substrate Depletion

For economy and legal reasons, the amount of ethanol present in industrially produced vinegar must not exceed prescribed levels depending on each specific type of vinegar. It is therefore important to identify the operating conditions optimizing acetic acid production ($P_{m\,est}$ in Equation (8)) under a constraint on the ethanol concentration at the time the second bioreactor is unloaded ($E_{u2\ est}$ in Equation (9)). The optimization problem is stated in Equations (20) and (21).

$$\max \ P_{m\ est} \tag{20}$$
$$\text{s.t.}$$

$$
\begin{aligned}
4.25 &\leq V_{u1} \leq 5.75 \\
4 &\leq E_{l1} \leq 6 \\
2 &\leq E_{u1} \leq 4 \\
28 &\leq T_1 \leq 32 \\
2.5 &\leq E_{l2} \leq 4.5 \\
28 &\leq T_2 \leq 32 \\
E_{u2\ est} &= constant
\end{aligned}
\tag{21}
$$

where *constant* is the ethanol concentration at unloading time in the second reactor to be imposed.

Table 13 shows the values of the operational variables providing the solution to the previous problem with $E_{u2\ est}$ values from 0.2% to 1.5% (v/v).

Table 13. Values of the operational variables needed to maximize $P_{m\ est}$ with an $E_{u2\ est}$ value of 0.2%, 0.5%, 1.0% or 1.5% (v/v).

V_{u1}, L	E_{l1}, % (v/v)	E_{u1}, % (v/v)	T_1, °C	E_{l2}, % (v/v)	T_2, °C	$P_{m\ est}$, g Acetic Acid·h^{-1}	$E_{u2\ est}$, % (v/v)
4.76	5.2	2.3	32.0	4.5	32.0	34.6	0.2
4.53	5.2	2.4	32.0	4.5	32.0	34.9	0.5
4.25	5.2	2.7	32.0	4.5	32.0	35.4	1.0
4.25	5.2	3.2	32.0	4.5	32.0	35.5	1.5

The solution was obtained by using the MATLAB script of Section 3.2 (see file "Optimal_Prod.m" in Supplementary Materials) as expanded with the constraint on $E_{u2\ est}$. The previous range of $E_{u2\ est}$ was used to consider virtually complete and less marked depletion of the substrate because some vinegars (particularly those under a designation of origin or aged in casks) are allowed to contain higher levels of residual ethanol.

As expected, the greater $E_{u2\ est}$ was (i.e., the less markedly the substrate was depleted in the second bioreactor), the more favored was AAB activity and the higher was acetic acid productivity as a result.

3.4. Comparison of the Performance of Serial and Parallel Bioreactors

This was one other point of especial interest. For this purpose, we compared the results obtained with the two bioreactors working in series here and those previously reported for others operating in parallel [8–10,22,23,31,32]. Because the previous studies were conducted at 31 °C, we estimated P_m and E_{u2} at that temperature. As can be seen from Table 14, there were no substantial differences in performance between the serial and parallel configurations when P_m was maximized under no constraint on E_{u2}—not even at the optimum T_1 and T_2 values found here. However, maximizing P_m with $E_{u2} = 0.5\%$ (v/v) resulted in the serial system being more productive than the parallel one. Specifically, at 31 °C the former system reached a P_m 12.2% higher than did the latter. It was impossible to directly compare productivity with the two systems at 32 °C and $E_{u2} = 0.5\%$ (v/v) because no productivity data for the parallel system at that temperature were available. In any case, the results of Table 14 suggest that they cannot be too different from those obtained at 31 °C.

Table 14. Comparison of acetic acid production and final ethanol concentration at the time of unloading with two bioreactors operating in series and in parallel.

Working mode	Optimizing Production			Optimizing Production with a Specific Final E_{u2} Value		
	Parallel (31 °C)	Series		Parallel (31 °C)	Series	
		$T_1 = 31$ °C $T_2 = 31$ °C	$T_1 = 32$ °C $T_2 = 28$ °C		$T_1 = 31$ °C $T_2 = 31$ °C	$T_1 = 32$ °C $T_2 = 32$ °C
P_m g acetic acid·h^{-1}	35.2 ± 0.5	34.8 ± 0.7	35.9 ± 0.7	29.6 ± 0.5	33.2 ± 0.7	34.2 ± 0.7
E_{u2} % (v/v)	3.0 ± 0.2	3.7 ± 0.3	4.3 ± 0.3	0.5 ± 0.2	0.5 ± 0.3	0.5 ± 0.3

Finally, it should be noted that using two serial bioreactors results in very high acetic acid productivity and virtually complete depletion of ethanol. On the other hand, as shown in a previous work [16], depleting the second bioreactor in a system operating in parallel would slow down the process as a result of the time needed for acetic acid bacteria to regain full activity after a period of substrate scarcity and a high acidity in the medium—all of which considerably detract from productivity [16].

4. Conclusions

The two serially arranged bioreactors system for the acetification process has been optimized in various respects using previous quadratic polynomial models. The values of the operational variables maximizing or minimizing the mean rate of acetic acid production were obtained: a maximum value of $(r_A)_{global\ est}= 0.27$ g acetic acid·$(100\ mL·h)^{-1}$ was achieved when $V_{u1} = 4.25$ L, $E_{l1} = 6\%$ (v/v), $E_{u1} = 3.07\%$ (v/v), $T_1 = 32$ °C, $E_{l2} = 4.5\%$ (v/v) and a minimum value of $(r_A)_{global\ est} = 0.11$ g acetic acid·$(100\ mL·h)^{-1}$ was achieved when $V_{u1} = 5.75$ L, $E_{l1} = 4\%$ (v/v), $E_{u1} = 2\%$ (v/v), $T_1 = 28$ °C, $E_{l2} = 4.5\%$ (v/v).

In addition, the conditions leading to the maximum and minimum productivity were: $P_{m\ est} = 36.6$ g acetic acid·h^{-1} was achieved when $V_{u1} = 4.25$ L, $E_{l1} = 5.1\%$ (v/v), $E_{u1} = 4\%$ (v/v), $T_1 = 32$ °C, $E_{l2} = 4.5\%$ (v/v), $T_2 = 28$ °C and $P_{m\ est} = 14.7$ g acetic acid·h^{-1} was achieved when $V_{u1} = 5.75$ L, $E_{l1} = 4\%$ (v/v), $E_{u1} = 4\%$ (v/v), $T_1 = 31.9$ °C, $E_{l2} = 2.5\%$ (v/v), $T_2 = 32$ °C.

The operating conditions needed to maximize productivity while ensuring ethanol depletion to a preset degree were also identified: $4.25 \leq V_{u1} \leq 4.76$ L, $E_{l1} = 5.2\%$ (v/v), $2.3 \leq E_{u1} \leq 3.2\%$ (v/v), $T_1 = 32$ °C, $E_{l2} = 4.5\%$ (v/v), $T_2 = 32$ °C to achieve $34.6 \leq P_m \leq 35.5$ g acetic acid·h^{-1} with $0.2 \leq E_{u2} \leq 1.5\%$ (v/v). Such conditions allowed near-maximum productivity to be obtained.

One other aim fulfilled by modelling the system and using results of previous research work was to compare the performance of the system with the two reactors operating in series and in parallel. P_m can be maximized with no restrictions on E_{u2} by using either configuration. However, if the substrate is to be depleted [e.g., $E_{u2} = 0.5\%$ (v/v)], then the serial configuration will be more productive.

Supplementary Materials: The following are available online at http://www.mdpi.com/xxx/s1. "Optimal_rA.m" and "Optimal_Prod.m": MATLAB scripts for optimizing $(r_A)_{global\ est}$ and $P_{m\ est}$, respectively. File "S1.pdf": Combinations of operational variables to analyze the maximum and minimum of $(r_A)_{global\ est}$ and $P_{m\ est}$.

Author Contributions: Conceptualization, I.G.-G.; methodology, C.M.Á.-C., I.M.S.-D. and I.G.-G.; formal analysis, C.M.Á.-C., I.M.S.-D., J.E.J.-H. and I.G.-G.; data curation, C.M.Á.-C. and I.M.S.-D.; writing—original draft preparation, J.E.J.-H. and I.G.-G.; writing—review and editing, J.E.J.-H., I.G.-G. and I.M.S.-D.; funding acquisition, I.G.-G. All authors have read and agreed to the published version of the manuscript.

Funding: This research was funded by "XXIII Programa Propio de Fomento de la Investigación 2018" (MOD 4.2. SINERGIAS, Ref XXIII. PP Mod 4.2) from University of Córdoba (Spain) and by "Programa PAIDI" from Junta de Andalucía (RNM-271).

Conflicts of Interest: The authors declare no conflict of interest.

References

1. Rončević, Z.; Grahovac, J.; Dodić, S.; Vučurović, D.; Dodić, J. Utilisation of winery wastewater for xanthan production in stirred tank bioreactor: Bioprocess modelling and optimisation. *Food Bioprod. Process.* **2019**, *117*, 113–125. [CrossRef]
2. El-Naggar, N.E.A.; Haroun, S.A.; El-Weshy, E.M.; Metwally, E.A.; Sherief, A.A. Mathematical modeling for bioprocess optimization of a protein drug, uricase, production by *Aspergillus welwitschiae* strain 1–4. *Sci. Rep.* **2019**, *9*, 12971. [CrossRef] [PubMed]
3. Beagan, N.; O'Connor, K.E.; Del Val, I.J. Model-based operational optimisation of a microbial bioprocess converting terephthalic acid to biomass. *Biochem. Eng. J.* **2020**, *158*, 107576. [CrossRef]
4. Adeetunji, A.I.; Olaniran, A.O. Statistical modelling and optimization of protease production by an autochthonous Bacillus aryabhattai Ab15-ES: A response surface methodology approach. *Biocatal. Agric. Biotechnol.* **2020**, *24*, 101528. [CrossRef]
5. Román-Camacho, J.J.; Santos-Dueñas, I.M.; García-García, I.; Moreno-García, J.; García-Martínez, T.; Mauricio, J.C. Metaproteomics of microbiota involved in submerged culture production of alcohol wine vinegar: A first approach. *Int. J. Food Microbiol.* **2020**, *333*, 108797. [CrossRef] [PubMed]
6. García-García, I.; Cantero-Moreno, D.; Jiménez-Ot, C.; Baena-Ruano, S.; Jiménez-Hornero, J.; Santos-Duenas, I.; Bonilla-Venceslada, J.L.; Barja, F. Estimating the mean acetification rate via on-line monitored changes in ethanol during a semi-continuous vinegar production cycle. *J. Food Eng.* **2007**, *80*, 460–464. [CrossRef]
7. García-García, I.; Jiménez-Hornero, J.E.; Santos-Dueñas, I.M.; González-Granados, Z.; Cañete-Rodríguez, A.M. Modelling and optimization of acetic acid fermentation (Chapter 15). In *Advances in Vinegar Production*; Bekatorou, A., Ed.; CRC Press (Taylor & Francis Group): Boca Raton, FL, USA, 2019; pp. 299–325. [CrossRef]
8. Jiménez-Hornero, J.E.; Santos-Dueñas, I.M.; García-García, I. Optimization of biotechnological processes. The acetic acid fermentation. Part I: The proposed model. *Biochem. Eng. J.* **2009**, *45*, 1–6. [CrossRef]
9. Jiménez-Hornero, J.E.; Santos-Dueñas, I.M.; Garcia-Garcia, I. Optimization of biotechnological processes. The acetic acid fermentation. Part II: Practical identifiability analysis and parameter estimation. *Biochem. Eng. J.* **2009**, *45*, 7–21. [CrossRef]
10. Jiménez-Hornero, J.E.; Santos-Dueñas, I.M.; Garcia-Garcia, I. Optimization of biotechnological processes. The acetic acid fermentation. Part III: Dynamic optimization. *Biochem. Eng. J.* **2009**, *45*, 22–29. [CrossRef]
11. Jiménez-Hornero, J.E.; Santos-Dueñas, I.M.; García-García, I. Modelling acetification with artificial neural networks and comparison with alternative procedures. *Processes* **2020**, *8*, 749. [CrossRef]
12. García-García, I.; Santos-Dueñas, I.M.; Jiménez-Ot, C.; Jiménez-Hornero, J.E.; Bonilla-Venceslada, J.L. Vinegar engineering. In *Vinegars of the World*; Solieri, L., Giudici, P., Eds.; Springer: Milano, Italy, 2009; Chapter 9; pp. 97–120. [CrossRef]
13. Jiménez-Ot, C.; Santos-Dueñas, I.M.; Jiménez-Hornero, J.; Baena-Ruano, S.; Martín-Santos, M.A.; Bonilla-Venceslada, J.L.; García-García, I. Influencia de la graduación total de un vino Montilla-Moriles sobre la velocidad de acetificación en el proceso de elaboración de vinagre. In *XVI Congreso Nacional de Microbiología de los Alimentos, Córdoba, Spain*; Fernández-Salguero, J., García-Jimeno, R., Medina-Canalejo, L., Cabezas Redondo, L., Eds.; Publication Services of Diputación de Córdoba: Córdoba, Spain, 2008; pp. 255–256.
14. Baena-Ruano, S.; Jiménez-Ot, C.; Santos-Dueñas, I.M.; Cantero-Moreno, D.; Barja, F.; García-García, I. Rapid method for total, viable and non-viable acetic acid bacteria determination during acetification process. *Process Biochem.* **2006**, *41*, 1160–1164. [CrossRef]
15. Baena-Ruano, S.; Jiménez-Ot, C.; Jiménez-Hornero, J.; Santos-Dueñas, I.M.; Bonilla-Venceslada, J.L.; Cantero-Moreno, D.; García-García, I. *In Proceedings of the Second Symposium on Research+Development+Innovation for Vinegars Production, Córdoba, Spain, 26–28 April 2006*; García-García, I., Ed.; University of Córdoba: Córdoba, Spain, 2006; pp. 180–183.
16. Baena-Ruano, S.; Jiménez-Ot, C.; Santos-Dueñas, I.M.; Jiménez-Hornero, J.E.; Bonilla-Venceslada, J.L.; Álvarez-Cáliz, C.; García-García, I. Influence of the final ethanol concentration on the acetification and production rate in the wine vinegar process. *J. Chem. Technol. Biotechnol.* **2010**, *85*, 908–912. [CrossRef]
17. Álvarez-Cáliz, C.; Santos-Dueñas, I.M.; Cañete-Rodríguez, A.M.; García-Martínez, T.; Maurico, J.C.; García-García, I. Free amino acids, urea and ammonium ion contents for submerged wine vinegar production: Influence of loading rate and air-flow rate. *Acetic Acid Bact.* **2012**, *1*, e1. [CrossRef]
18. Emde, F. State of the art technologies in submersible vinegar production. In Proceedings of the Second Symposium on Research+Development+Innovation for Vinegars Production, Córdoba, Spain, 26–28 April 2006; García-García, I., Ed.; University of Córdoba: Córdoba, Spain, 2006; pp. 101–109.
19. Sellmer, S. New strategies in process control for the production of wine vinegar. In Proceedings of the Second Symposium on Research+Development+Innovation for Vinegars Production, Córdoba, Spain, 26–28 April 2006; García-García, I., Ed.; University of Córdoba: Córdoba, Spain, 2006; pp. 127–132.
20. González-Sáiz, J.M.; Pizarro, C.; Garrido-Vidal, D. Evaluation of kinetic models for industrial acetic fermentation: Proposal of a new model optimized by genetic algorithms. *Biotechnol. Prog.* **2003**, *19*, 599–611. [CrossRef] [PubMed]
21. Garrido-Vidal, D.; Pizarro, C.; Gonzalez-Saiz, J.M. Study of process variables in industrial acetic fermentation by a continuous pilot fermentor and response surfaces. *Biotechnol. Prog.* **2003**, *19*, 1468–1479. [CrossRef] [PubMed]
22. Jiménez-Hornero, J.E.; Santos-Dueñas, I.M.; Garcia-Garcia, I. Structural identifiability of a model for the acetic acid fermentation process. *Math. Biosci.* **2008**, *216*, 154–162. [CrossRef] [PubMed]

23. Santos-Dueñas, I.M.; Jimenez-Hornero, J.E.; Cañete-Rodríguez, A.M.; García-García, I. Modeling and optimization of acetic acid fermentation: A polynomial-based approach. *Biochem. Eng. J.* **2015**, *99*, 35–43. [CrossRef]
24. Nguyen, N.K.; Borkowski, J.J. New 3-level response surface designs constructed from incomplete block designs. *J. Stat. Plan Inference* **2008**, *138*, 294–305. [CrossRef]
25. Abilov, A.G.; Aliev, V.S.; Rustamov, M.I.; Aliev, N.M.; Lutfaliev, K.A. Problems of control and chemical engineering experiment, Part 1 1D: 45. In Proceedings of the IFAC 6th Triennal World Congress, Boston/Cambridge, MA, USA, 24–30 August 1975; pp. 1–7.
26. Box, G.E.P.; Hunter, J.S.; Hunter, W.G. *Statistics for Experimenters: Design, Innovation and Discovery*, 2nd ed.; John Wiley & Sons, Inc.: Hoboken, NJ, USA, 2008; pp. 1–672. ISBN 978-0-471-71813-0.
27. Álvarez-Cáliz, C.; Santos-Dueñas, I.M.; Jiménez-Hornero, J.E.; García-García, I. Modelling of the acetification stage in the production of wine vinegar by use of two serial bioreactors. *Appl. Sci.* **2020**, *10*, 9064. [CrossRef]
28. Karush, W. Minima of Functions of Several Variables with Inequalities as Side Constraints. Ph.D. Thesis, Department of Mathematics, University of Chicago, Chicago, IL, USA, 1939.
29. Kuhn, H.W.; Tucker, A.W. Nonlinear programming. In Proceedings of the Second Berkeley Symposium on Mathematical Statistics and Probability, University of California, Berkeley, CA, USA, 31 July–12 August 1950; Neyman, J., Ed.; University of California Press: Berkeley, CA, USA, 1951; pp. 481–492.
30. Mathworks Inc. *MATLAB Version 9.4*; Mathworks Inc.: Natick, MA, USA, 2018. Available online: www.mathworks.com (accessed on 28 January 2021).
31. Jiménez-Hornero, J.E. Contribuciones al Modelado y Optimización del Proceso de Fermentación Acética. Ph.D. Thesis, Universidad Nacional de Educación a Distancia, Madrid, Spain, 2007.
32. Santos-Dueñas, I.M. Modelización Polinominal y Optimización de la Acetificación de Vino. Ph.D. Thesis, Universidad de Córdoba, Córdoba, Spain, 2009.

Article

Effect of Time and Temperature on Physicochemical and Microbiological Properties of Sous Vide Chicken Breast Fillets

Hossein Haghighi [1,*,†], Anna Maria Belmonte [1,†], Francesca Masino [1,2], Giovanna Minelli [1,2], Domenico Pietro Lo Fiego [1,2] and Andrea Pulvirenti [1,2]

[1]   Department of Life Sciences, University of Modena and Reggio Emilia, 42122 Reggio Emilia, Italy; 71483@studenti.unimore.it (A.M.B.); francesca.masino@unimore.it (F.M.); giovanna.minelli@unimore.it (G.M.); domenicopietro.lofiego@unimore.it (D.P.L.F.); andrea.pulvirenti@unimore.it (A.P.)

[2]   Interdepartmental Research Centre BIOGEST-SITEIA, University of Modena and Reggio Emilia, 42124 Reggio Emilia, Italy

*   Correspondence: hossein.haghighi@unimore.it

†   Share first authorship.

Abstract: Temperature and time are two critical parameters in sous vide cooking which directly affect eating quality characteristics and food safety. This study aimed to evaluate physicochemical and microbiological properties of sous vide chicken breast fillets cooked at twelve different combinations of temperature (60, 70, and 80 °C) and time (60, 90, 120, and 150 min). The results showed that cooking temperature played a major role in the moisture content, cooking loss, pH, a* color value, shear force, and thiobarbituric acid reactive substances (TBARS). Increasing cooking temperature caused an increase in cooking loss, lipid oxidation, TBARS, and pH, while moisture content was reduced ($p < 0.05$). Cooking time played a minor role and only moisture content, cooking loss, and a* color value were affected by this parameter ($p < 0.05$). Total mesophilic aerobic bacteria, *Psychrotrophic* bacteria, and *Enterobacteriaceae* were not detected during 21 days of storage at 4 °C. Cooking at 60 °C for 60 min showed the optimum combination of temperature and time for sous vide cooked chicken breast fillets. The result of this study could be interesting for catering, restaurants, ready-to-eat industries, and homes to select the optimum combination of temperature and time for improving the eating quality characteristics and ensuring microbiological safety.

Keywords: chicken breast fillets; color; cooking loss; cooking temperature; cooking time; microbiological safety; shear force; sous vide cooking; TBARS

Citation: Haghighi, H.; Belmonte, A.M.; Masino, F.; Minelli, G.; Lo Fiego, D.P.; Pulvirenti, A. Effect of Time and Temperature on Physicochemical and Microbiological Properties of Sous Vide Chicken Breast Fillets. *Appl. Sci.* **2021**, *11*, 3189. https://doi.org/10.3390/app11073189

Academic Editor: Isidoro Garcia-Garcia

Received: 25 February 2021
Accepted: 30 March 2021
Published: 2 April 2021

1. Introduction

Meat plays a key role in human nutrition and evolution thanks to its components, including proteins and essential micronutrients such as Zn, Se, Fe, vitamin A, vitamin B_{12}, and folate [1,2]. Most often, raw meat is subjected to various cooking methods such as boiling in water, grilling, steaming, microwave radiation, and sous vide to enhance its digestibility, sensory characteristic, and to improve its hygienic quality [3–5]. In each type of cooking method, several changes occur as a consequence of heating, such as denaturation, aggregation, and degradation of proteins, fiber shrinkage, and collagen solubilization [5–7]. The bio-accessibility of nutrients also can be affected during the cooking process mainly due to the degradation of vitamins, amino acids, and minerals [8]. Therefore, selecting an appropriate cooking method is a critical step before consumption which directly affects physicochemical, textural, and microbiological properties. Among different cooking methods, sous vide cooking has received considerable attention from catering, restaurants, ready-to-eat industries, and homes [9,10]. This technique provides more efficient heat transfer from water to food compared to other cooking methods [11], resulting improvement in eating quality characteristics such as texture, tenderness, juiciness, color, flavor, and also provides high nutritional value [6,12,13]. Besides, this technique is simple to apply for cooking different kinds of food (e.g., meat, cereals, legumes, etc.) [14]. The term "sous vide" is a French word that refers to the uniform cooking of food inside the food grade and heat-stable vacuumed pouches incubated in a circulating water bath with monitored conditions of temperature and time followed by chilled storage [15,16]. Sous vide cooking has been reported to enhance the quality attributes, inhibiting off-flavors from lipid oxidation, reducing aerobic bacteria and the risk of post-cooking contamination during storage [9,17–19]. Besides, it is beneficial for preserving vitamins, antioxidant compounds, essential amino acids, and unsaturated fatty acids during solubilization, volatilization, and high-temperature application [11,15].

Selecting the right temperature and time combinations plays an important role in sous vide cooking to reduce the risk of overcooking, loss of volatile compounds, and heat-sensitive nutrients [8]. In this context, the effect of cooking temperature and time in sous vide has been reported on the physicochemical properties and eating quality of pork [20–23], lamb [7], beef [24,25], turkey [26,27], and chicken [28–30]. Sánchez del Pulgar et al. [23] found that sous vide pork cheeks cooked at 60 °C had lower water losses, more moisture content, more lightness (L*), and redness (a*) compared to those cooked at 80 °C. Roldán et al. [7] reported that sous vide lamb loins cooked at 60 °C had the highest lightness and redness compared to those cooked at 70 and 80 °C. Besides, increasing cooking temperature caused an increase in cooking loss and a decrease in moisture content. However, the interaction between time and temperature was only effective on microstructural properties. Bıyıklı et al. [26] found that sous vide turkey cutlet cooked at 65 °C had a lower cooking loss, thiobarbituric acid reactive substances (TBARS), and pH compared to those cooked at 70 °C and 75 °C. Besides, the cooking loss, fat content, and pH were increased by increasing cooking time from 20 min to 60 min.

According to the United Nations Food and Agriculture Organization (FAO), poultry meat is the second most widely eaten meat in the world after pork. It is estimated that global poultry consumption will reach 133 million tons by 2024. This is mainly due to the high consumer demands for a healthier diet with high protein content, good amino acid composition, low levels of fat and cholesterol, as well as lower selling price [31]. Because of these features, poultry meat, including chicken breast fillets, has received much attention recently. To the best of our knowledge, literature concerning the combinations of temperature and time on physicochemical and microbiological properties of sous vide chicken breast fillets is still limited. Therefore, the focus of this research was to evaluate the effect of these parameters on eating quality characteristics such as moisture content, cooking loss, lipid oxidation, pH, shear force, color, and microbial safety of sous vide chicken breast fillets.

2. Materials and Methods

2.1. Experimental Design

Fresh skinless and boneless raw chicken breasts were purchased from the local market (Reggio Emilia, Italy) supplied by the same producer within 24 h postmortem and transported to the Department of Life Sciences, University of Modena and Reggio Emilia, Italy using a thermocol box filled with ice and used immediately. Surface fat was trimmed off and samples were cut into pieces with 125 ± 5 g weight and 2.5 ± 0.2 cm thickness. Samples were randomly assigned into the 13 groups. Twelve groups were vacuum-sealed in the food-grade nylon-polyethylene plastic pouches (150×200 mm^2) using a vacuum sealer (La Grandispensa, Elegen, Reggio Emilia, Italy) with a pump flow rate of 30 L per minute to create 98% vacuum degree inside the pouches. Plastic pouches had wide thermal stability (-40 °C–+120 °C) with O_2 permeability of 9 cm^3/day m^2 (4 °C/80% relative humidity), and water vapor permeability of 1.2 g/day m^2 (Joelplas SL, Barcelona, Spain). As a control group, chicken breast fillets sealed in plastic pouches without a vacuum (0% vacuum degree) were boiled at 100 °C for 60 min. The samples were cooked in a sous vide cooker (Elegen, Reggio Emilia, Italy). Three independent replicate trials with two repeats based on different combinations of temperature (60, 70, and 80 °C) and time (60, 90, 120, and 150 min) were analyzed (Table 1). Overall, a total of 78 chicken breast fillets were analyzed (13 groups of samples × 3 independent replicate × 2 repeats). The sous vide chicken breast fillets were cooled in an ice bath for one hour and overnight in the fridge at 2–4 °C. Moisture content, cooking loss, pH, color, TBARS, and shear force were measured the day after the cooking process [7].

Table 1. Temperature, Time, and Vacuum Conditions Applied in This Study for Cooking Chicken Breast Fillets.

Group	Temperature (°C)	Time (min)	Vacuum Degree (%)
Control	100	60	0
1	60	60	98
2	60	90	98
3	60	120	98
4	60	150	98
5	70	60	98
6	70	90	98
7	70	120	98
8	70	150	98
9	80	60	98
10	80	90	98
11	80	120	98
12	80	150	98

2.2. Moisture Content and Cooking Loss

The moisture content and cooking loss were determined according to the AOAC International 950.46 method [32]. The moisture content of the chicken fillets (5 g) was calculated as the percentage of weight loss to a constant weight (M_d) after drying in an oven at 105 ± 2 °C and the initial weight (M_i) according to Equation (1):

$$\text{Moisture content (\%): } (M_i\text{-}M_d)/(M_i) \times 100 \tag{1}$$

The cooking loss was measured by the weight difference of meat samples (5 g) before (W_1) and after cooking (W_2) according to Equation (2):

$$\text{Cooking loss (\%): } (W_1\text{-}W_2)/(W_1) \times 100 \tag{2}$$

Moisture content and cooking loss measurements were performed in triplicate.

2.3. pH

The pH value was measured before and after cooking according to the AOAC 981.12 method [32] using a pH meter equipped with a Xerolite electrode (Crison Instrument, Allela, Spain). The pH was determined by blending a 10 g sample with 50 mL distilled water for 60 s in a homogenizer (IKA, Labortechnik, Staufen, Germany). The analysis was performed in triplicate.

2.4. Color

The color of meat samples before and after cooking was measured on the external surface of each fillet with a colorimeter (CR-400, Konica Minolta, Osaka, Japan) equipped with a standard illuminant D65 and 10° observer angle [33]. The results are reported as L* (lightness), a* (redness/greenness), and b* (yellowness/blueness). The instrument was calibrated with a white standard (L* = 99.36, a* = −0.12, b* = −0.06) before each measurement [34]. The average of six measurements at different positions was calculated.

2.5. Warner-Bratzler Shear Force (WBSF)

The WBSF was performed according to Honikel [35] with slight modification. Texture analyzer (Z1.0, Zwick/Roell, Ulm, Germany) with loading cell of 1000 N and crosshead speed 250 mm/min was used to perform shear force analysis on cooked chicken breast fillets ($3 \times 1.5 \times 1$ cm^3) using a Warner-Bratzler blade [22]. The data was obtained from TestXpert® II 161 (V3.31) software (Zwick/Roell, Ulm, Germany). The maximum peak force (kg) to shear the sample was reported as a shear force. The average of five measurements was recorded.

2.6. Thiobarbituric Acid Reactive Substances (TBARS)

TBARS measurement was carried out based on Siu and Draper [36]. A total of 2.5 g of minced meat sample and 12.5 mL distilled water were homogenized at 9500 rpm for 120 s using an ultra-turrax homogenizer (IKA, Labortechnik, Staufen, Germany). The homogenized sample mixed with 12.5 mL of 10% trichloroacetic acid (TCA) (CAS Number: 76-03-09, Sigma-Aldrich, Milan, Italy) and centrifuged for 20 min at 2000 rpm at 4 °C. The supernatant was filtered by a filter paper (Whatman No. 1). A total of 4 mL of the filtrate aliquots was mixed with 1 mL of 0.06 M 2-thiobarbituric acid (TBA) (CAS Number: 504-17-6, Sigma-Aldrich, Milan, Italy) and the solution was heated in a water bath at 80 °C for 90 min. A distilled water-TCA-TBA reagent was also prepared and presented as a blank. The absorbance at 532 nm was measured in duplicate by a spectrophotometer (Jasco Corporation, Tokyo, Japan). Results were expressed as mg of malondialdehyde (MDA) equivalents/kg sample. The average of three measurements was recorded.

2.7. Microbiological Analyses

Microbiological analysis was performed during the storage at 4 °C for 21 days [20]. For each day (0, 5, 10, 15, and 21 days) of analysis, 10 g sliced chicken breast fillets were collected aseptically, and 90 mL sterile saline solution (0.9% NaCl) was added and homogenized for 2 min in a stomacher (Lab blenders Stomacher 400, Instrument Lab Control, Reggio Emilia, Italy). Appropriate dilutions were made with sterile saline solution and 1 mL was plated onto the culture media. Total mesophilic aerobic bacteria counts were determined after aerobic incubation at 30 °C for 48 h using Plate Count Agar (Biolife, Milan, Italy) in accordance with ISO 4833-1: 2013 [37]. Total *Psychrotrophic* counts were determined after aerobic incubation at 4 °C for 10 days using Plate Count Agar (Biolife, Milan, Italy) in accordance with ISO 17410: 2019 [38]. *Enterobacteriaceae* were counted on Violet Red Bile Glucose Agar (Biolife, Milan, Italy) after aerobic incubation at 37 °C for 24 h in accordance with ISO 21528-1: 2017 [39]. The average of three measurements was recorded.

2.8. Statistical Analysis

The experiment was performed in three independent replicates and the number of repeats varied from one analysis to another and was reported in each subsection. The data were analyzed through two-way analysis of variance (ANOVA). The differences between means were compared by Tukey's post-hoc test ($p < 0.05$). A principal component analysis (PCA) was then performed to establish the variations and relationships among physicochemical properties of sous vide chicken breast fillets cooked at twelve different combinations of temperature and time. All the analysis was performed in SPSS software (IBM SPSS 20, New York, NY, USA).

3. Results and Discussion

3.1. Moisture Content, Cooking Loss, and pH

Moisture content is one of the important physicochemical characteristics in meat which plays a basic role in the palatability of meat. Moisture content of raw chicken breast fillet 24 h post-mortem is presented in Table 2. Raw meat showed a moisture content of 72.4%. These results were consistent with those obtained by Sanchez Brambila et al. [40].

Table 2. Moisture Content, Thiobarbituric Acid Reactive Substances (TBARS), Color Parameters (L*: Lightness, a*: Redness/Greenness, and b*: Yellowness/Blueness), and pH of the Raw Chicken Breast Fillet 24 h Post-Mortem.

Parameters	Results
Moisture (%)	72.4 ± 1.02
TBARS (mg/Kg)	0.08 ± 0.011
Weight (g)	125 ± 5
L*	58.4 ± 1.7
a*	0.8 ± 0.1
b*	9.1 ± 0.9
pH	5.8 ± 0.03

Values are presented as means ± standard deviations (n = 3).

The moisture content of sous vide chicken breast fillets cooked at different temperature and time combinations ranged from 68.25% to 71.89% (Table 3). Moisture content was affected by cooking temperature, cooking time, and interaction between temperature and time ($p < 0.05$). As expected, there was a reduction in moisture content by increasing temperature from 60 °C to 80 °C. Control treatment cooked at 100 °C for 60 min showed the lowest moisture content with 68.25% ($p < 0.05$). Increasing cooking time from 60 min to 150 at higher temperatures (70 and 80 °C) caused a reduction in moisture content ($p < 0.05$). During cooking, the fluid is released as water and other ingredients such as fat and soluble proteins. Releasing the sarcoplasmic fluid from the muscle fibers results in lower water content at higher temperatures [6,8,41]. Murphy et al. [42] reported that the denaturation of myosin and actin at higher temperatures caused structural changes and changes in porosity of the chicken breast patties which can directly affect the moisture content. This result is in accordance with those obtained for chicken and beef [28,43].

Cooking loss is an important factor to consider because it is directly related to juiciness which could influence the consumer's perception of the final product [25]. The cooking loss is defined as total liquid and soluble matter lost from the meat during cooking and it is influenced by different factors such as the quality of the raw meat, genetics of the animals, and cooking conditions. This loss relies on the mass transfer process during heat treatment [44]. In this study, cooking loss ranged from 10.23% to 28.08%. Control samples cooked at 100 °C showed the highest cooking loss ($p < 0.05$). Cooking loss was affected by both cooking temperature and cooking time and it was increased by increasing cooking temperature and time ($p < 0.05$). Increasing temperature causes denaturation of myofibrillar proteins and the actomyosin complex, resulting in shrinkage of the muscle fiber. Thus, less water can be captured within the protein structures kept by capillary forces [41,45]. Our result is in agreement with previous studies on sous vide cooking on chicken [28], beef [25,43], pork [23], and lamb [7]. According to Purslow et al. [46], the cooking loss is mainly determined by the shrinkage of myofibrillar proteins (40–60 °C), shrinkage of collagen (60–70 °C), and denaturation of actin (70–80 °C). Denaturation of proteins occurs with increasing temperature which causes structural changes and the release of fluid from muscle fiber leading to a decrease in the water holding capacity and higher cooking loss [47].

The pH of raw chicken breast fillet 24 h post-mortem was 5.8 (Table 2). The pH value of sous vide chicken breast fillets cooked at different temperature and time combinations slightly increased and ranged from 6.07 to 6.3. The pH was affected by temperature and the interaction between temperature and time. Increasing temperature from 60 °C to 80 °C caused an increase in pH value. Similarly, Bıyıklı et al. [26] reported that increasing cooking temperature from 65 °C to 75 °C and cooking time from 20 min to 60 min caused an increase in the pH of sous vide turkey cutlet. Becker et al. [20] reported that increasing temperature caused an increase in pH mainly due to the protein denaturation and the change in protein charge.

Table 3. Moisture Content, Cooking Loss, Shear Force, TBARS, Color Parameters (L*: Lightness, a*: Redness/Greenness, and b*: Yellowness/Blueness), and pH of Sous vide Chicken Breast Fillets Cooked at Different Temperature and Time Combinations.

Temp (°C)	60				70				80				100				
Time (min)	60	90	120	150	60	90	120	150	60	90	120	150	60	SEM	Temp	Time	Temp × Time
Moisture (%)	71.41 f,g	71.30 e,f,g	71.72 f,g	71.89 g	71.71 f,g	70.86 e,f,g	69.97 c,d,e	70.46 d,e,f	70.43 d,e,f	69.76 b,c,d	69.47 a,b,c	69.02 a,b	68.25 a	0.21	*	*	*
Cooking loss (%)	10.23 a	11.02 a	12.42 a,b	12.47 a,b	14.01 a,b,c	16.88 b,c,d	18.38 c,d,e	18.69 c,d,e	17.86 c,d,e	21.77 d,e,f	22.77 e,f	24.23 f,g	28.08 g	3.11	*	*	N.S
Shear force (kg)	0.75 a	0.83 a,b	0.76 a	0.62 a	0.66 a	0.73 a	0.62 a	0.63 a	0.88 a,b	0.97 b	0.79 a	0.88 a,b	1.37 c	0.02	*	N.S	N.S
TBARS (mg/kg)	0.29 a	0.77 a,b	0.92 a,b	0.94 a,b	1.50 b	1.47 b	1.63 b	1.71 b	2.31 c	2.42 c	2.54 c	2.60 c	2.91 d	0.12	*	N.S	N.S
L*	80.94 a	81.71 a	80.11 a	79.63 a	80.82 a	81.72 a	82.27 a	82.43 a	81.39 a	81.19 a	80.85 a	81.15 a	80.75 a	2.65	N.S	N.S	N.S
a*	1.95 a	1.95 a	1.81 a	1.71 a	1.73 a	1.71 a	1.74 a	1.50 ab	1.44 b	1.39 b	1.33 b	1.29 b	1.29 b	0.05	*	*	N.S
b*	14.71 a	14.65 a	14.95 a	15.15 a	14.91 a	14.83 a	14.95 a	15.40 a	15.64 a	15.55 a	15.60 a	15.36 a	14.82 a	0.41	N.S	N.S	N.S
pH	6.17 a,b,c	6.14 a	6.07 a	6.08 a	6.11 a	6.14 a,b	6.07 a	6.13 a	6.15 a,b	6.25 b,c	6.30 d	6.27 c,d	6.17 a,b,c	0.93	*	N.S	*

Means value with different superscripts letters (a–g) in the same row indicate a significant difference ($p < 0.05$). Values are presented as means (n = 3). SEM: Standard error of the mean. N.S: not significant; *: indicate a significant difference ($p < 0.05$).

3.2. Color

The L*, a*, b* values of raw chicken breast fillet 24 h post-mortem were 58.4, 0.8, and 9.1, respectively (Table 2). Color parameters are usually used as an indicator of the doneness of cooked meat which directly impacts the appearance and attractiveness of the product [8,19]. The color is mainly affected inside the muscle by myoglobin content, oxidative state of myoglobin, muscle fiber orientation, space between the muscle fibers, packaging conditions, Millard reactions, and pH [48,49]. Color parameters of chicken breast fillets cooked at different temperature-time combinations are presented in Table 3. The L* value (lightness) was not affected by cooking temperature, time, and their interaction ($p > 0.05$). A similar result was reported by Park et al. [29] in sous vide chicken breast cooked at different combinations of temperature (60 and 70 °C) and time (60, 120, and 180 min). In contrast to this result, Sánchez del Pulgar et al. [23] reported that sous vide pork cheeks cooked at 60 °C had a higher L* compared to those cooked at 80 °C. The authors concluded that samples cooked at lower temperatures preserved more water during cooking which might be released to the surface during the slicing process before color measurement. On the other hand, the chicken breast color can be classified into pale (L* > 53), dark (L* < 46), and normal (46 < L* < 53) based on the L* value [33]. In our study, sous vide chicken breast fillets in all combinations of temperature and time showed a pale appearance.

The a* value (redness/greenness) ranged from 1.29 to 1.95. The low a*value in poultry meat is mainly due to the presence of white muscle fibers with low myoglobin content [50]. In this study, a* value was affected by cooking temperature and cooking time ($p < 0.05$). Control samples cooked at 100 °C revealed a lower a* value than those cooked at 60 °C and 70 °C ($p < 0.05$). A similar result was reported by Naveena et al. [28] and García-Segovia et al. [51]. The pink color in poultry meat is evidence of a poorly cooked product. Holownia et al. [52] defined a subjective pink threshold at a* = 3.8 in chicken breast fillets. In our study, a* values were under this threshold level at all different temperatures and time combinations. In a general context, a* value is conversely linked to the degree of myoglobin thermal denaturation in cooked meat [23]. Myoglobin thermal denaturation happens quickly with increasing temperature which can directly interact with by-products of lipid oxidation leading to a reduction in a* value [50].

The b* value (yellowness/blueness) ranged from 14.65 to 15.64. The b* value was not affected by cooking temperature, time, and their interaction ($p > 0.05$). In contrast to our result, Park et al. [29] reported that b* value was affected by cooking temperature in sous vide chicken samples cooked at different combinations of temperature (60 and 70 °C) and time (60, 120, and 180 min).

3.3. Lipid Oxidation

A thiobarbituric acid reactive substances (TBARS) test was used to determine secondary lipid oxidation products (e.g., aldehydes) as an indicator of oxidative deterioration [45], off-flavors, and rancidity [13]. Raw meat showed a TBARS value of 0.08 mg/Kg (Table 2). TBARS values of chicken breast fillets cooked at different temperature-time combinations ranged from 0.29 to 2.60 mg/Kg (Table 3). This parameter was only affected by the cooking temperature ($p < 0.05$). Chicken breast fillets cooked at 60 °C at every time point showed TBARS values below one. Akoğlu et al. [27] reported that oxidative rancidity cannot be detected by a sensory panel under a threshold level of one (mg/kg). TBARS value was increased by increasing temperature up to 80 °C ($p < 0.05$). Control treatment cooked at 100 °C showed a similar value to sous vide chicken cooked at 80 °C. In contrast to our result, Sánchez del Pulgar et al. [23] reported that time (5 and 12 h) and temperature (60 and 80 °C) and their interaction were affected by the TBARS of sous vide pork cheeks.

3.4. Warner-Bratzler Shear Force (WBSF)

The WBSF is commonly used for evaluating tenderness. It is an important eating quality character due to the impact on texture and consumer acceptance [53]. The WBSF values of chicken breast fillets cooked at different temperature-time combinations are presented in Table 3. The WBSF was only affected by cooking temperature and it was increased by increasing temperature ($p < 0.05$). This parameter ranged between 0.62 and 1.37 kg. The lowest shear force was found in sous vide chicken treatment cooked at 60 °C and 70 °C. This result might be associated with higher moisture content and lower cooking loss of samples cooked at lower temperatures [29,41]. Cooking at low temperatures reduces the protein–protein association and gelation and increases water retention [6,22]. On the other hand, the control sample cooked at 100 °C showed the highest WBSF, which could be attributed to higher cooking loss, lower moisture content, and formation of gelatin due to the collagen denaturation and myofibrillar hardening [54]. Barbanti and Pasquini [55] reported that the enhancement of tenderness is mainly caused by the solubilization of connective tissues, while denaturation of myofibrillar proteins led to toughening. Overall, from previous studies it was suggested that solubilization of the connective tissue [51,56], aggregation of sarcoplasmic proteins [6,9], and water retention inside the muscles [25,53,57] are three major factors contributing to the increase in tenderness.

3.5. Principal Component Analysis (PCA) Analysis

Figure 1 reports the loading plot of the PCA model computed on the physicochemical variables considered in this study. The analysis showed that about 59.84% of the total variation is explained by the first principal component (PC1) and 14.58% by the second principal component (PC2). These two PCs account for about 74.42% of the total data variance. PC1 correlated positively with shear force, cooking loss, TBARS, and pH, while it had a negative correlation with moisture and a* color value. This tendency confirms the opposite relationship between moisture content and shear force, cooking loss, TBARS, and pH. PC2 was only correlated positively with L* color value. A similar result was reported by Fabre et al. [53].

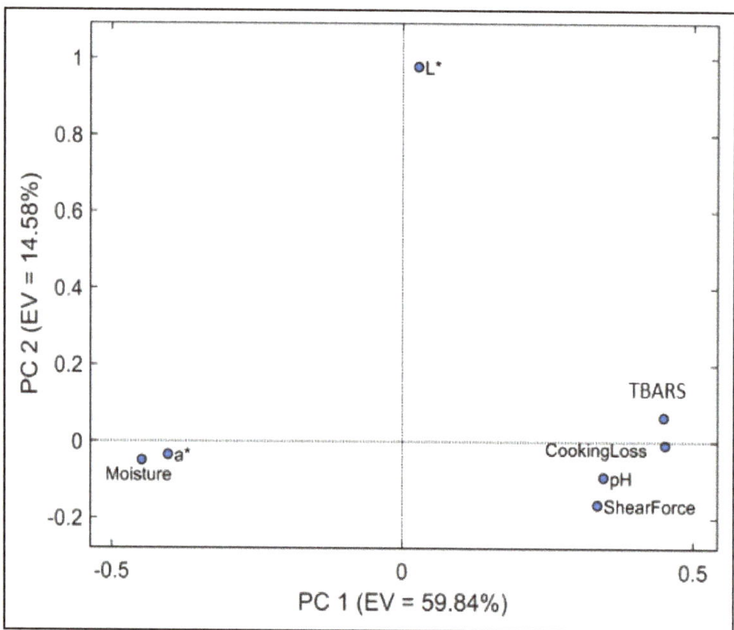

Figure 1. PC1 vs. PC2: Loading plot of physicochemical variables studied in sous vide chicken fillet breasts cooked at different combinations of temperature (60, 70, 80, and 100 °C) and time (60, 90, 120, and 150 min).

Figure 2 reports the score plot. The colors on the plot refer to the different temperatures (60, 70, 80, and 100 °C) while the numbers indicate the cooking time (60, 90, 120, and 150 min). Chicken breast samples cooked at 60 °C and 70 °C were at negative values of PC1. Conversely, the chicken breast samples cooked at 80 and 100 °C were at positive values of PC1. By comparing the score plot with the corresponding loading plot, it is possible to interpret the relationships between samples and variables [58]. The score plot in conjugation with the loading plot demonstrated that increasing cooking temperature caused an increase in cooking loss, lipid oxidation, TBARS, and pH. Comparing the results in Table 3 with the PCA model allowed us to conclude that the cooking temperature played a major role in measured variables while the effect of cooking times seemed to be negligible.

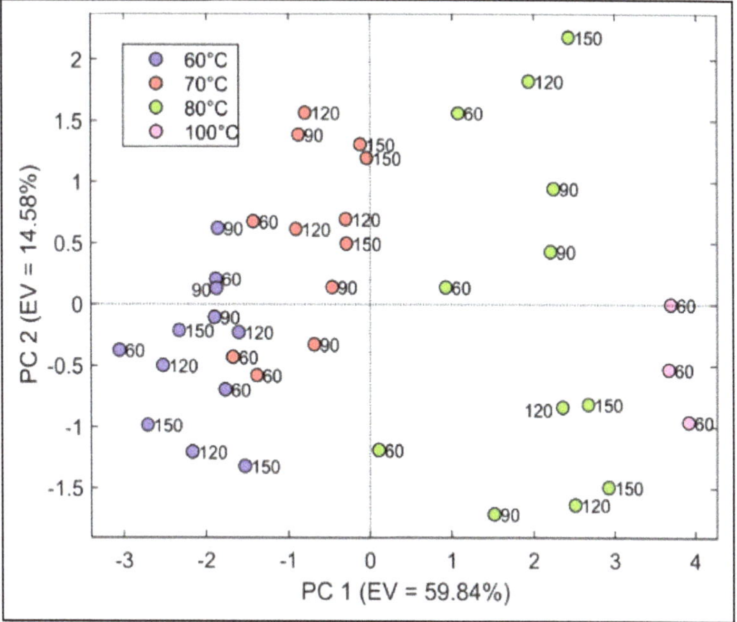

Figure 2. PC1 vs. PC2: Score plot of physicochemical variables studied in sous vide chicken fillet breasts cooked at different combinations of temperature (60, 70, 80, and 100 °C) and time (60, 90, 120, and 150 min). The symbol's color corresponds to cooking temperature (60, 70, 80, and 100 °C) and the number beside the symbol corresponds to cooking time (60, 90, 120, and 150 min).

3.6. Microbiology

The microbial load before and after sous vide cooking was analyzed to verify different temperature and time combinations applied in this study guarantee microbiological safety. The microbial counts of mesophilic aerobic bacteria, *Psychrotrophic* bacteria, and *Enterobacteriaceae* at raw chicken breast fillets are presented in Table 4. The selection of these three groups was based on their significant importance in food quality and safety. Raw chicken breast fillets showed 2.8 and 2.3 log CFU/g counts for total mesophilic aerobic bacteria and *Enterobacteriaceae*, respectively. The *Psychrotrophic* bacteria count was lower than 1 log CFU/g. These results are below reference values recommended by food quality standards for fresh poultry meat (EC No. 2073/2005). The counts of all microbial groups were not detectable at 4 °C for 21 days in sous vide chicken breast fillets confirming that even the lowest temperature and time combinations (60 °C–60 min) were enough to pasteurize meat. This result might be due to the growth inhibition of microorganisms under anaerobic conditions caused by vacuum packaging together with heat treatment and storage at a low temperature (4 °C) [17,27]. In accordance with these results, Can and Harun [31] reported that total mesophilic aerobic bacteria, *Psychrotrophic* bacteria, and *Enterobacteriaceae* counts were for sous vide chicken meatballs cooked at 90 °C for 20 min. In contrast to our results, Akoğlu et al. [27] found that total mesophilic aerobic bacterial counts exceeded 5 log CFU/g for sous vide turkey cutlet cooked at 45 °C for 60 min and stored at 4 and 12 °C, respectively. The presence of total mesophilic aerobic bacteria might be due to the low temperature (45 °C) applied in this study which was not enough to inhibit the growth of microorganisms.

Table 4. Microbiological Counts of the Raw Chicken Breast Fillet 24 h Post-Mortem (Day 0) and Sous Vide Chicken Breast Fillets (Cooked at All Different Combinations of Temperature and Time) during 21 Days of Storage at 4 °C.

Table	Treatment	Total Mesophilic Aerobic Log (CFU/g)	*Enterobacteriaceae* Log (CFU/g)	*Psychrotrophic* Aerobic Log (CFU/g)
0	Raw meat	2.8 ± 0.6	2.3 ± 0.4	<1
0	Sous vide	n.d	n.d	n.d
5	Sous vide	n.d	n.d	n.d
10	Sous vide	n.d	n.d	n.d
15	Sous vide	n.d	n.d	n.d
21	Sous vide	n.d	n.d	n.d

Values are presented as means ± standard deviations (n = 3). n.d: not detected.

4. Conclusions

Sous vide cooking is gaining more and more attention from catering, restaurants, ready-to-eat industries, and homes recently mainly due to the improvement in eating quality characteristics, extended shelf lives, and reduced risk of post-cooking contamination compared to other cooking methods. Temperature and time are two critical parameters in sous vide cooking that directly affect eating quality and safety. The finding of this study showed that cooking temperature played a major role in the moisture content, cooking loss, shear force, TBARS, a*, and pH value. Increasing cooking time from 60 min to 150 min caused a reduction in moisture content and a* value while cooking loss increased. Chicken breast fillets cooked at 60 °C revealed less cooking loss, lipid oxidation, shear force, and a more intense red color compared to those cooked at 70 and 80 °C. Total mesophilic aerobic bacteria, *Psychrotrophic* bacteria, and *Enterobacteriaceae* were not detected during storage at 4 °C for 21 days, ensuring microbiological safety for consumers. Overall, the optimum condition obtained in this study for chicken breast fillets was cooking at 60 °C for 60 min. Future studies need to be carried out to assess the sensory quality parameters and palatability of sous vide chicken breast fillets during the storage time to determine shelf life and consumer acceptability. Besides, it is necessary to perform inoculum studies targeting specific pathogenic and spoilage microorganisms to assess the effectiveness of selected temperature and time combinations on microbiological quality.

Author Contributions: Conceptualization, H.H., and A.M.B.; methodology, H.H., and A.M.B.; software, H.H., A.M.B., and F.M.; validation, H.H., and A.M.B.; formal analysis, H.H., and A.M.B.; investigation, H.H., and A.M.B.; resources, D.P.L.F., and A.P.; data curation, H.H., and A.M.B.; writing—original draft preparation, H.H., and A.M.B.; writing—review and editing, H.H; and A.M.B.; visualization, H.H., and A.M.B.; supervision, D.P.L.F., G.M., and A.P.; project administration, H.H.; funding acquisition, A.P. All authors have read and agreed to the published version of the manuscript.

Funding: This research was funded "FAR" INTERDISCIPLINARI-LS 2018 PROT.161714 Titolo: "DEL TRATTAMENTO TERMICO SOUS-VIDE: QUALITÀ REALE E PERCEPITA.

Institutional Review Board Statement: Not applicable.

Informed Consent Statement: Not applicable.

Data Availability Statement: The data presented in this study are available on request from the corresponding author.

Conflicts of Interest: The authors declare no conflict of interest.

References

1. Biesalski, H.-K. Meat as a component of a healthy diet—Are there any risks or benefits if meat is avoided in the diet? *Meat Sci.* **2005**, *70*, 509–524. [CrossRef] [PubMed]
2. Czarnowska-Kujawska, M.; Draszanowska, A.; Gujska, E. Effect of Different Cooking Methods on Folate Content in Chicken Liver. *Foods* **2020**, *9*, 1431. [CrossRef] [PubMed]
3. Davey, C.L.; Gilbert, K.V. Temperature-dependent cooking toughness in beef. *J. Sci. Food Agric.* **1974**, *25*, 931–938. [CrossRef]
4. Gök, V.; Uzun, T.; Tomar, O.; Çağlar, M.Y.; Çağlar, A. The effect of cooking methods on some quality characteristics of gluteus medius. *Food Sci. Technol.* **2019**, *39*, 999–1004. [CrossRef]
5. Nithyalakshmi, V.; Preetha, R. Effect of cooking conditions on physico-chemical and textural properties of emu (Dromaius novaehollandiae) meat. *Int. Food Res. J.* **2015**, *22*, 1924–1930.
6. Tornberg, E. Effects of heat on meat proteins—Implications on structure and quality of meat products. *Meat Sci.* **2005**, *70*, 493–508. [CrossRef]
7. Roldán, M.; Antequera, T.; Martín, A.; Mayoral, A.I.; Ruiz, J. Effect of different temperature–time combinations on physicochemical, microbiological, textural and structural features of sous-vide cooked lamb loins. *Meat Sci.* **2013**, *93*, 572–578. [CrossRef] [PubMed]
8. Dominguez-Hernandez, E.; Salaseviciene, A.; Ertbjerg, P. Low-temperature long-time cooking of meat: Eating quality and underlying mechanisms. *Meat Sci.* **2018**, *143*, 104–113. [CrossRef]
9. Baldwin, D.E. Sous vide cooking: A review. *Int. J. Gastron. Food Sci.* **2012**, *1*, 15–30. [CrossRef]
10. Zavadlav, S.; Blažić, M.; Van De Velde, F.; Vignatti, C.; Fenoglio, C.; Piagentini, A.M.; Pirovani, M.E.; Perotti, C.M.; Kovačević, D.B.; Putnik, P. *Sous-Vide* as a Technique for Preparing Healthy and High-Quality Vegetable and Seafood Products. *Foods* **2020**, *9*, 1537. [CrossRef]
11. Albistur, A.R.-; Gámbaro, A. Consumer perception of a non-traditional market on sous-vide dishes. *Int. J. Gastron. Food Sci.* **2018**, *11*, 20–24. [CrossRef]
12. Christensen, L.; Gunvig, A.; Tørngren, M.A.; Aaslyng, M.D.; Knøchel, S.; Christensen, M. Sensory characteristics of meat cooked for prolonged times at low temperature. *Meat Sci.* **2012**, *90*, 485–489. [CrossRef]
13. Cho, D.K.; Lee, B.; Oh, H.; Lee, J.S.; Kim, Y.S.; Choi, Y.M. Effect of Searing Process on Quality Characteristics and Storage Stability of Sous-Vide Cooked Pork Patties. *Foods* **2020**, *9*, 1011. [CrossRef]
14. Rondanelli, M.; Daglia, M.; Meneghini, S.; Di Lorenzo, A.; Peroni, G.; Faliva, M.A.; Perna, S. Nutritional advantages of sous-vide cooking compared to boiling on cereals and legumes: Determination of ashes and metals content in ready-to-eat products. *Food Sci. Nutr.* **2017**, *5*, 827–833. [CrossRef]
15. Schellekens, M. New research issues in sous-vide cooking. *Trends Food Sci. Technol.* **1996**, *7*, 256–262. [CrossRef]
16. Jeong, S.-H.; Kim, E.-C.; Lee, D.-U. The Impact of a Consecutive Process of Pulsed Electric Field, Sous-Vide Cooking, and Reheating on the Properties of Beef Semitendinosus Muscle. *Foods* **2020**, *9*, 1674. [CrossRef]
17. Hong, G.-E.; Kim, J.-H.; Ahn, S.-J.; Lee, C.-H. Changes in Meat Quality Characteristics of the Sous-vide Cooked Chicken Breast during Refrigerated Storage. *Food Sci. Anim. Resour.* **2015**, *35*, 757–764. [CrossRef]
18. Rizzo, V.; Amoroso, L.; Licciardello, F.; Mazzaglia, A.; Muratore, G.; Restuccia, C.; Lombardo, S.; Pandino, G.; Strano, M.G.; Mauromicale, G. The effect of sous vide packaging with rosemary essential oil on storage quality of fresh-cut potato. *LWT* **2018**, *94*, 111–118. [CrossRef]
19. Głuchowski, A.; Czarniecka-Skubina, E.; Wasiak-Zys, G.; Nowak, D.; Skubina, C.-; Zys, W. Effect of Various Cooking Methods on Technological and Sensory Quality of Atlantic Salmon (Salmo salar). *Foods* **2019**, *8*, 323. [CrossRef]
20. Becker, A.; Boulaaba, A.; Pingen, S.; Krischek, C.; Klein, G. Low temperature cooking of pork meat—Physicochemical and sensory aspects. *Meat Sci.* **2016**, *118*, 82–88. [CrossRef]
21. Christensen, L.; Ertbjerg, P.; Aaslyng, M.D.; Christensen, M. Effect of prolonged heat treatment from 48 °C to 63°C on toughness, cooking loss and color of pork. *Meat Sci.* **2011**, *88*, 280–285. [CrossRef]
22. Jeong, K.; Hyeonbin, O.; Shin, S.Y.; Kim, Y.-S. Effects of sous-vide method at different temperatures, times and vacuum degrees on the quality, structural, and microbiological properties of pork ham. *Meat Sci.* **2018**, *143*, 1–7. [CrossRef] [PubMed]
23. del Pulgar, J.S.; Gázquez, A.; Ruiz-Carrascal, J. Physico-chemical, textural and structural characteristics of sous-vide cooked pork cheeks as affected by vacuum, cooking temperature, and cooking time. *Meat Sci.* **2012**, *90*, 828–835. [CrossRef] [PubMed]
24. Botinestean, C.; Keenan, D.F.; Kerry, J.P.; Hamill, R.M. The effect of thermal treatments including sous-vide, blast freezing and their combinations on beef tenderness of M. semitendinosus steaks targeted at elderly consumers. *LWT* **2016**, *74*, 154–159. [CrossRef]
25. Gómez, I.; Ibañez, F.C.; Beriain, M.J. Physicochemical and sensory properties of sous vide meat and meat analog products marinated and cooked at different temperature-time combinations. *Int. J. Food Prop.* **2019**, *22*, 1693–1708. [CrossRef]
26. Bıyıklı, M.; Akoğlu, A.; Kurhan, Ş.; Akoğlu, İ.T. Effect of different Sous Vide cooking temperature-time combinations on the physicochemical, microbiological, and sensory properties of turkey cutlet. *Int. J. Gastron. Food Sci.* **2020**, *20*, 100204. [CrossRef]

27. Akoğlu, I.; Bıyıklı, M.; Akoglu, A.; Kurhan, Ş.; Biyikli, M. Determination of the Quality and Shelf Life of Sous Vide Cooked Turkey Cutlet Stored at 4 and 12 °C. *Braz. J. Poult. Sci.* **2018**, *20*, 1–8. [CrossRef]
28. Naveena, B.M.; Khansole, P.S.; Kumar, M.S.; Krishnaiah, N.; Kulkarni, V.V.; Deepak, S.J. Effect of sous vide processing on physicochemical, ultrastructural, microbial and sensory changes in vacuum packaged chicken sausages. *Food Sci. Technol. Int.* **2016**, *23*, 75–85. [CrossRef]
29. Park, C.; Lee, B.; Oh, E.; Kim, Y.; Choi, Y. Combined effects of sous-vide cooking conditions on meat and sensory quality characteristics of chicken breast meat. *Poult. Sci.* **2020**, *99*, 3286–3291. [CrossRef]
30. Karpińska-Tymoszczyk, M.; Draszanowska, A.; Danowska-Oziewicz, M.; Kurp, L. The effect of low-temperature thermal processing on the quality of chicken breast fillets. *Food Sci. Technol. Int.* **2020**, *26*, 563–573. [CrossRef]
31. Can, Ö.; Harun, F. Shelf Life of Chicken Meat Balls Submitted to Sous Vide Treatment. *Braz. J. Poult. Sci.* **2015**, *17*, 137–144. [CrossRef]
32. AOAC. *AOAC Official Methods of Analysis*, 18th ed.; AOAC International: Rockville, MD, USA, 2005; ISBN 0935584870.
33. Da Silva-Buzanello, R.A.; Schuch, A.F.; Gasparin, A.W.; Torquato, A.S.; Scremin, F.R.; Canan, C.; Soares, A.L. Quality parameters of chicken breast meat affected by carcass scalding conditions. *Asian-Australas. J. Anim. Sci.* **2019**, *32*, 1186–1194. [CrossRef] [PubMed]
34. Haghighi, H.; Gullo, M.; La China, S.; Pfeifer, F.; Siesler, H.W.; Licciardello, F.; Pulvirenti, A. Characterization of bio-nanocomposite films based on gelatin/polyvinyl alcohol blend reinforced with bacterial cellulose nanowhiskers for food packaging applications. *Food Hydrocoll.* **2021**, *113*, 106454. [CrossRef]
35. Honikel, K.O. Reference methods for the assessment of physical characteristics of meat. *Meat Sci.* **1998**, *49*, 447–457. [CrossRef]
36. Siu, G.M.; Draper, H.H. A Survey of the Malonaldehyde Content of Retail Meats and Fish. *J. Food Sci.* **1978**, *43*, 1147–1149. [CrossRef]
37. ISO. *ISO 4833-1 Microbiology of the Food Chain—Horizontal Method for the Enumeration of Microorganisms, Part 1, Count at 30 °C by the Pour Plate Technique*; ISO: Geneva, Switzerland, 2013; p. 9.
38. ISO. *ISO 17410 Microbiology of the Food Chain—Horizontal Method for the Enumeration of Psychrotrophic Microorganisms*; ISO: Geneva, Switzerland, 2019; p. 10.
39. ISO. *ISO 21528-1 Microbiology of the Food Chain—Horizontal Method for the Detection and Enumeration of Enterobacteriaceae—Part 1: Detection of Enterobacteriaceae*; ISO: Geneva, Switzerland, 2017; p. 17.
40. Brambila, G.S.; Chatterjee, D.; Bowker, B.; Zhuang, H. Descriptive texture analyses of cooked patties made of chicken breast with the woody breast condition. *Poult. Sci.* **2017**, *96*, 3489–3494. [CrossRef] [PubMed]
41. Murphy, R.Y.; Marks, B.P. Effect of meat temperature on proteins, texture, and cook loss for ground chicken breast patties. *Poult. Sci.* **2000**, *79*, 99–104. [CrossRef]
42. Murphy, R.Y.; Johnson, E.R.; Duncan, L.K.; Clausen, E.C.; Davis, M.D.; March, J.A. Heat Transfer Properties, Moisture Loss, Product Yield, and Soluble Proteins in Chicken Breast Patties During Air Convection Cooking. *Poult. Sci.* **2001**, *80*, 508–514. [CrossRef]
43. Ismail, I.; Hwang, Y.-H.; Bakhsh, A.; Joo, S.-T. The alternative approach of low temperature-long time cooking on bovine semitendinosus meat quality. *Asian Australas. J. Anim. Sci.* **2019**, *32*, 282–289. [CrossRef] [PubMed]
44. Aaslyng, M.D.; Bejerholm, C.; Ertbjerg, P.; Bertram, H.C.; Andersen, H.J. Cooking loss and juiciness of pork in relation to raw meat quality and cooking procedure. *Food Qual. Prefer.* **2003**, *14*, 277–288. [CrossRef]
45. Roldán, M.; Antequera, T.; Hernández, A.; Ruiz, J.; Ruiz-Carrascal, J. Physicochemical and microbiological changes during the refrigerated storage of lamb loins sous-vide cooked at different combinations of time and temperature. *Food Sci. Technol. Int.* **2014**, *21*, 512–522. [CrossRef]
46. Purslow, P.; Oiseth, S.; Hughes, J.; Warner, R. The structural basis of cooking loss in beef: Variations with temperature and ageing. *Food Res. Int.* **2016**, *89*, 739–748. [CrossRef] [PubMed]
47. Li, C.; Wang, D.; Xu, W.; Gao, F.; Zhou, G. Effect of final cooked temperature on tenderness, protein solubility and microstructure of duck breast muscle. *LWT* **2013**, *51*, 266–274. [CrossRef]
48. Wideman, N.; O'Bryan, C.; Crandall, P. Factors affecting poultry meat colour and consumer preferences—A review. *World's Poult. Sci. J.* **2016**, *72*, 353–366. [CrossRef]
49. Cobos, Á.; Diaz, O. Chemical Composition of Meat and Meat Products. In *Handbook of Food Chemistry*; Cheung, P.C.K., Ed.; Springer: Berlin/Heidelberg, Germany, 2014; pp. 1–32.
50. Khan, A.; Allen, K.; Wang, X. Effect of Type I and Type II Antioxidants on Oxidative Stability, Microbial Growth, pH, and Color in Raw Poultry Meat. *Food Nutr. Sci.* **2015**, *6*, 1541–1551. [CrossRef]
51. García-Segovia, P.; Andrés-Bello, A.; Martínez-Monzó, J. Effect of cooking method on mechanical properties, color and structure of beef muscle (M. pectoralis). *J. Food Eng.* **2007**, *80*, 813–821. [CrossRef]
52. Holownia, K.; Chinnan, M.S.; Reynolds, A.E.; Koehler, P.E. Evaluation of induced color changes in chicken breast meat during simulation of pink color defect. *Poult. Sci.* **2003**, *82*, 1049–1059. [CrossRef] [PubMed]
53. Fabre, R.; Dalzotto, G.; Perlo, F.; Bonato, P.; Teira, G.; Tisocco, O. Cooking method effect on Warner-Bratzler shear force of different beef muscles. *Meat Sci.* **2018**, *138*, 10–14. [CrossRef]
54. Turner, B.E.; Larick, D.K. Palatability of Sous Vide Processed Chicken Breast. *Poult. Sci.* **1996**, *75*, 1056–1063. [CrossRef]

55. Barbanti, D.; Pasquini, M. Influence of cooking conditions on cooking loss and tenderness of raw and marinated chicken breast meat. *LWT* **2005**, *38*, 895–901. [CrossRef]
56. Warner, R.; Ha, M.; Sikes, A.; Vaskoska, R. Cooking and Novel Postmortem Treatments to Improve Meat Texture. In *New Aspects of Meat Quality*; Elsevier BV: Amsterdam, The Netherlands, 2017; pp. 387–423.
57. Laakkonen, E.; Wellington, G.H.; Sherbon, J.N. Low-Temperature, Long-Time Heating of Bovine Muscle 1. Changes in Tenderness, Water-Binding Capacity, pH and Amount of Water-Soluble Components. *J. Food Sci.* **1970**, *35*, 175–177. [CrossRef]
58. Bigi, F.; Haghighi, H.; De Leo, R.; Ulrici, A.; Pulvirenti, A. Multivariate exploratory data analysis by PCA of the combined effect of film-forming composition, drying conditions, and UV-C irradiation on the functional properties of films based on chitosan and pectin. *LWT* **2021**, *137*, 110432. [CrossRef]

applied sciences

MDPI

Article

Improvement of Paper Resistance against Moisture and Oil by Coating with Poly(-3-hydroxybutyrate-co-3-hydroxyvalerate) (PHBV) and Polycaprolactone (PCL)

Emanuela Lo Faro [1,†], **Camilla Menozzi** [1,†], **Fabio Licciardello** [1,2,*] **and Patrizia Fava** [1,2]

[1] Department of Life Sciences, University of Modena and Reggio Emilia, 42122 Reggio Emilia, Italy; emanuela.lofaro@unimore.it (E.L.F.); 214138@studenti.unimore.it (C.M.); patrizia.fava@unimore.it (P.F.)

[2] Interdepartmental Research Centre for the Improvement of Agro-Food Biological Resources (BIOGEST-SITEIA), University of Modena and Reggio Emilia, 42122 Reggio Emilia, Italy

* Correspondence: fabio.licciardello@unimore.it

† Shared first authorship.

Abstract: Surface hydrophobicity and grease resistance of paper may be achieved by the application of coatings usually derived from fossil-oil resources. However, poor recyclability and environmental concerns on generated waste has increased interest in the study of alternative paper coatings. This work focuses on the study of the performances offered by two different biopolymers, poly(3-hydroxybutyrate-co-3hydroxyvalerate) (PHBV) and polycaprolactone (PCL), also assessing the effect of a plasticizer (PEG) when used as paper coatings. The coated samples were characterized for the structural (by scanning electron microscopy, SEM), diffusive (water vapor and grease barrier properties), and surface properties (affinity for water and oil, by contact angle measurements). Samples of polyethylene-coated and fluorinated paper were used as commercial reference. WVTR of coated samples generally decreased and PHBV and PCL coatings with PEG at 20% showed interesting low wettability, as inferred from the water contact angles. Samples coated with PCL also showed increased grease resistance in comparison with plain paper. This work, within the limits of its lab-scale, offers interesting insights for future research lines toward the development of cellulose-based food contact materials that are fully recyclable and compostable.

Keywords: compostable bioplastics; coatings; contact angle; grease resistance; paper; WVTR

Citation: Lo Faro, E.; Menozzi, C.; Licciardello, F.; Fava, P. Improvement of Paper Resistance against Moisture and Oil by Coating with Poly(-3-hydroxybutyrate-co-3-hydroxyvalerate) (PHBV) and Polycaprolactone (PCL). *Appl. Sci.* **2021**, *11*, 8058. https://doi.org/10.3390/app11178058

Academic Editors: Isidoro Garcia-Garcia, Jesus Simal-Gandara and Maria Gullo

Received: 3 July 2021
Accepted: 26 August 2021
Published: 31 August 2021

Publisher's Note: MDPI stays neutral with regard to jurisdictional claims in published maps and institutional affiliations.

1. Introduction

Paper is a light, flexible, biodegradable, highly recyclable and compostable material, derived from renewable resources and its appreciable environmental compatibility often confirms it as the first choice by several food industries, both as primary and secondary packaging [1–3]. Nonetheless, the characteristic porous structure and hydrophilicity, lead to low barrier properties against possible food deterioration sources (moisture, gases, aromas), which are often not adequate to satisfy food products needs [3]. To overcome these weaknesses and improve final product properties, thus widening possible applications, different paper treatment technologies have been developed over time. Paper coatings with functional materials, and in particular, polyethylene and fluorinated compounds are among the most performing, cost-effective and hence widely used solutions because of the excellent barrier properties they offer. However, concerns related to the impact of plastic materials and fluorinated compounds (and derivatives) on the ecosystem and on human health have urged the search for alternative solutions. Moreover, food-grade paper contamination with food residues makes recycling often impossible. In the case of polyethylene coated papers, if the recycling of such a composite product already requires specialized facilities and systems and additional costs (the plastic layer can break down in flakes which tend to clog the fine screens, and by melting from roller heat, provoke paper breaking, meaning

177

downtimes and production losses), further issues are encountered at the time when food solid residues could remain anchored to the packaging: food residues can bypass filtration and fiber separation systems. This means they can interfere, by contamination, with paper sheet formation, or increase the organic load that treatment systems then have to handle (mechanical, physical, biological or physicochemical wastewater treatment). That is why paper mills are unwilling to receive such contaminated products [4,5]. According to CONAI (2020) [4], composting can be an alternative disposal for cellulosic materials contaminated with food, at least as long as the percentages entering in the organic waste stream remain at the actual level. At the same time, research on the development of bioplastics, suitable both as stand-alone film and as paper coatings, has moved forward: these materials may offer interesting functionalities contributing to the food products protection and shelf life extension and to sustainability [6]. The use of compostable bioplastics faces drawbacks at the industrial upscale. Traditional, fossil-based materials are often more cost-effective compared to bioplastics, and this represents an important challenge in a market that is particularly cost-conscious. Moreover, the compostable bioplastic end-of-waste management lacks standardization and awareness [5]. For this reason, the need to harmonize the test to verify the actual compostability of materials is urgent, especially in response to the development of new possible solutions. The UNI EN 13432 [7] standard establishes the criteria a material must comply to be defined as compostable. In 2021, the ASTM D6868 standard [8] was revised and updated and it defines the requirements for labeling of materials and products (packaging included), entirely designed to be composted in municipal and industrial aerobic composting facilities, wherein a biodegradable plastic film or coating is attached, either through lamination or extrusion directly onto the paper, to compostable substrates. It does not describe product contents nor their compostability or biodegradability performance. A satisfactorily compostable product must demonstrate proper disintegration during composting, adequate level of inherent biodegradation, and no adverse impacts on the ability of composts to support plant growth.

This work aims to develop and characterize the properties of paper coated with compostable bioplastics: in particular, poly(3-hydroxybutyrate-co-3-hydroxyvalerate) (PHBV) and polycaprolactone (PCL) have been assessed. PHBV belongs to the polyhydroxyalkanoates (PHA) category, an aliphatic biogenic polyesters family, produced by bacterial fermentation and it is an isotactic copolyester obtained by copolymerization of hydroxybutyrate and hydroxyvalerate [9]. PCL is a semicrystalline biodegradable synthetic polyester obtained by a ring-opening polymerization (in presence of catalysts) of ε-caprolactone, which represents the monomeric unit [10,11]. Their biodegradability has been extensively studied and many different degradation times can be found; however, biodegradability is a material property that is influenced by many factors. A recent review [12] provides an overview of the main environmental conditions in which biodegradation occurs and then presents the degradability of numerous polymers. For example, strips of PHBV with a thickness of 1 mm showed a total mass loss of 100% after 70 days at 20–24 °C in industrial composting facilities; a complete degradation via composting can be observed for PCL within a few weeks. Figures on biodegradation of PHA and PHB can be found in another review [13] and the biodegradability in marine environment of PHA and PCL has been extensively discussed [14].

Many studies concerning the processing of these materials and the comparison with traditional plastic materials are available, but they are mainly focused on the development and evaluation of films, while the paper coating application still remains a little-explored research field. First, structural, diffusional and surface (water and oil affinity) properties of paper samples coated with PHBV at different thicknesses (in single and double layer) were evaluated; then, the optimization of PHBV and PCL coating solutions was attempted by addition of polyethylene glycol (PEG) (at 5%, 10% and 20% on biopolymer dry weight) to improve coating uniformity and spreadability. PEG may be defined as a natural-based plasticizer, whose ability to improve the mechanical properties of different biopolymers is well known [9,15]. PEG was used in the production of film via chloroform casting of

20% PEG and 80% PHB mixtures [16], reducing stiffness and increasing the elongation at break; PHB added with 10% of PEG (different MW) was used to produce film by compression molding [17–19]; similar results have been obtained for PCL films prepared by solvent casting and added with 10% PEG 400 and PEG 1500 [20]. In this work, different concentrations of PEG have been used, considering a range already investigated by other authors for different application: production of standalone film via solvent casting [21], via extrusion [22] or for the extrusion of PHBV on paper [23].

2. Materials and Methods

2.1. Materials

In this work a calendered bleached paper (Advantage MG White High Gloss, Mondi Group, Addlestone, UK) was used as reference paper. According to technical sheet, it is obtained from a long-fiber sulphate pulp, with a grammage of 35 g/m^2, a thickness of 50 μm, a tensile strength of 3.5 and 1.5 kN/m, a tear resistance of 280 and 420 mN (respectively in machine and cross direction). Commercial polyethylene-coated paper and fluorinated paper were used as commercial references. Polyethylene coated paper (CCM Soc. Coop. Group, Modena, Italy) has a thickness of 73 μm and a grammage of 54 g/m^2, 9 g/m^2 of which represent the polymeric coating. Fluorinated paper (ACL, Azienda Cartaria Lombarda s.p.a., Malagnino, Italy) is superficially treated with fluorinated compounds, to grant an elevated grease resistance, and with epichlorohydrin, a polyamide polymer which cross-links during paper drying, partially hindering water penetration. Chloroform was used because of the good solubility of PHBV, PCL and PEG [24] in this solvent and to avoid paper modification during the coating. Only few solvents are suitable candidates for PHB, of which chloroform is one of the most compatible and most commonly used [24,25]. The good solubility of PCL in chloroform has been demonstrated in previous studies [26–28].

Coating paper solutions were prepared dissolving in chloroform two biopolymers: poly(-3-hydroxybutyrate-co-3-hydroxyvalerate) (PHBV) and polycaprolactone (PCL). In the first case, the product used was PHI 003 (NaturePlast$^®$, Ifs, France). It is 94% biobased according to ASTM D6866 standard and industrially compostable according to ASTM D6400 standard. The product is a powder with a melt flow rate of 15–30 g/10 min (tested at 190 °C and with 2.16 Kg). PCL (Sigma-Aldrich, Mw ~80.000) has a melt flow rate in the range from 2.01 to 4.03 (tested at 160 °C and with 5.00 Kg) and a water content lower than 0.5%; melting point at 60 °C and relative density of 1.145 g/mL at 25 °C. As a plasticizer, in the second phase of the study, polyethylene glycol (PEG) 200 (Mw ~190–210; Fluka Analytical) was used.

2.2. Methods

2.2.1. Coating Solution Preparation

This work was organized in two sequential operational phases, starting from the preparation of the solutions. In the first set, coating solutions were prepared dissolving 5% w/v PHBV in previously heated chloroform, under continuous stirring in water bath for about 2 h at 55 °C. In the second set, in the purpose to obtain clearer solutions and a better solubilization of polymers, solvent was heated, always in water bath and under magnetic stirring, at 50 °C; then, the polymer was added (5% w/v of PHBV or PCL) and dissolved in chloroform under magnetic stirring at 60 °C for 50 min, and subsequently, at about 75 °C for 10 min. The solutions were cooled at room temperature under continuous stirring. A recovery system for the vapors condensation was used, to minimize the solvent evaporation losses. Any loss of solvent by evaporation was, however, restored to the initial volume. Solution of pure PHBV and PCL and solutions of each biopolymer added with plasticizer (PEG) at 5%, 10%, and 20% (on polymer dry weight basis) were prepared. A PEG stock solution in chloroform was previously prepared [29], which was added in due amounts to the bioplastic solutions, followed by stirring at room temperature for 10 min.

2.2.2. Sheets Coating

Paper sheets (210 × 297 mm) were coated via bar coating with a Compact AB3650 (TQC Sheen) automatic film applicator, working under a fume hood. The instrument allows different coating rates, which could affect continuity and uniformity of the coatings. Once coated, sheets were partially dried under fume hood, to remove the chloroform, and then placed in an oven at 40 °C for about 20 min. The samples were kept for 24 h at room temperature before testing. If a double layer was applied, the drying phase in the oven was performed only after the deposition of the second layer. The first set of samples was realized varying coating thickness with an application rate of 80 mm/s, except one sample coated at 50 mm/s. In Table 1 the different coating conditions and resulting samples are listed: nominal thickness refers to the thickness of the solution (wet coating), 3 replicate coated sheets were prepared for each formulation.

Table 1. Coating conditions and codification of samples from the first and second set.

Sample Code	Layers	Nominal Coating Thickness (µm)	Application Speed (mm/s)	Bioplastic	PEG (% *w/w*)
		SET 1			
80s80	1	80	80	PHBV	-
80s50	1	80	50	PHBV	-
80d	2	80 + 80	80	PHBV	-
100s	1	100	80	PHBV	-
100d	2	100 + 100	80	PHBV	-
		SET 2			
PHBV	2	80 + 80	80	PHBV	-
PHBVpeg5	2	80 + 80	80	PHBV	5
PHBVpeg10	2	80 + 80	80	PHBV	10
PHBVpeg20	2	80 + 80	80	PHBV	20
PCL	2	80 + 80	80	PCL	-
PCLpeg5	2	80 + 80	80	PCL	5
PCLpeg10	2	80 + 80	80	PCL	10
PCLpeg20	2	80 + 80	80	PCL	20

The second set of coated sheets were all obtained with double layer coatings, using the 80 µm application bar at 80 mm/s but with two different biopolymers, PHBV and PCL, also considering the addition of PEG as plasticizer at different concentrations (5, 10 and 20% *w/w*). In Table 1 the different samples are listed; 2 replicate coated sheets were prepared for each formulation.

2.2.3. Coating Grammage Determination

The coatings grammage was obtained by mass difference of 20 cm^2—coated paper samples and uncoated paper samples of the same size.

Results reported in Table 2 are the mean ± standard deviation of 3 determinations.

2.2.4. Scanning Electron Microscopy (SEM)

The surface and structure of the samples were analyzed by scanning electron microscopy using a Nova NanoSEM 450 (FEI, Hillsboro, OR, USA) (available at CIGS, Centro Interdipartimentale Grandi Strumenti, of the University of Modena and Reggio Emilia) provided with LVD detector, under low vacuum conditions (80 KPa) and with an acceleration of 10 kV. Images were captured with different magnifications (500×–3000×) and tilts (0°–40°), thus enabling the visualization of both surface and cross section.

Table 2. Coating grammages.

Code		Coating Grammage (g/m^2)
	SET 1	
80s80		2.82 ± 0.97
80s50		4.81 ± 1.14
80d		7.72 ± 1.31
100s		4.60 ± 0.88
100d		9.65 ± 0.82
	SET 2	
PHBV		8.48 ± 1.15
PHBVpeg5		7.42 ± 1.71
PHBVpeg10		7.92 ± 1.16
PHBVpeg20		8.57 ± 1.45
PCL		8.30 ± 1.01
PCLpeg5		6.73 ± 1.32
PCLpeg10		5.42 ± 2.83
PCLpeg20		9.77 ± 0.79

2.2.5. Measurement of Water Vapor Transmission Rate (WVTR)

The measurement of the water vapor transmission rate (WVTR) of the different samples was performed in triplicate according to the ASTM E96 standard, with modifications. Two grams of anhydrous $CaCl_2$ were placed in 25 mL glass bottles (55 mm high, 10 mm internal diameter mouth) to achieve a 0% internal RH. They were positioned in an oven at 45 °C, to remain completely dry and to be progressively sealed with the paper samples glued on the top of the vials, with the coated part inwards to prevent water vapor tangential diffusion. The vials were weighed at zero time, then placed in a desiccator with 90% RH obtained by $BaCl_2$ saturated solution at 45 °C.

The vials were weighted daily for 4 days of storage. The WVTR value (g 24 h^{-1} m^{-2}) was calculated using the following formula:

$$WVTR = \Delta W / (\Delta t \times A) \tag{1}$$

where: "$\Delta W / \Delta t$" represents the weight gain as a function of time (g 24 h^{-1}), obtained as the slope of the linear regression of the mass gain versus time; "A" corresponds to the exposed surface of the film (7.85×10^{-5} m^2). The reported WVTR data are the average of three replicates.

2.2.6. Oil and Grease Resistance

Oil and grease resistance was determined according to the ASTM F119-82 standard, with modifications [30]. Test specimens (60×60 mm) were placed on ground-glass plate (50×50 mm). Then, a cotton flannel (20 mm in diameter) was placed on specimens with 50 g weights (20 mm in diameter at the base) on them, and the assembly was conditioned in an oven at 40 °C for 30 min. It was subsequently removed, and after removing the weights, 150 µL of sunflower oil was pipetted on the flannel piece. Then, the weights were re-assembled and conditioned again in the oven at 40 °C. At convenient time intervals, the bottom of the ground glass plate was checked with the assembly in the oven. The time at which the first trace of wetting is visible at the position of the weight was recorded. Three specimens for each sample were used.

2.2.7. Water and Oil Contact Angle Measurement

Contact angle values were determined by means of an OCA 15EC contact angle meter and using OCA 20 (Dataphysics) software by the sessile drop method. For each type of sample, 1×3 cm paper strips were positioned on a film holder. Contact angle measurements were taken depositing 3 µL of water, and 7 µL of oil on the sample surface. For the contact angle measurement of each type of paper, 9 replicates were taken for water,

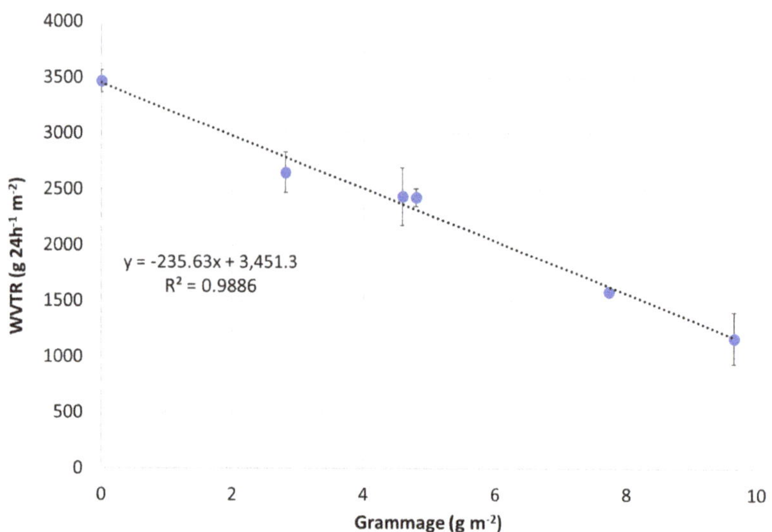

Figure 5. Linear correlation between coating grammage and WVTR of first set samples. Grammage equal to 0 corresponds to uncoated paper sample.

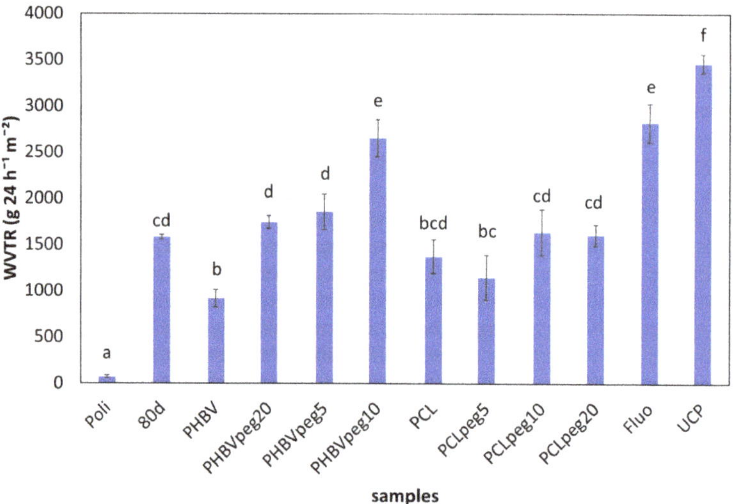

Figure 6. WVTR values ($g\ 24\ h^{-1}\ m^{-2}$) of PHBV and PCL coated samples both pure and with PEG addition at different concentrations (set 2). Bars represents the average value of three determinations, error bars report standard deviation. Different letters indicate statistically significant differences ($p < 0.05$). "UCP" refers to base paper, "Fluo" to fluorinated paper, while "Poli" to polyethylene-coated paper. See Table 1 for other samples codification.

Concerning the addition of PEG to PHBV, a general increase of WVTR in presence of plasticizer can be observed. These results could be explained considering the hydrophilic nature of PEG and its plasticizing action, according with previous studies that attribute this behavior to a free volume increase (which promotes molecular diffusion) and to the PEG hydrophilic nature, which tends to increase sorption capacity, solubility and water permeability of plasticized films [17]. In particular, WVTR values significantly increased in presence of plasticizer up to 10%, and then decreased at 20% PEG concentration, while maintaining a value higher than PHBV not plasticized and not significantly different compared with PHBV with 5% PEG. This trend agrees with a study about PHBV extruded films incorporated with PEG1000 at 5%, 10%, 20%. The WVTR decrease was explained in the study assuming a demixing between plasticizer and PHBV, which could, given the relatively low PHBV permeability, allow a reduction in water vapor permeability [22]. This hypothesis is not necessarily in contrast with the SEM structure observed for these samples, which appeared as more uniform compared to non-plasticized films and could be explained as the result of a beginning demixing between the two materials, not always detectable by SEM images. According to Figure 6, pure PHBV coating showed lower, although not significantly, WVTR values compared to PCL. It is possible that discontinuities observed at SEM on PCL coatings could facilitate water vapor permeation. In accordance with a previous study [41], this result could be further explained considering both that water vapor diffusion coefficient in an amorphous or semicrystalline polymer depends on molecular dynamic or amorphous regions segmental movements, and the different glass transition temperatures of the biopolymers (PCL has a lower Tg compared to PHBV), but the presence and arrangement of substituent groups (such as methyl or ethyl groups) as in PHBV case, which tend to have a minor chain flexibility on account of steric effect, and therefore, a lower WVTR, but the lack of information does not allow to elaborate additional considerations. Compared to PHBV-based coatings, plasticizer addition did not significantly affect WVTR in PCL coatings: the ones with 10% and 20% resulted to be more permeable (though not significantly) to water vapor compared to pure PCL. In general, coated samples, in particular PCL ones, showed an improvement of barrier to vapor compared to base paper; nevertheless, barrier properties can be further improved using other application techniques: processing method plays a key role in film production and can do the same in the case of coatings. Long drying required by solvent casting process can contribute to the formation of more open microstructures and channels for easy diffusion of water vapor molecules [42]. Biopolymers hot processing may allow to achieve denser and more resistant structures.

3.3. Oil and Grease Resistance

The determination of oil and grease resistance of food packaging materials, cellulosic included, is important in many applications, e.g., retail packing of fresh and cured meat, cheese, fish, but also when exchange of food components with packaging materials occurs, occasionally affecting the performances of the material itself (delamination). There is a discrete number of available and known tests to assess this performance, but they (including ASTM F119 applied in this work) are largely based on subjective statements, often not allowing to obtain measurements with a statistically meaningful physical value. Especially when grease barrier properties differences are low, a more precise and reproducible measurement method will be required: for instance, mixing oils used in the tests with coloring agent and determining colorimetric coordinates using a spectrophotometer), with the aim of making results independent from individual subjectivity [43].

Table 3 shows grease resistance times of all samples of this work. All the coatings improved base paper resistance (resistance ranging between 0 and 10 min), but performances were not comparable with fluorinated and polyethylene-coated papers, which did not show any failure after 72 h. Resistance times recorded for PHBV coated papers were generally lower than 20 min; only PHBVpeg5 showed resistance ranging between 40 and 60 min. The few studies available in literature report variable grease resistance data and ascribe this variability to the presence of cracks or to lack of uniformity in the material [23]. PCL coated samples showed in all cases a resistance of more than 4 h, but less than 12 h, except one sample showing a higher resistance between 12 and 24 h.

Table 3. Grease resistance times.

Sample Code	Time (minutes)		
UCP	10	-	0
Fluo	>4320	>4320	>4320
Poli	>4320	>4320	>4320
80s80	20	20	20
80s50	10	30	20
80d	10	10	10
100s	10	10	40
100d	10	20	20
PHBV	10	20	20
PHBVpeg5	40	60	40
PHBVpeg10	20	20	20
PHBVpeg20	20	10	40
PCL	240–720	240–720	240–720
PCLpeg5	240–720	240–720	240–720
PCLpeg10	240–720	720–1440	240–720
PCLpeg20	240–720	240–720	-

3.4. Contact Angle Measurements

3.4.1. Water

As shown in Figures 7 and 8, fluorinated paper resulted far more hydrophobic than other samples, possibly because of the epichlorohydrin treatment. Contact angle values of first set samples resulted to decrease as number of layers and coating amount increase. This result could be explained by the Wenzel theory [44], assuming the effect of the relative roughness of coatings, and in particular, assuming a raise of the measured hydrophilicity as roughness of the material increases: as noted in SEM images, double layer coatings showed a higher roughness compared with single layers and resulted to be the most hydrophilic.

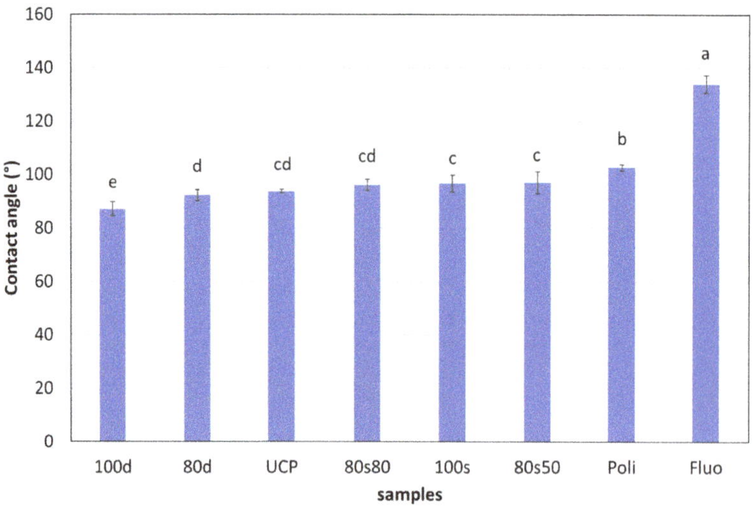

Figure 7. Average water contact angle values for first set samples. Error bars report standard deviation. Different letters indicate statistically significant differences (*p* < 0.05). "UCP" refers to base paper, "Fluo" to fluorinated paper, while "Poli" to polyethylene-coated paper. See Table 1 for other samples codification.

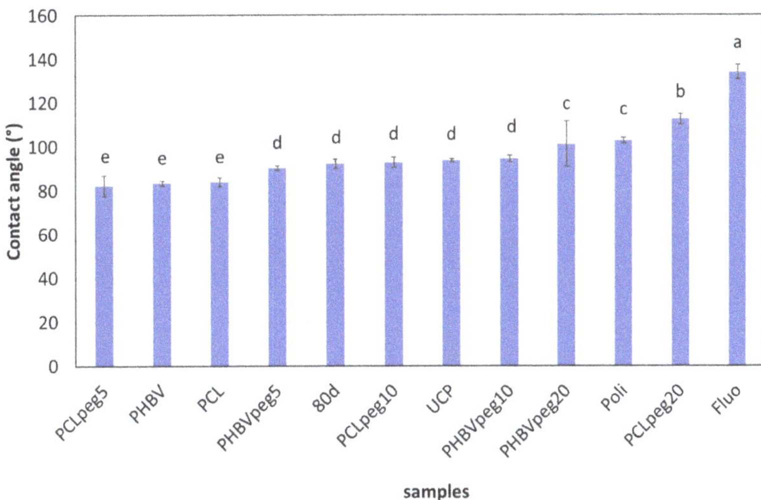

Figure 8. Average water contact angle values for the second set samples. Error bars report standard deviation. Different letters indicate statistically significant differences ($p < 0.05$). "UCP" refers to base paper, "Fluo" to fluorinated paper, while "Poli" to polyethylene-coated paper. See Table 1 for other samples codification.

Calendered paper showed a good hydrophobicity. The relatively high value detected for this paper could be related to a behavior that comes closer to the Cassie-Baxter model, assuming the contribution of air bubbles entrapped between interfiber gaps, given the high porosity of this material, but it is difficult to verify such a hypothesis (Figure 8) [44,45].

It is possible to notice a general upward trend of the contact angle (or a downward trend of wettability), by adding plasticizer. Given the relative hydrophilicity of PHBV and PEG, this result could be explained considering the Wenzel theory and the effect of surface roughness [44]. The use of plasticizer in the coating solution formulation could be associated with a better solution spreadability and a more uniform coating distribution. The SEM images seemed to support the effectiveness of PEG in this regard, as the coated surface appeared more uniform and less rough with increasing plasticizer concentrations. PCL contact angle values were higher than the ones found in literature for films based on the same PCL type used in this work (always at 5% w/v, but in trichloromethane and differently processed), equal to 64.5° [46].

Even if the trend is not as linear as for PHBV, the Wenzel theory could explain the significantly higher contact angle values detected for PCL with PEG at 10% and 20% samples, considering the progressive increase of smoothness observed for these coatings by the SEM images. Although PCL with PEG at 5% sample showed a contact angle value lower than pure PCL, the difference was not statistically significant. PHBV and PCL both with PEG at 20% samples showed the lowest wettability (highest contact angle values) among coated samples. In particular, PCLpeg20 reached a wettability level significantly lower compared to reference polyethylene-coated paper (Figure 9).

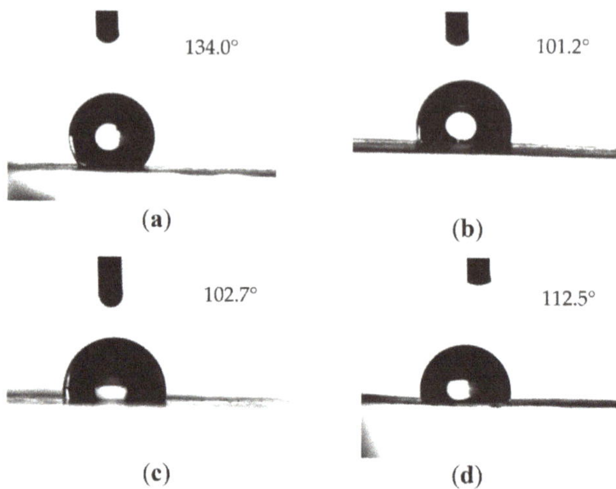

Figure 9. Water contact angles of (**a**) fluorinated paper (Fluo); (**b**) paper coated with PHBV with PEG 20% (PHBVpeg20); (**c**) polyethylene-coated paper (Poli); (**d**) paper coated with PCL with PEG 20% (PCLpeg20).

Furthermore, it is important to remember that unexpected results may occur, as the hydrophobic and hydrophilic surface chemistry physics is more complex than Wenzel and Cassie-Baxter models, which represent simplification of what is really going on. Rough surfaces contact angles study is still a vibrant research field [44].

3.4.2. Oil

Tables 4 and 5 show oil contact angle values of, respectively, first and second set and commercial reference samples. Oil contact angle values in both sets showed an elevated wettability (oleophilicity) of the coatings, expressed by low contact angle values (in the range from 11.3 to 37.3°), and therefore difficult to determine. As expected, given the low surface energy, fluorinated paper showed high contact angle values (average value of 101.9°) and particularly reduced wettability (high oleophobicity) (Figure 10).

Polyethylene-coated paper showed a low oil contact angle value (average value of 16.9°). Several coatings of the second set showed a higher oleophobicity than this commercial reference, with special regards for PCLpeg20, showing an average measured value of 37.3°.

Table 4. Oil contact angle values of first set samples.

Sample Code	Mean Contact Angle Value
UCP	15.1 ± 3.8 [bc]
Fluo	101.9 ± 1.5 [a]
Poli	16.9 ± 0.2 [b]
80s80	14.0 ± 1.6 [bc]
80s50	11.3 ± 0.7 [c]
80d	12.6 ± 1.8 [bc]
100s	15.5 ± 3.2 [bc]
100d	12.4 ± 1.5 [bc]

Different superscript letters indicate statistically significant differences ($p < 0.05$).

Table 5. Oil contact angle values of second set samples.

Sample Code	Mean Contact Angle Value
UCP	15.1 ± 3.8 [de]
Fluo	101.9 ± 1.5 [a]
Poli	16.9 ± 0.2 [d]
80d	12.6 ± 1.8 [e]
PHBV	24.7 ± 1.6 [c]
PHBVpeg5	19.3 ± 1.0 [d]
PHBVpeg10	17.2 ± 1.7 [d]
PHBVpeg20	17.6 ± 2.3 [d]
PCL	18.9 ± 1.9 [d]
PCLpeg5	16.6 ± 0.8 [de]
PCLpeg10	27.0 ± 2.3 [c]
PCLpeg20	37.3 ± 1.9 [b]

Different superscript letters indicate statistically significant differences ($p < 0.05$).

Figure 10. Oil contact angle of fluorinated paper (Fluo).

4. Conclusions

To conclude, this work, within the limits of its lab-scale, offers insights for future research lines toward the development of cellulose-based food contact materials, with special attention to a greater sustainability compared to actual commercial solutions. The SEM analysis of the developed coated samples showed the disappearance of the typical fiber network of paper and allowed to observe a continuous layer of coating. The smoothness of the coating surface was used as an indicator of good biopolymer solubilization. Coated samples developed in this work showed a significant improvement of water vapor barrier compared to uncoated paper and also to commercial fluorinated paper. The significant reduction of WVTR is a promising feature of the developed coated paper samples, according to the importance of this parameter for food quality preservation. Biopolymer-coated paper samples showed improved grease resistance, still not comparable with fluorinated and polyethylene-coated commercial samples. Among samples, PCL-coated offered the best resistance, ranging from 4 to 12 h. PHBV and PCL coatings with PEG at 20% showed good water contact angles, thus a low wettability. The measured oil contact angles were generally much lower compared to fluorinated paper, however certain coating formulations (i.e., PCLpeg20, PCLpeg10 and PHBV) showed significantly higher surface oleophobicity compared to commercial polyethylene-coated paper.

Chloroform utilized in this study is a poorly biocompatible solvent; therefore, future developments should include the evaluation of more eco-friendly solvents, when possible, or other coating techniques, such as extrusion coating. The increasing availability and commercial variety of compostable biopolymers allow a more concrete technical feasibility. The real compostability of biopolymer-coated paper should, however, be verified.

Author Contributions: Conceptualization, F.L. and P.F.; methodology, F.L., P.F., E.L.F., and C.M.; formal analysis, E.L.F. and C.M.; investigation, E.L.F. and C.M.; data curation, F.L. and C.M; writing—original draft preparation, C.M. and E.L.F.; writing—review and editing, F.L. and P.F.; supervision, F.L. and P.F. All authors have read and agreed to the published version of the manuscript.

Funding: This research received no external funding.

Conflicts of Interest: The authors declare no conflict of interest.

References

1. Khwaldia, K.; Arab-Tehrany, E.; Desobry, S. Biopolymer Coatings on Paper Packaging Materials. *Compr. Rev. Food Sci. Food Saf.* **2010**, *9*, 82–91. [CrossRef]
2. Deshwal, G.K.; Panjagari, N.R.; Alam, T. An Overview of Paper and Paper Based Food Packaging Materials: Health Safety and Environmental Concerns. *J. Food Sci. Technol.* **2019**, *56*, 4391–4403. [CrossRef]
3. Nechita, P.; Roman (Iana-Roman), M. Review on Polysaccharides Used in Coatings for Food Packaging Papers. *Coatings* **2020**, *10*, 566. [CrossRef]
4. Marinelli, A.; Santi, R.; Del Curto, B. Guidelines for Facilitating the Recycling of Packaging Made Predominantly from Paper, CONAI. 2020. Available online: http://www.progettarericiclo.com/en/docs/guidelines-facilitating-recycling-packaging-made-predominantly-paper (accessed on 2 April 2021).
5. Triantafillopoulos, N.; Koukoulas, A.A. The Future of Single-Use Paper Coffee Cups: Current Progress and Outlook. *BioResources* **2020**, *15*, 7260–7287. [CrossRef]
6. Nicu, R.; Lupei, M.; Balan, T.; Bobu, E. Alkyl-Chitosan as Paper Coating to Improve Water Barrier Properties. *Cellul. Chem. Technol.* **2013**, *47*, 623–630.
7. UNI EN 13432. *Packaging—Requirements for Packaging Recoverable through Composting and Biodegradation—Test Scheme and Evaluation Criteria for the Final Acceptance of Packaging*; European Committee for Standardisation: Brussels, Belgium, 2000.
8. ASTM D6868-21. *Standard Specification for Labeling of End Items That Incorporate Plastics and Polymers as Coatings or Additives with Paper and Other Substrates Designed to Be Aerobically Composted in Municipal or Industrial Facilities*; ASTM International: West Conshohocken, PA, USA, 2021.
9. Bugnicourt, E.; Cinelli, P.; Lazzeri, A.; Alvarez, V. Polyhydroxyalkanoate (PHA): Review of Synthesis, Characteristics, Processing and Potential Applications in Packaging. *Express Polym. Lett.* **2014**, *8*, 791–808. [CrossRef]
10. Rydz, J.; Musioł, M.; Zawidlak-Węgrzyńska, B.; Sikorska, W. Present and future of biodegradable polymers for food packaging applications. In *Handbook of Food Bioengineering, Biopolym.Food Design*; Alexandru, M.G., Alina, M.H., Eds.; Academic Press: Cambridge, MA, USA, 2018; Volume 20, pp. 431–467. ISBN 978-0-12-811449-0.
11. McKeen, L.W. Environmentally friendly polymers. In *Permeability Properties of Plastics and Elastomers*, 3rd ed.; William Andrew Publishing: Norwich, NY, USA, 2012; pp. 287–304. ISBN 978-1-4377-3469-0.
12. Kliem, S.; Kreutzbruck, M.; Bonten, C. Review on the Biological Degradation of Polymers in Various Environments. *Materials* **2020**, *13*, 4586. [CrossRef] [PubMed]
13. Meereboer, K.W.; Misra, M.; Mohanty, A.K. Review of Recent Advances in the Biodegradability of Polyhydroxyalkanoate (PHA) Bioplastics and Their Composites. *Green Chem.* **2020**, *22*, 5519–5558. [CrossRef]
14. Suzuki, M.; Tachibana, Y.; Kasuya, K. Biodegradability of Poly(3-Hydroxyalkanoate) and Poly(ε-Caprolactone) via Biological Carbon Cycles in Marine Environments. *Polym. J.* **2021**, *53*, 47–66. [CrossRef]
15. Vieira, M.G.A.; da Silva, M.A.; dos Santos, L.O.; Beppu, M.M. Natural-Based Plasticizers and Biopolymer Films: A Review. *Eur. Polym. J.* **2011**, *47*, 254–263. [CrossRef]
16. Gamba, A.M.; Fonseca, J.S.; Méndez, D.A.; Viloria, A.C.; Fajardo, D.; Moreno, N.C.; Rojas, I.C. Assessment of Different Plasticizer Polyhydroxyalkanoate Mixtures to Obtain Biodegradable Polimeric Films. *Chem. Eng. Trans.* **2017**, *57*, 1363–1368. [CrossRef]
17. Requena, R.; Jiménez, A.; Vargas, M.; Chiralt, A. Effect of Plasticizers on Thermal and Physical Properties of Compression-Moulded Poly[(3-Hydroxybutyrate)-Co-(3-Hydroxyvalerate)] Films. *Polym. Test.* **2016**, *56*, 45–53. [CrossRef]
18. Requena, R.; Jiménez, A.; Vargas, M.; Chiralt, A. Poly[(3-Hydroxybutyrate)-*Co*-(3-Hydroxyvalerate)] Active Bilayer Films Obtained by Compression Moulding and Applying Essential Oils at the Interface: Antimicrobial PHBV Active Bilayer Films Obtained by Compression Moulding. *Polym. Int.* **2016**, *65*, 883–891. [CrossRef]
19. Requena, R.; Vargas, M.; Chiralt, A. Release Kinetics of Carvacrol and Eugenol from Poly(Hydroxybutyrate-Co-Hydroxyvalerate) (PHBV) Films for Food Packaging Applications. *Eur. Polym. J.* **2017**, *92*, 185–193. [CrossRef]
20. Rosa, D.S.; Guedes, C.G.F.; Casarin, F.; Bragança, F.C. The Effect of the Mw of PEG in PCL/CA Blends. *Polym. Test.* **2005**, *24*, 542–548. [CrossRef]
21. Parra, D.F.; Fusaro, J.; Gaboardi, F.; Rosa, D.S. Influence of Poly (Ethylene Glycol) on the Thermal, Mechanical, Morphological, Physical–Chemical and Biodegradation Properties of Poly (3-Hydroxybutyrate). *Polym. Degrad. Stab.* **2006**, *91*, 1954–1959. [CrossRef]
22. Jost, V.; Langowski, H.-C. Effect of Different Plasticisers on the Mechanical and Barrier Properties of Extruded Cast PHBV Films. *Eur. Polym. J.* **2015**, *68*, 302–312. [CrossRef]
23. Sängerlaub, S.; Brüggemann, M.; Rodler, N.; Jost, V.; Bauer, K.D. Extrusion Coating of Paper with Poly(3-Hydroxybutyrate-Co-3-Hydroxyvalerate) (PHBV)—Packaging Related Functional Properties. *Coatings* **2019**, *9*, 457. [CrossRef]
24. Anbukarasu, P.; Sauvageau, D.; Elias, A. Tuning the Properties of Polyhydroxybutyrate Films Using Acetic Acid via Solvent Casting. *Sci. Rep.* **2016**, *5*, 17884. [CrossRef]
25. Xu, Y.; Zou, L.; Lu, H.; Kang, T. Effect of Different Solvent Systems on PHBV/PEO Electrospun Fibers. *RSC Adv.* **2017**, *7*, 4000–4010. [CrossRef]

26. Tang, Z.G.; Black, R.A.; Curran, J.M.; Hunt, J.A.; Rhodes, N.P.; Williams, D.F. Surface Properties and Biocompatibility of Solvent-Cast Poly[ε-Caprolactone] Films. *Biomaterials* **2004**, *25*, 4741–4748. [CrossRef] [PubMed]
27. Zhang, S.; Campagne, C.; Salaün, F. Preparation of Electrosprayed Poly(Caprolactone) Microparticles Based on Green Solvents and Related Investigations on the Effects of Solution Properties as Well as Operating Parameters. *Coatings* **2019**, *9*, 84. [CrossRef]
28. Zhang, S.; Campagne, C.; Salaün, F. Influence of Solvent Selection in the Electrospraying Process of Polycaprolactone. *Appl. Sci.* **2019**, *9*, 402. [CrossRef]
29. Catoni, S.E.M.; Trindade, K.N.; Gomes, C.A.T.; Schneider, A.L.S.; Pezzin, A.P.T.; Soldi, V. Influence of Poly(Ethylene Grycol)—(PEG) on the Properties of Influence of Poly(3-Hydroxybutyrate-CO-3-Hydroxyvalerate)—PHBV. *Polímeros* **2013**, *23*, 320–325. [CrossRef]
30. ASTM F119-82. *Standard Test Method for Rate of Grease Penetration of Flexible Barrier Materials (Rapid Method)*; ASTM International: West Conshohocken, PA, USA, 2015.
31. Bedane, A.H.; Xiao, H.; Eić, M.; Farmahini-Farahani, M. Structural and Thermodynamic Characterization of Modified Cellulose Fiber-Based Materials and Related Interactions with Water Vapor. *Appl. Surf. Sci.* **2015**, *351*, 725–737. [CrossRef]
32. Bota, J.; Vukoje, M.; Brozović, M.; Hrnjak-Murgić, Z. Reduced Water Permeability of Biodegradable PCL Nanocomposite Coated Paperboard Packaging. *Chem. Biochem. Eng. Q.* **2017**, *31*, 417–424. [CrossRef]
33. Helanto, K.; Matikainen, L.; Talja, R.; Rojas, O.J. Bio-Based Polymers for Sustainable Packaging and Biobarriers: A Critical Review. *BioResources* **2019**, *14*, 4902–4951. [CrossRef]
34. Reichert, C.L.; Bugnicourt, E.; Coltelli, M.-B.; Cinelli, P.; Lazzeri, A.; Canesi, I.; Braca, F.; Martínez, B.M.; Alonso, R.; Agostinis, L.; et al. Bio-Based Packaging: Materials, Modifications, Industrial Applications and Sustainability. *Polymers* **2020**, *12*, 1558. [CrossRef]
35. Din, M.I.; Ghaffar, T.; Najeeb, J.; Hussain, Z.; Khalid, R.; Zahid, H. Potential Perspectives of Biodegradable Plastics for Food Packaging Application-Review of Properties and Recent Developments. *Food Addit. Contam. Part A* **2020**, *37*, 665–680. [CrossRef]
36. Wu, F.; Misra, M.; Mohanty, A.K. Challenges and New Opportunities on Barrier Performance of Biodegradable Polymers for Sustainable Packaging. *Prog. Polym. Sci.* **2021**, *117*, 101395. [CrossRef]
37. Jost, V. Packaging Related Properties of Commercially Available Biopolymers—An Overview of the Status Quo. *Express Polym. Lett.* **2018**, *12*, 429–435. [CrossRef]
38. Cyras, V.P.; Commisso, M.S.; Mauri, A.N.; Vázquez, A. Biodegradable Double-Layer Films Based on Biological Resources: Polyhydroxybutyrate and Cellulose. *J. Appl. Polym. Sci.* **2007**, *106*, 749–756. [CrossRef]
39. Han, J.; Salmieri, S.; Le Tien, C.; Lacroix, M. Improvement of Water Barrier Property of Paperboard by Coating Application with Biodegradable Polymers. *J. Agric. Food Chem.* **2010**, *58*, 3125–3131. [CrossRef] [PubMed]
40. Rastogi, V.; Samyn, P. Bio-Based Coatings for Paper Applications. *Coatings* **2015**, *5*, 887–930. [CrossRef]
41. Shogren, R. Water Vapor Permeability of Biodegradable Polymers. *J. Environ. Polym. Degrad.* **1997**, *5*, 91–95. [CrossRef]
42. Fabra, M.J.; López-Rubio, A.; Cabedo, L.; Lagaron, J.M. Tailoring Barrier Properties of Thermoplastic Corn Starch-Based Films (TPCS) by Means of a Multilayer Design. *J. Colloid Interface Sci.* **2016**, *483*, 84–92. [CrossRef]
43. Gietl, M.L.; Schmidt, H.-W.; Giesa, R.; Terrenoire, A.; Balk, R. Semiquantitative Method for the Evaluation of Grease Barrier Coatings. *Prog. Org. Coat.* **2009**, *66*, 107–112. [CrossRef]
44. Ryan, B.J.; Poduska, K.M. Roughness Effects on Contact Angle Measurements. *Am. J. Phys.* **2008**, *76*, 1074–1077. [CrossRef]
45. Kubiak, K.J.; Wilson, M.C.T.; Mathia, T.G.; Carval, P. Wettability versus Roughness of Engineering Surfaces. *Wear* **2011**, *271*, 523–528. [CrossRef]
46. Barabaszová, K.Č.; Holešová, S.; Hundáková, M.; Mohyla, V. Vermiculite in Polycaprolactone Films Prepared with the Used of Ultrasound. *Mater. Today Proc.* **2021**, *37*, 13–20. [CrossRef]

Article

Kombucha Tea as a Reservoir of Cellulose Producing Bacteria: Assessing Diversity among *Komagataeibacter* Isolates

Salvatore La China, Luciana De Vero, Kavitha Anguluri, Marcello Brugnoli, Dhouha Mamlouk and Maria Gullo *

Department of Life Sciences, University of Modena and Reggio Emilia, 42122 Reggio Emilia, Italy;
salvatore.lachina@unimore.it (S.L.C.); luciana.devero@unimore.it (L.D.V.); kavitha.anguluri@unimore.it (K.A.);
marcello.brugnoli96@gmail.com (M.B.); mamloukdhouha@gmail.com (D.M.)
* Correspondence: maria.gullo@unimore.it

Abstract: Bacterial cellulose (BC) is receiving a great deal of attention due to its unique properties such as high purity, water retention capacity, high mechanical strength, and biocompatibility. However, the production of BC has been limited because of the associated high costs and low productivity. In light of this, the isolation of new BC producing bacteria and the selection of highly productive strains has become a prominent issue. Kombucha tea is a fermented beverage in which the bacteria fraction of the microbial community is composed mostly of strains belonging to the genus *Komagataeibacter*. In this study, Kombucha tea production trials were performed starting from a previous batch, and bacterial isolation was conducted along cultivation time. From the whole microbial pool, 46 isolates were tested for their ability to produce BC. The obtained BC yield ranged from 0.59 g/L, for the isolate K2G36, to 23 g/L for K2G30—which used as the reference strain. The genetic intraspecific diversity of the 46 isolates was investigated using two repetitive-sequence-based PCR typing methods: the enterobacterial repetitive intergenic consensus (ERIC) elements and the (GTG)$_5$ sequences, respectively. The results obtained using the two different approaches revealed the suitability of the fingerprint techniques, showing a discrimination power, calculated as the D index, of 0.94 for (GTG)$_5$ rep-PCR and 0.95 for ERIC rep-PCR. In order to improve the sensitivity of the applied method, a combined model for the two genotyping experiments was performed, allowing for the ability to discriminate among strains.

Keywords: kombucha tea; microbial diversity; bacterial cellulose; *Komagataeibacter xylinus*; repetitive elements sequence-based rep-PCR; typing

Citation: La China, S.; De Vero, L.; Anguluri, K.; Brugnoli, M.; Mamlouk, D.; Gullo, M. Kombucha Tea as a Reservoir of Cellulose Producing Bacteria: Assessing Diversity among *Komagataeibacter* Isolates. *Appl. Sci.* **2021**, *11*, 1595. https://doi.org/10.3390/app11041595

Academic Editor: Maria Kanellaki
Received: 31 December 2020
Accepted: 6 February 2021
Published: 10 February 2021

Publisher's Note: MDPI stays neutral with regard to jurisdictional claims in published maps and institutional affiliations.

1. Introduction

The attraction of bacterial cellulose (BC) as a natural biopolymer, produced by microorganisms, arises from its main functional properties and subsequent applications, especially in the biomedical, food, and engineering fields [1–3]. Bacteria able to synthesize cellulose can be considered ubiquitous since they have been found to inhabit different ecosystems. They can be isolated from environmental sources (such as soil and plants), from insects, humans, and from food sources [4].

Within the family *Acetobacteraceae*, the genus *Komagataeibacter* includes species, such as *K. xylinus*, *K. europaeus* and *K. hansenii*, that have been described as cellulose producing organisms [5]. The main studied strains belong to the *K. xylinus* species, which is recognized as the model organism for studying the mechanism of BC synthesis [6,7].

One of the most important reservoirs in which it is possible to detect strains belonging to the *Komagataeibacter* genus is kombucha tea, a fermented beverage traditionally produced in Asia, probably originating from the northeast of China [8]. Kombucha tea is characterized by a microbial community in which different microbial groups, mainly yeasts and bacteria, live in a symbiotic lifestyle [9]. Among bacteria groups, acetic acid bacteria are the main functional organisms, although lactic acid bacteria are also found; moreover, among yeasts, strains of the genera *Saccharomyces* and *Zygosaccharomyces* have

195

been described [10]. Kombucha tea production consists of two fermentation reactions operated by yeasts and acetic acid bacteria [11]. The symbiotic relation between yeast and bacteria is established during the fermentation process, in which the yeasts are capable of hydrolysing sucrose, the main substrate for kombucha tea production, into glucose and fructose, and then ethanol. Acetic acid bacteria use glucose, fructose and ethanol to produce gluconic acid, glucuronic acid, acetic acid, and BC [12,13].

Kombucha tea can be considered as a dynamic environment, rapidly changing its composition, acting as an ecophysiological reservoir of very specialized organisms. Considering the occurrence and function of acetic acid bacteria within kombucha, previous studies have described *K. xylinus* as the main species.

The ability of *K. xylinus* strains to produce BC is highly variable, making the selection of the best ones a crucial step in order to obtain candidate strains for large scale BC production. In fact, the availability of selected organisms in producing BC is a requirement for exploiting them as industrial machinery. The production of BC is mainly influenced by two factors. The first one is represented by the carbon sources used in the medium [14,15]; pure carbon sources could increase the production cost, making the process unsustainable. The second issue is related to the ability of the organism to grow and produce BC in the selected conditions [16]. Some studies have described the abilities of strains belonging to the *Komagataeibacter* genus for producing BC at different yields [5].

In this work, acetic acid bacteria were collected from liquid and pellicle fraction of kombucha tea samples. In total, 46 isolates were screened for their ability to produce BC and for the main phenotypic characteristics. Moreover, they were typed by applying a fingerprint analysis based on $(GTG)_5$ repetitive elements and Enterobacterial Repetitive Intragenic Consensus (ERIC) elements in order to assay the intraspecific genetic diversity.

2. Materials and Methods

2.1. Kombucha Preparation, Sampling and Physico-Chemical Determinations

Two native local kombucha precultures (P1 and P2) were provided by the National Institute of Applied Sciences and Technology (INSAT, Tunisia), previously cultivated on two different substrates: black tea (P1) and green tea (P2). Each preculture was composed of kombucha cellulose pellicle with approximately 200 mL of starter culture. Duplicates of kombucha, referred to green tea kombucha (GTK) and black tea kombucha (BTK), were prepared and inoculated according to Jayabalan et al. [17]. Briefly, sucrose 10% (w/v) was added to demineralized water and allowed to boil. Afterwards, 1.2% (w/v) black or green tea (Les Jardins du thé, Office Tunisien du Commerce, Tunisia) was added and allowed to infuse for about 5 min. After filtration of tea leaves through a sterile sieve, 200 mL of tea was poured into 500 mL glass jars that had been previously sterilized at +121 °C for 20 min. Once the sucrose–tea solution had cooled at room temperature, 10% (v/v) of the fermented tea (of the previous fermentation brew from kombucha with the same origin, in this case black or green tea as substrate, corresponding to the aforementioned starter culture) was added. Then, 3% (w/v) of representative pellicle fragments cut from the preculture were placed in the culture light side up. The jars were covered with a sterile gauze and fixed with an elastic band. Fermentations were carried out in the dark for 12 days. All experiments were carried out aseptically.

Cell morphology and direct observation of samples were carried out using optical microscopy at $100\times$ of magnification, using C. Zeiss microscope apparatus (Axiolab).

Titratable acidity was determined by neutralizing samples at pH 7.2 with 0.1 N of NaOH (it was assumed that all media acidity was due to acetic acid) and expressed as % (wt/wt); pH was measured by CRISON, MicropH, 2002 pHmeter. Ethanol, expressed as % (v/v), was checked by densitometry measure using a hydrostatic balance after distillation.

2.2. Plating and Isolation

Plating was performed immediately after sampling. Therefore, 10 mL of 0.1% (w/v) peptone water was added to 1 g of black or green kombucha pellicle in a sterile stomacher bag that was vigorously shaken for 5 min in a laboratory blender STOMACHER® 400 (Seward, England) to obtain a uniform homogenate. Samples (1 mL each) of the homogenate and liquid phase were 10-fold serially diluted in 0.1% (w/v) peptone water, from which aliquots (0.1 mL) were plated on GYC medium (50 g/L glucose, 10 g/L yeast extract, 15 g/L CaCO$_3$, 9 g/L bacteriological agar) and ACB agar medium (30 g/L yeast extract, 1 mL of a 22 g/L bromocresol green solution, 20 g/L bacteriological agar). The media were previously autoclaved at 121 °C for 15 min and after 20 mL/L of filtered ethanol 95% (v/v) was added aseptically to the ACB medium. A total of 0.2 mL of cycloheximide hydroalcoholic solution at 25% m/v was added directly to petri dishes.

Plates were incubated at +28 °C for 72–120 h. Colonies were picked up from a suitable dilution of each sample on GYC and ACB agar media, grown for 3–5 days at +28 °C and purified through subculturing and plating. Pure kombucha isolates were named and strains were cultivated on the corresponding isolation medium for 3–5 days at +28 °C [18]. Long-term preservation methods of the strains were performed according to the standard procedures of the Microbial Resource Research Infrastructure—Italian Joint Research Unit (MIRRI-IT) [19]. Specifically, a seed lot of the strains was stored at −80 °C, adding glycerol 50% (v/v) to the liquid media. The strains with significant cellulose production were deposited in the University of Modena and Reggio Emilia (UNIMORE) Microbial Culture Collection (UMCC, www.umcc.unimore.it, accessed on 3 February 2020).

2.3. Phenotypic Characterization

Strains were phenotypically characterized by determination of cell morphology, gram staining, catalase activity, potassium hydroxide (KOH) test, consumption of calcium carbonate and oxidation of ethanol and acetic acid on ACB agar medium, as previously reported in Gullo et al. [20].

The BC production test was carried out by collecting the pellicles and boiling them in 4 mL of 5.0% NaOH solution for 2 h. Cellulose was confirmed when the pellicle did not dissolve after boiling [21]. DSMZ (German Collection of Microorganisms and Cell Cultures GmbH) strains *K. xylinus* (DSMZ 2004) and *G. oxydans* (DSMZ 3503) were used as positive and negative controls, respectively. Tests were conducted in triplicate.

The amount of BC was estimated, as reported by Gullo et al. [22]. Briefly, pellicles were washed with distilled water and then treated with 1% NaOH at 90 °C for 30 min. Treated BC was washed twice with distilled water, the pH of the residual water was determined; the washing step was repeated until reaching neutral pH. Drying was conducted at 80 °C. BC weight was measured by analytical balance (Gibertini E42S) [23]. The yield of BC produced was expressed in grams of dried BC per liter of broth.

2.4. Genomic DNA Extraction, RFLP and Amplification of (GTG)$_5$/REP-PCR and ERIC Elements

Genomic DNA extraction was performed on liquid cultures, which were previously incubated for 72 h at +28 °C. Flasks were vigorously shaken to remove the embedded cells from BC, the liquid medium was transferred to a sterile tube and centrifuged at 12,000× g for 5 min. gDNA was extracted using the method previously proposed by Gullo et al. [24]. gDNA was checked by 3% gel agarose in 1X TBE buffer and quantified by spectrophotometric measure (NanoDrop ND-1000). Band sizes were determined using a 100 bp DNA ladder (Invitrogen, Carlsbad, CA, USA).

RFLP analysis on the full-length 16S rRNA gene was performed using *Rsa*I and *Alu*I endonucleases (Fermentas, Hanover, ND, USA), incubating at +37 °C for 2 h. The restriction reaction was stopped by incubation at 80 °C for 20 min. The fragments were checked on 3% gel agarose in 1X TBE buffer and sizes were determined using 100 bp DNA ladder (Invitrogen).

(GTG)$_5$ rep-PCR fingerprinting was carried out according to the method of Versalovic et al. [25], with some minor modifications. Isolates were subjected to REP-PCR with a single oligonucleotide GTG$_5$ (5′- GTGGTGGTGGTGGTG-3′). Samples were incubated for 5 min at 94 °C and then cycled 35 times at 94 °C for 30 s, 40 °C for 1 min, and 72 °C for 4 min.

The samples were incubated for 7 min at 72 °C for final extension and kept at 4 °C. ERIC elements were initially amplified according to Versalovic et al. [26] using the primers ERIC 1R (5′-ATGTAAGCTCCTGGGGATTCAC-3′) and ERIC 2 (5′-AAGTAAGTGGGGTGAGCG-3′). Samples were incubated for 5 min at 94 °C and then cycled 30 times at 94 °C for 30 s, 57 °C for 30 s, and 65 °C for 4 min. The final extension was performed at 65 °C at 8 min and kept at 4 °C until tested. The reproducibility of ERIC/PCR was also tested by amplifying gDNA from randomly chosen strains several times. For both rep-PCR, pattern band lengths were determined by comparison against a 100 bp plus DNA ladder for the smallest bands and by 1 Kbp DNA ladder for the largest bands (Takara Bio, Inc., Otsu, Shiga, Japan).

2.5. Statistics and Clustering Analysis of $(GTG)_5$ and ERIC Rep-PCR Patterns

Patterns obtained from $(GTG)_5$ and ERIC rep-PCR runs were imported into BioNumerics tool, package version 8. The cluster analysis was performed calculating Dice coefficients using a tolerance of 1% and optimization of 1%. Clustering based on Dice coefficients was performed using the unpaired group method with arithmetic average (UPGMA) method. For each fingerprint assay, the cluster cut-off analysis was applied to define the most reliable clusters. The cophenetic correlation index was used as a statistical tool to evaluate the quality of branches. For each of the genotyping experiments, the discriminatory index (D), expressed as the probability that two strains consecutively taken from a sample would be placed into different clades, was calculated as described by Hunter and Gaston [27].

Correlation analysis was performed considering the most abundant clades (Clade 5, 6 and 7), in order to obtain statistically significant results. The Spearman correlation index was calculated using the Hmisc v4.4–2 R package [28] implemented in R v 4.0.3 [29]. The correlation plot was obtained using the Corrplot v 0.84 package [30].

3. Results

3.1. Kombucha Characteristics and Isolated Strains

Within 3 days of cultivation in 600 mL beakers, the two kombucha tea samples, GTK and BTK, demonstrated a thin exopolysaccharide layer that became thicker with time (from 2–3 mm to 10 mm at the end of the cultivation period). Optical microscopy observations showed a high number of free bacterial and yeast cells, as well as aggregates of cells within the matrix, making cell counting uncertain and not enlightening even from the first day of fermentation.

Titratable acidity reached a maximum of 12 g/L at the end of fermentation in the black kombucha trial (GTK) and 6 g/L in the green one (BTK). pH dropped from approximately 3.7 to 2.75 for both samples, as a result of acid formation. Although no inhibition compounds were determined in this study, the difference in the final amount of acetic acid of the GTK and BTK samples could be due to the occurrence of more antibacterial compounds in GTK [31]. Ethanol was nearly 0 at the beginning of the cultivation time and reached 0.28% (v/v) and 0.30% (v/v) at the final time, for samples GTK and BTK, respectively. The low value of observed ethanol is in agreement with the gradual increase in acetic acid during cultivation time.

The isolation was performed from preculture (P1 and P2) and samples at 0, 6, and 12 days of cultivation. A total of 122 isolates were selected based on the colony and/or cellular morphologies. Microscopically, presumptive acetic acid bacteria cells appeared in single or grouped pairs or short chains, Gram-negative, catalase-positive, and KOH positive. A total of 46 isolates, which oxidized both ethanol and acetic acid, as confirmed by the ethanol chalk–ethanol test on GYC and the colorimetric assay on ACB, and produced a well-defined cellulose layer and/or formed both a cellulose layer and cellulose particles dispersed in the surrounding liquid, were considered in this study (Table 1).

Table 1. Features of 46 isolates from Kombucha tea used in this study. Site of isolation and RFLP analysis. In bold are reported the strains deposited in UNIMORE Microbial Culture Collection (UMCC) collection; within the brackets is the assigned accession number.

* Strain	Sample	Time	BC (g/L)	Colony Morphology	Liquid Growth	*Alu*I (pb)	Group	*Rsa*I (bp)	Group	Reference
K1A18	liquid	6	4.0448 ± 0.0008	green cellulosic	cellulose layer	-	-	120, 400, 425, 550	A	This study
K1A34	liquid	12	5.6214 ± 0.0010	green with light cellulosic edge	cellulose layer	150, 450, 800	1	120, 400, 425, 550	A	This study
K1A6	liquid	0	3.0024 ± 0.0009	cellulosic small	cellulose layer	150, 450, 800	1	120, 400, 425, 550	A	This study
K1A7	liquid	0	1.5886 ± 0.0004	cellulosic	cellulose layer/deposit	150, 450, 800	1	120, 400, 425, 550	A	This study
K1A8	liquid	0	2.5886 ± 0.0002	cellulosic	cellulose layer	150, 450, 800	1	120, 400, 425, 550	A	This study
K1A9	liquid	0	3.4014 ± 0.0004	green cellulosic	cellulose layer	-	-	-	-	This study
K1G2 (=UMCC 2964)	liquid	P1	4.7895 ± 0.0014	white cellulosic edge	cellulose layer/deposit	150, 450, 800	1	120, 400, 425, 550	A	[20]
K1G22	liquid	12	0.7057 ± 0.0003	cellulose light	cellulose layer/deposit	150, 450, 800	1	120, 400, 425, 550	A	[20]
K1G23	liquid	12	4.8122 ± 0.0003	dark cellulose	cellulose layer	150, 450, 800	1	120, 400, 425, 550	A	[20]
K1G24	liquid	12	4.1210 ± 0.0001	white cellulose	only cellulose deposit	-	-	120, 400, 425, 550	A	[20]
K1G3	liquid	P1	2.1857 ± 0.0052	light beige	cellulose layer	150, 450, 800	1	120, 400, 425, 550	A	[20]
K1G4	liquid	0	1.1738 ± 0.0032	cellulosic	cellulose layer/deposit	150, 450, 800	1	120, 400, 425, 550	A	[20]
K1G5	liquid	0	4.9590 ± 0.0027	cellulosic	cellulose layer/deposit	150, 450, 800	1	120, 400, 425, 550	A	[20]
K1G6	liquid	0	5.3676 ± 0.0025	white cellulosic	cellulose layer/deposit	150, 450, 800	1	120, 400, 425, 550	A	[20]
K2A10 (=UMCC 2965)	liquid	0	5.0405 ± 0.0008	cellulosic green small	cellulose layer	150, 450, 800	1	120, 400, 425, 550	A	This study
K2A28	liquid	6	5.8962 ± 0.0007	green cellulosic	cellulose layer	150, 450, 800	1	120, 400, 425, 550	A	This study
K2A32	pellicle	6	5.6143 ± 0.0001	green cellulosic	cellulose layer	-	-	-	-	This study
K2A33	liquid	6	0.6414 ± 0.0005	green cellulosic	cellulose layer	150, 450, 800	1	120, 400, 425, 550	A	This study
K2A44	liquid	12	5.1971 ± 0.0002	small green	cellulose layer/deposit	150, 450, 800	1	120, 400, 425, 550	A	This study

Table 1. *Cont.*

* Strain	Sample	Time	BC (g/L)	Colony Morphology	Liquid Growth	*Alu*I (pb)	Group	*Rsa*I (bp)	Group	Reference
K2A45	pellicle	12	5.2133 ± 0.0004	light edge	cellulose layer	150, 450, 800	1	120, 400, 425, 550	A	This study
K2A46	liquid	12	5.7833 ± 0.0001	dark center light edge	cellulose layer	150, 450, 800	1	120, 400, 425, 550	A	This study
K2A47	pellicle	12	5.8019 ± 0.0001	blue medium displayed	cellulose layer	150, 450, 800	1	120, 400, 425, 550	A	This study
K2A7	liquid	0	4.6076 ± 0.0007	green cellulosic small	cellulose layer	150, 450, 800	1	400, 450, 550	B	This study
K2A8	liquid	0	4.4781 ± 0.0004	green cellulosic small	cellulose layer	-	-	-	-	This study
K2G1	liquid	P2	6.8843 ± 0.0004	cellulosic	cellulose layer/deposit	150, 450, 800	1	120, 400, 425, 550	A	This study
K2G10	liquid	0	4.9752 ± 0.0016	cellulosic small	cellulose layer/deposit	150, 450, 800	1	400, 450, 550	B	[20]
K2G11	liquid	0	1.0638 ± 0.0003	white cellulosic	cellulose layer/deposit	150, 450, 800	1	120, 400, 425, 550	A	[20]
K2G12	liquid	0	6.2805 ± 0.0001	small cellulosic	cellulose layer/deposit	150, 450, 800	1	120, 400, 425, 550	A	[20]
K2G14	liquid	0	7.7414 ± 0.0005	white cellulosic	cellulose layer/deposit	150, 450, 800	1	120, 400, 425, 550	A	[20]
K2G15	liquid	0	7.5448 ± 0.0001	creamy	cellulose layer/deposit	150, 450, 800	1	400, 450, 550	A	[20]
K2G2	pellicle	P2	6.2548 ± 0.0004		cellulose layer/deposit	150, 450, 800	1	120, 400, 425, 550	A	[20]
K2G29 (=UMCC 2967)	pellicle	6	7.5848 ± 0.0002	cellulosic	cellulose layer	150, 450, 800	1	400, 450, 550	B	[20]
K2G30	pellicle	6	23.1829 ± 0.0001	cellulosic	cellulose layer/deposit	150, 450, 800	1	400, 450, 550	B	[20]
K2G31	pellicle	6	1.1267 ± 0.0007	cellulosic	cellulose layer/deposit	-	-	-	-	[20]
K2G32 (=UMCC 2968)	liquid	6	9.9976 ± 0.0003	cellulosic	cellulose layer/deposit	150, 450, 800	1	120, 400, 425, 550	A	[20]
K2G34	liquid	6	6.6614 ± 0.0001	cellulosic	cellulose layer	-	-	-	-	[20]
K2G35	liquid	6	1.9314 ± 0.0005	cellulosic	cellulose layer/deposit	150, 450, 800	1	120, 400, 425, 550	A	[20]
K2G36	liquid	6	0.5929 ± 0.0004	yellow cellulosic	cellulose layer	150, 450, 800	1	400, 450, 550	B	[20]

Table 1. *Cont.*

* Strain	Sample	Time	BC (g/L)	Colony Morphology	Liquid Growth	*Alu*I (pb)	Group	*Rsa*I (bp)	Group	Reference
K2G38 (=UMCC 2969)	liquid	6	9.6957 ± 0.0009	cellulosic	cellulose layer	150, 450, 800	1	120, 400, 425, 550	A	[20]
K2G39 (=UMCC 2970)	liquid	6	10.3957 ± 0.0001	cellulosic	cellulose	150, 450, 800	1	400, 450, 550	B	[20]
K2G41 (=UMCC 2971)	liquid	6	9.1443 ± 0.0003	cellulosic	cellulose layer	150, 450, 800	1	400, 450, 550	B	[20]
K2G44 (=UMCC 2972)	pellicle	12	5.9500 ± 0.0005	dark brown cellulose	cellulose layer/deposit	150, 450, 800	1	120, 400, 425, 550	A	[20]
K2G46	liquid	12	9.1195 ± 0.0007	brown cellulose	cellulose layer/deposit	-	-	-	-	[20]
K2G5 (=UMCC 2966)	liquid	P2	8.0605 ± 0.0003	white cellulosic	cellulose layer	-	-	120, 400, 425, 550	A	[20]
K2G6	liquid	P2	7.7533 ± 0.0003	white cellulosic	cellulose layer/deposit	150, 450, 800	1	120, 400, 425, 550	A	[20]
K2G8	liquid	0	6.8343 ± 0.0002	white cellulosic	cellulose layer	150, 450, 800	1	120, 400, 425, 550	A	[20]

(-) not detected; * Strain labeling: K (Kombucha tea); 1 (BTK); 2 (GTK); A (ACB medium); G (GYC medium); Number: progressive number. BC: bacterial cellulose; BTK: black tea kombucha; GTK: green tea kombucha.

Qualitative cellulose tests confirmed their ability to produce BC. Significant macroscopic diversity in the cellulosic material was observed, coupled with high variability in terms of amount of BC, as reported in Figure 1. The strains with a BC production level of at least higher than 4 g/L were deposited in the UMCC culture collection with the code reported in Table 1.

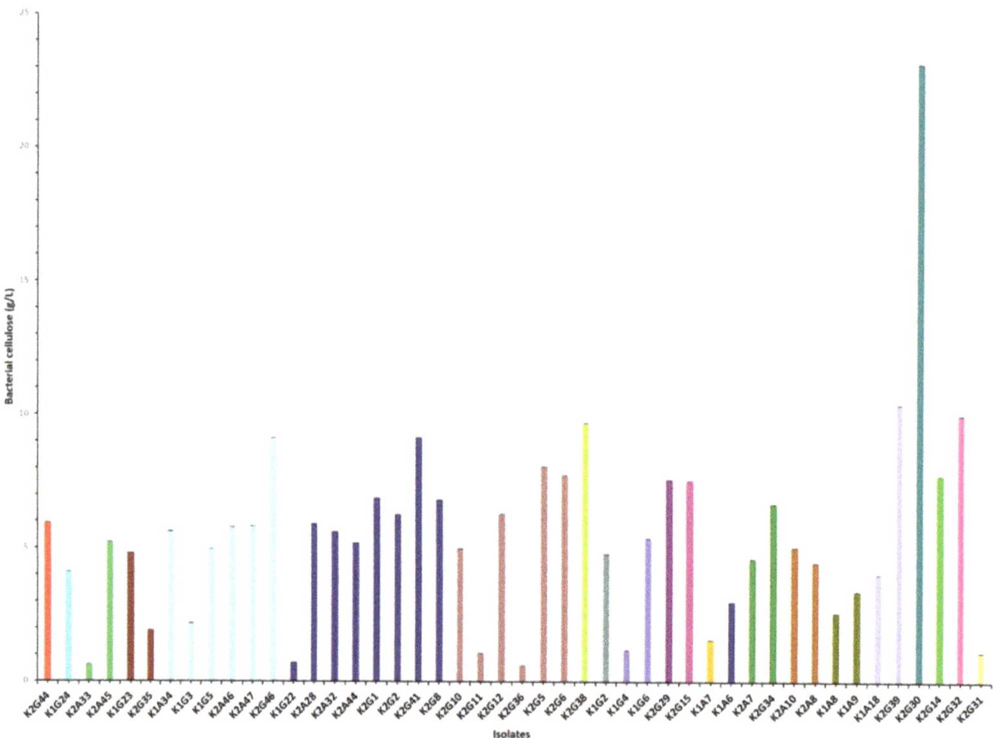

Figure 1. Bacterial cellulose yields of selected strains isolated from Kombucha tea. Each value is the mean of three parallel replicates ± standard deviation.

3.2. Diversity of Acetic Acid Bacteria from Kombucha Tea

The RFLP analysis on 16S rRNA, using *Rsa*I and *Alu*I restriction endonucleases, was conducted as a preliminary test to investigate the diversity of isolates recovered from kombucha tea samples. The mapping analysis based on the size of the fragments obtained by *Rsa*I and *Alu*I showed high homogeneous results, indicating the isolates as members of the same species. This evidence was in agreement with a previous study [16] which grouped *Komagataeibacter* strains according to *Rsa*I and *Alu*I restriction enzymes and observed similarities between the pattern of *K. xylinus* DSMZ 2004 with the *Komagataeibacter* strains investigated in this study. The low variability of the fingerprinting pattern is also consistent with previous studies on the high similarity of the 16S rRNA gene among the *Komagataeibacter* genus and especially within the *K. xylinus* species [32,33].

From the genotyping of strains performed by using (GTG)$_5$ and ERIC fingerprinting techniques, two dendrograms were generated (Figure 2). Based on the cluster cut-off analysis, profiles obtained using the amplification of interspersed tandem repeats GTG (Figure 2) were capable of discriminating between the strains considering a minimum percentage of similarity of 27.9%. A total of five major clades were created with a discrimination power of 0.42, calculated using the D index (Simpson index) [27]. In order to improve the discrimination power of the analysis, we considered a minimum percentage of similarity of 70%. At this percentage of similarity, the resulted discrimination power was 0.94, defining 20 different biotypes. Among the detected clades, eight of them included just one isolate while seven were formed by two isolates. It was possible to distinguish two major clades, of which, one included nine isolates (clade 18) and another included five isolates (clade 19). The two reference strains, represented by K1G4 (=UMCC 2947) and K2G30 (=UMCC 2756) clustered into two different clades, represented by clades 19 and 7, respectively. The similarity percentage among the two references was 40.2%.

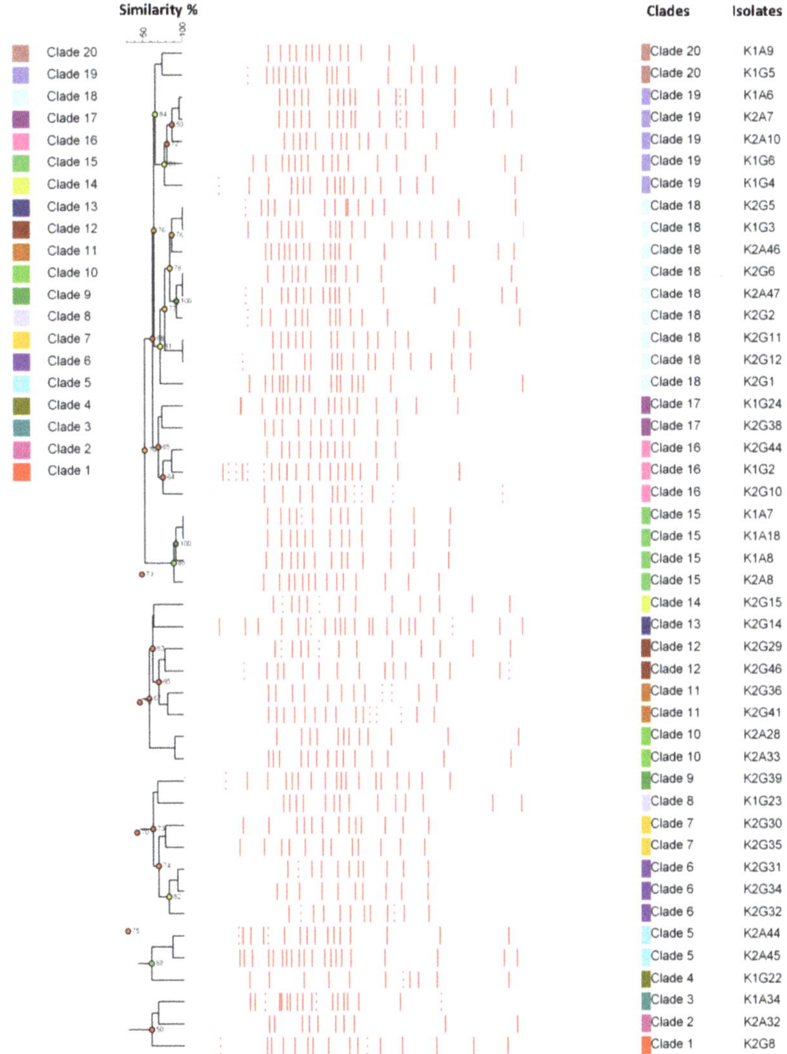

Figure 2. Dendrograms obtained from UPGMA analysis, using Dice's coefficient, of the digitized patterns obtained from (GTG)₅ rep-PCR. The similarity threshold for biotypes discrimination was 70%. The cophenetic coefficient is represented by numbers and dots colored as red–orange–yellow–green, based on the branch quality.

The phylogenetic tree obtained from the digitized pattern profile using ERIC rep-PCR is represented in Figure 3. Based on the clustering cut-off analysis, ERIC was able to discriminate using a minimum similarity percentage of 40%, identifying five major clades, as in the case of $(GTG)_5$. The clustering cut-off analysis discriminated a total of 19 biotypes, with a discrimination power of 0.95, slightly higher compared to $(GTG)_5$. A total of six clades were represented by just one isolate, while the remaining isolates were well distributed among the detected clades. Three clades were represented by five isolates (clade 7, 16, and 17) and two clusters by four isolates (clade 8 and 9). The remaining clades were represented by three and two isolates. The number of detected clades was not so different compared to $(GTG)_5$ and, also in this case, the reference strains (K1G4 and K2G30) were found to be separated into two different groups (clade 15 for K1G4 and clade 12 for K2G30). The similarity of Dice indexes based on the patterns obtained from ERIC rep-PCR was 56.7%; quite similar to that obtained using $(GTG)_5$ rep-PCR (40.2%).

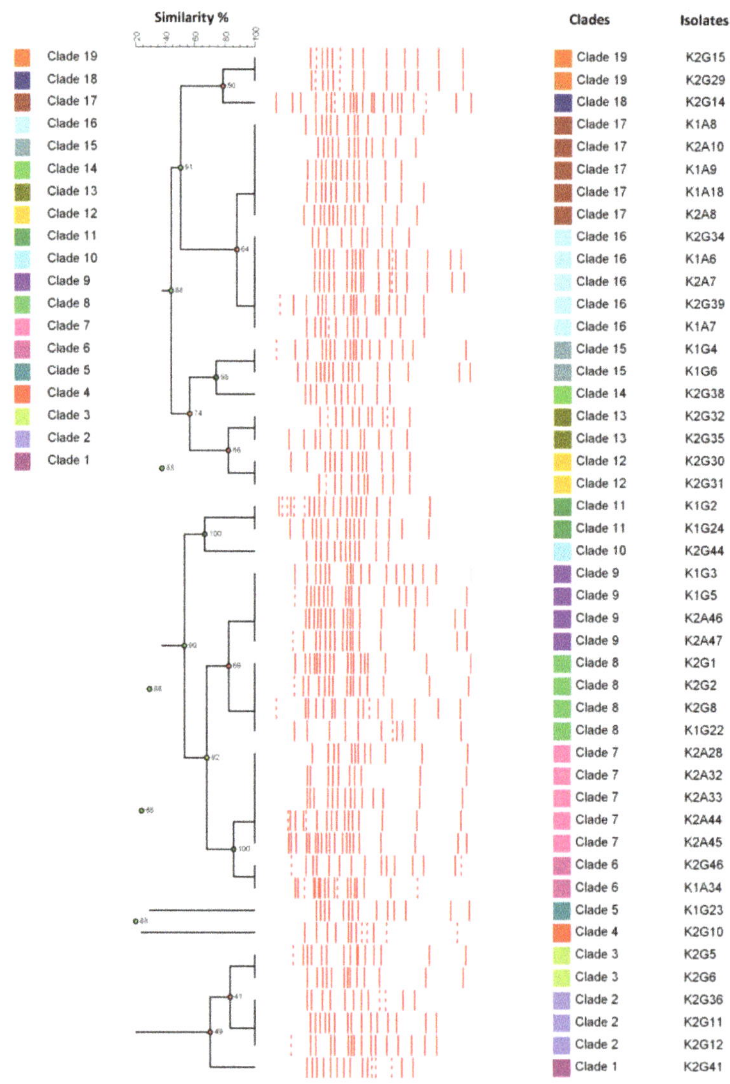

Figure 3. Dendrograms obtained from UPGMA analysis, using Dice's coefficient, of the digitized patterns obtained from ERIC rep-PCR. The similarity threshold for biotypes discrimination was 70%. The cophenetic coefficient is represented by numbers and dots colored as red–orange–yellow–green, based on the branch quality.

3.3. Improving the Fingerprinting of Komagataeibacter Strains

In order to improve the sensitivity of the intraspecific discrimination, the data obtained from both rep-PCR assays were combined. The clustering analysis is reported in Figure 4. Based on the cluster cut-off analysis, the combination of (GTG)$_5$ and ERIC rep-PCR allowed for discriminating the isolates, grouping them into 22 clades, considering a similarity percentage threshold of 94%. The discrimination power, calculated as the Simpson index, was 0.94, mining a high intraspecies diversity. Most of the isolates were grouped into three clusters (clade 5, 6, and 7) representing 43.5% of the isolates ($n = 20$). The remaining isolates ($n = 26$) were dispersed in 19 clades, of which, most of them consisted of just one isolate ($n = 12$). The reference strains were clustered into two different clades, as observed in both the case of (GTG)$_5$ and ERIC rep-PCR. Clade 10, in which K1G4 was clustered with K1G6, and clade 19, where K2G30 clustered alone, had a similarity percentage of Dice index of 84.9%, higher considering the separated results from (GTG)$_5$ and ERIC rep-PCR. This result is in agreement with the meaning of the combined analysis, since the two reference strains belong to the species of *K. xylinus*, as previously stated [14,34].

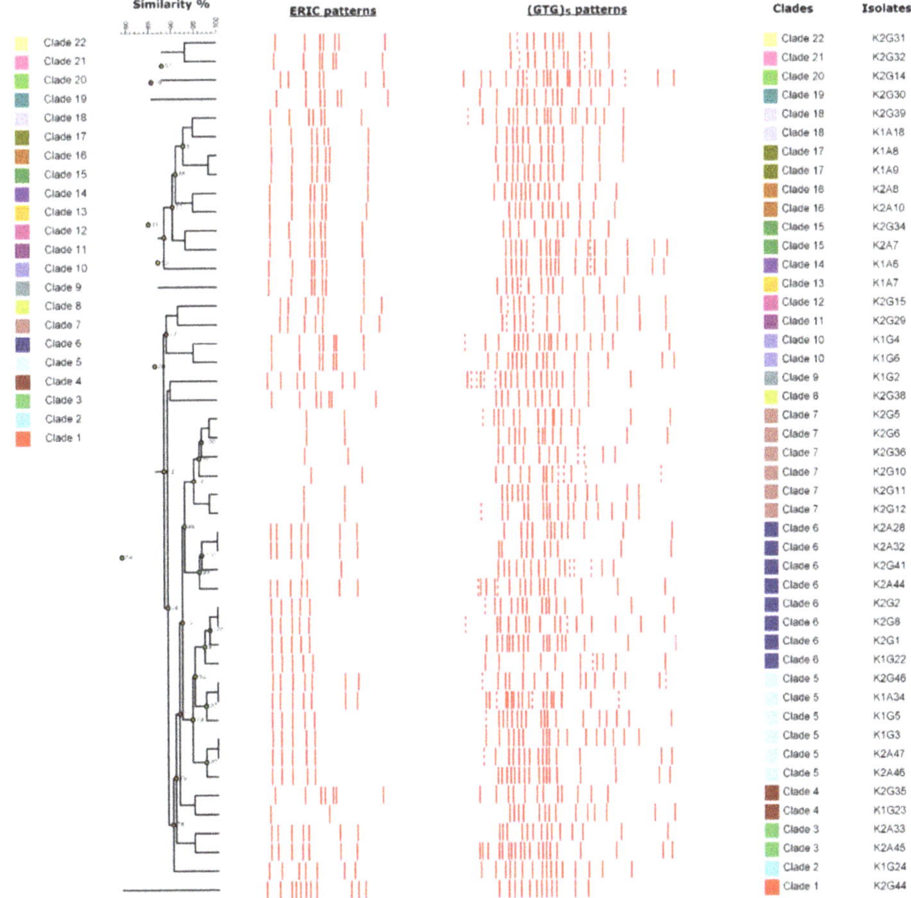

Figure 4. Dendrogram obtained combining the digitalized patterns from both (GTG)5 rep-PCR and ERIC rep-PCR. The dendrogram was drawn from UPGMA analysis using Dice's coefficient. The discrimination of biotypes was performed considering a similarity threshold of 94%. The cophenetic coefficient is represented by numbers and dots colored as red–orange–yellow–green, based on the branch quality. Cluster colors were defined in agreement with the Figure 1.

3.4. Correlation between Strains and Cellulose Yield

In order to understand putative correlation of the BC yield obtained from isolates with time and environmental factors, such as sampling fraction (pellicle or liquid) and clustering methods applied in this study, a Spearman correlation index was calculated. At the beginning all clades that resulted from the combined fingerprinting approach (Figure 4) were compared with all variables (Figure 5a). All factors seemed to strongly influence the BC yields. Only the time of sampling seemed to slightly influence BC production, but with a low value (R2 = -0.38, $p = 0.009$). In order to understand how the sampling time could affect BC yield, the three most abundant clades, represented by clade 5, 6 and 7 were evaluated by calculating the Spearman coefficient, considering both time and sampling fraction. Regarding clade 5 ($n = 6$) (Figure 5b), the time seems to correlate positively (R2 = 0.84, $p = 0.03$). Indeed, the amount of BC (Figure 1) was higher for isolates sampled at day 12 (K2A46, K2A47, K2G46 and K1A34) and lower for isolates sampled at day 0 and from the preculture (K1G5 and K1G3 respectively). In clade 6 ($n = 8$) (Figure 5c), BC yield seems to be not statistically correlated by the sampling time (R2 = -0.70, $p > 0.05$). Regarding clade 7 ($n = 6$) (Figure 5d), the sampling time negatively affects the BC amount (R2 = -0.92, $p = 0.009$).

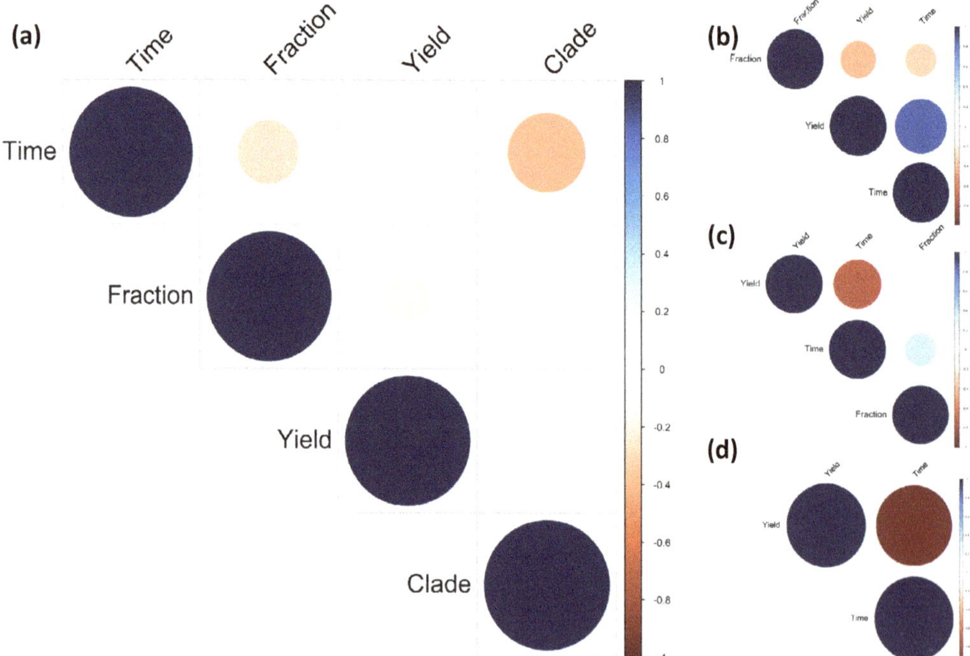

Figure 5. Correlation plot of the most abundant clades. Spearman correlation among time, environmental variables and obtained yields. (**a**) All retrieved clades were compared with time and environmental variables. Yield obtained within (**b**) clade 5, (**c**) clade 6 were compared with time and fraction variables. For clade 7 (**d**), includes only isolates sampled from liquid fraction.

4. Discussion

Bacterial cellulose is receiving a great deal of attention due to several features that make it suitable for many biotechnological applications, such as biomedical devices, food packaging components and engineering materials [1,2,35].

It is well known from the literature that kombucha tea represents one of the most abundant reservoirs of BC producer bacteria, in particular, strains belonging to the *Komagataeibacter* genus [10,12].

In this study, kombucha tea was chosen as a selective source for recovering cellulose producing bacteria. A total of 46 isolates were selected as the main components of the liquid and pellicle fractions of GTK and BTK kombucha tea samples and their phenotypic and genotypic variability was evaluated. RFLP of 16S rRNA gene and an analysis of the interspecific variability of isolates was conducted, and the results were discussed, considering the available genome sequences of strains K1G4 and K2G30.

The data obtained by RFLP analysis showed a very low variability in terms of the number of different species isolated from kombucha tea. This result is in agreement with previous studies reflecting both the low species diversity in kombucha tea [10], as well as the difficulty of taxonomic assignment of *Acetobacteraceae* by using 16S rRNA gene, due to its similarity in high phylogenetic related organisms, as extensively reported in the literature [36].

Considering the analysis of the intraspecific variability within bacteria, usually, genotyping is performed using the amplification of highly conserved tandem repeat regions that are widely dispersed in the genome. The amplification of repeated elements of genomic DNA, such as $(GTG)_5$ and ERIC sequences, offers some advantages in terms of costs and rapidity, and is well developed for several Gram-negative and Gram-positive bacterial species, such as *Lactobacilli* [37] and *Enterococci* [38] and acetic acid bacteria [39].

In this study, the clustering analysis using the digitized patterns of $(GTG)_5$ elements revealed the clustering of the 46 isolates into five large clusters considering a cluster cut-off value of 50% in similarity. The discrimination power of this analysis was 0.94 considering a similarity threshold of 70%. At this similarity value, a total of 20 different biotypes were detected. Similar results were obtained using the amplification of ERIC repeated elements. Considering the cluster cut-off analysis, the comparison of the digitized patterns of ERIC rep-PCR showed a clustering of the 46 isolates with a similarity percentage of 50%, the same result obtained using $(GTG)_5$ rep-PCR. The discrimination power of ERIC fingerprint was 0.95, very similar to the $(GTG)_5$ rep-PCR, considering a similarity percentage of 70. Based on these results, a total of 19 different clades were identified, compared to the 20 obtained using $(GTG)_5$ rep-PCR. In both the fingerprint approaches, K1G4 and K2G30, previously identified as *K. xylinus* [14,34], were considered as reference strains and they were clustered into two different clades. Among the two reference strains, the percentage of similarity was 44.2% in the case of $(GTG)_5$ rep-PCR and 56.7% for ERIC rep-PCR. These results support the hypothesis that most of the strains analyzed belong to the *K. xylinus* species and that the discrimination was performed at the strain level.

In order to improve the genotyping analysis, a combined model of the two fingerprint assays was performed (Figure 3). The resulting analysis showed a clusterization of the isolates into seven clades based on the cluster cut-off value of 88% of pattern similarity. The discrimination power of this analysis was 0.94 at 94% of similarity. The number of biotypes defined by the combined model was 22; meaning that an improvement of the discrimination of the isolates was achieved compared to the single results obtained by the two approaches. Additionally, concerning the reference strains (K1G4 and K2G30) in this case, we observed a grouping into two different clades, with a similarity percentage of 84.9%, meaning that all the isolates with equal or major than that value possibly belong to the species of *K. xylinus*. Based on this consideration, all of the isolates could be assigned to the *K. xylinus* except one isolate, the K2G44, clustering out of the group, with a similarity percentage of 78.8%. The data were in agreement with previous studies of the fingerprinting analysis of acetic acid bacteria [39,40], in which the differences of strains among the same species (as *Komagataeibacter xylinus*) were about 80%. In this case, combining the two methods, the detection of different strains was improved.

The correlation analysis deriving from the combined strain clustering allowed us to link the BC yield with isolation time and environmental variables. A possible explanation of the obtained results may lie in the genetic plasticity features of acetic acid bacteria. It is well known in the literature that acetic acid bacteria are able to adapt themselves to environmental stressors, such as temperature and high concentration of organic acids, by acquiring genetic material from the environment or losing it. Indeed, acetic acid bacteria possess a large number of transposons genes and plasmids [41]. One example is provided by *Acetobacter* species which strains adapt themselves to environmental conditions along cultivation and fermentation cycles [42,43]. In *Komagataeibacter* species the genetic plasticity has not yet been evaluated. However, we believe this aspect should be elucidated in the light to better understand the variability of BC yield within strains of *Komagataeibacter* genus.

The kombucha tea samples analyzed in this study allowed us to recover a pool of acetic acid bacteria strains, which have been characterized and deposited into the UMCC culture collection as promising candidates for BC production.

Author Contributions: M.G. conceptualized the research, provided resources and supervised the work; D.M. performed experiments; S.L.C. performed genotyping analysis and wrote part of the original draft; L.D.V. reviewed and edited the manuscript; K.A. wrote part of the original draft; M.B. participated in the investigation, reviewed and edited the manuscript. All authors have read and agreed to the published version of the manuscript.

Funding: Part of this research was granted by The Research Doctorate School in Food and Agricultural Science, Technology and Biotechnology (STEBA).

Institutional Review Board Statement: Not applicable.

Informed Consent Statement: Not applicable.

Acknowledgments: The National Institute of Applied Sciences and Technology (INSAT, Tunisia), is acknowledged for providing Kombucha samples. A special thanks to The Research Doctorate School in Food and Agricultural Science, Technology and Biotechnology (STEBA) is acknowledged for assigning the *"Michela Stanca"* award. The JRU MIRRI-IT is greatly acknowledged for technical and scientific support.

Conflicts of Interest: The authors declare no conflict of interest.

References

1. Robotti, F.; Bottan, S.; Fraschetti, F.; Mallone, A.; Pellegrini, G.; Lindenblatt, N.; Starck, C.; Falk, V.; Poulikakos, D.; Ferrari, A. A micron-scale surface topography design reducing cell adhesion to implanted materials. *Sci. Rep.* **2018**. [CrossRef]
2. Haghighi, H.; Gullo, M.; La China, S.; Pfeifer, F.; Siesler, H.W.; Licciardello, F.; Pulvirenti, A. Characterization of bio-nanocomposite films based on gelatin/polyvinyl alcohol blend reinforced with bacterial cellulose nanowhiskers for food packaging applications. *Food Hydrocoll.* **2020**, 106454. [CrossRef]
3. Barbi, S.; Taurino, C.; La China, S.; Anguluri, K.; Gullo, M.; Montorsi, M. Mechanical and structural properties of environmental green composites based on functionalized bacterial cellulose. *Cellulose* **2021**, 1–7. [CrossRef]
4. Valera, M.J.; Torija, M.J.; Mas, A. Detrimental effects of acetic acid bacteria in foods. In *Acetic Acid Bacteria: Fundamentals and Food Applications*, 1st ed.; Sengun, I., Ed.; CRC Press: Boca Raton, FL, USA; Taylor & Francis Group: Boca Raton, FL, USA, 2017; Volume 10, pp. 299–320.
5. Gullo, M.; La China, S.; Falcone, P.M.; Giudici, P. Biotechnological production of cellulose by acetic acid bacteria: Current state and perspectives. *Appl. Microbiol. Biotechnol.* **2018**, 1–14. [CrossRef] [PubMed]
6. Son, H.J.; Heo, M.S.; Kim, Y.G.; Lee, S.J. Optimization of fermentation conditions for the production of bacterial cellulose by a newly isolated *Acetobacter* sp. A9 in shaking cultures. *Biotechnol. Appl. Biochem.* **2001**, *33*, 1. [CrossRef] [PubMed]
7. La China, S.; Zanichelli, G.; De Vero, L.; Gullo, M. Oxidative fermentations and exopolysaccharides production by acetic acid bacteria: A mini review. *Biotechnol. Lett.* **2018**, *40*, 1289–1302. [CrossRef]
8. Coton, M.; Pawtowski, A.; Taminiau, B.; Burgaud, G.; Deniel, F.; Coulloumme-Labarthe, L.; Fall, A.; Daube, G.; Coton, E. Unraveling microbial ecology of industrial-scale Kombucha fermentations by metabarcoding and culture-based methods. *FEMS Microbiol. Ecol.* **2017**, *93*. [CrossRef]
9. Laureys, D.; Britton, S.J.; De Clippeleer, J. Kombucha tea fermentation: A review. *J. Am. Soc. Brew. Chem.* **2020**, *78*, 165–174. [CrossRef]
10. Sievers, M.; Lanini, C.; Weber, A.; Schuler-Schmid, U.; Teuber, M. Microbiology and fermentation balance in a Kombucha beverage obtained from a tea fungus fermentation. *Syst. Appl. Microbiol.* **1995**, *18*, 590–594. [CrossRef]
11. Arıkan, M.; Mitchell, A.L.; Finn, R.D.; Gürel, F. Microbial composition of Kombucha determined using amplicon sequencing and shotgun metagenomics. *J. Food Sci.* **2020**, *85*, 455–464. [CrossRef]

12. Villarreal-Soto, S.A.; Beaufort, S.; Bouajila, J.; Souchard, J.P.; Taillandier, P. Understanding Kombucha tea fermentation: A review. *J. Food Sci.* **2018**, *83*, 580–588. [CrossRef]
13. Gaggìa, F.; Baffoni, L.; Galiano, M.; Nielsen, D.; Jakobsen, R.; Castro-Mejía, J.; Bosi, S.; Truzzi, F.; Musumeci, F.; Dinelli, G.; et al. Kombucha beverage from green, black and rooibos teas: A comparative study looking at microbiology, chemistry and antioxidant activity. *Nutrients* **2018**, *11*, 1. [CrossRef]
14. Gullo, M.; La China, S.; Petroni, G.; Di Gregorio, S.; Giudici, P. Exploring K2G30 genome: A high bacterial cellulose producing strain in glucose and mannitol based media. *Front. Microbiol.* **2019**, *10*, 1–12. [CrossRef]
15. Singhsa, P.; Narain, R.; Manuspiya, H. Physical structure variations of bacterial cellulose produced by different *Komagataeibacter xylinus* strains and carbon sources in static and agitated conditions. *Cellulose* **2018**, *25*, 1571–1581. [CrossRef]
16. Mamlouk, D.; Gullo, M. Acetic acid bacteria: Physiology and carbon sources oxidation. *Indian J. Microbiol.* **2013**, *53*, 377–384. [CrossRef] [PubMed]
17. Jayabalan, R.; Marimuthu, S.; Swaminathan, K. Changes in content of organic acids and tea polyphenols during kombucha tea fermentation. *Food Chem.* **2007**, *102*, 392–398. [CrossRef]
18. Mamlouk, D. Insight Into Physiology and Functionality of Acetic Acid Bacteria Through a Multiphasic Approach. Ph.D. Thesis Dissertation, University of Modena and Reggio Emilia, Reggio Emilia, Italy, 2012.
19. De Vero, L.; Boniotti, M.B.; Budroni, M.; Buzzini, P.; Cassanelli, S.; Comunian, R.; Gullo, M.; Logrieco, A.F.; Mannazzu, I.; Musumeci, R.; et al. Preservation, characterization and exploitation of microbial biodiversity: The perspective of the Italian Network of Culture Collections. *Microorganisms* **2019**, *7*, 685. [CrossRef]
20. Gullo, M.; Mamlouk, D.; De Vero, L.; Giudici, P. *Acetobacter pasteurianus* strain AB0220: Cultivability and phenotypic stability over 9 years of preservation. *Curr. Microbiol.* **2012**, *64*, 576–580. [CrossRef] [PubMed]
21. Navarro, R.; Uchimura, T.; Komagata, K. Taxonomic heterogeneity of strains comprising *Gluconacetobacter hansenii*. *J. Gen. Appl. Microbiol.* **1999**, *45*, 295–300. [CrossRef] [PubMed]
22. Gullo, M.; Sola, A.; Zanichelli, G.; Montorsi, M.; Messori, M.; Giudici, P. Increased production of bacterial cellulose as starting point for scaled-up applications. *Appl. Microbiol. Biotechnol.* **2017**, *101*, 8115–8127. [CrossRef] [PubMed]
23. Hwang, J.W.; Yang, Y.K.; Hwang, J.K.; Pyun, Y.R.; Kim, Y.S. Effects of pH and dissolved oxygen on cellulose production by *Acetobacter xylinum* BRC5 in agitated culture. *J. Biosci. Bioeng.* **1999**, *88*, 183–188. [CrossRef]
24. Gullo, M.; De Vero, L.; Giudici, P. Succession of selected strains of *Acetobacter pasteurianus* and other acetic acid bacteria in traditional balsamic vinegar. *Appl. Environ. Microbiol.* **2009**, *75*, 2585–2589. [CrossRef] [PubMed]
25. Versalovic, J.; Schneider, M.; De Bruijn, F.J.; Lupski, J.R. Genomic fingerprinting of bacteria using repetitive sequence-based polymerase chain reaction. *Methods Mol. Cell. Biol.* **1994**, *5*, 25–40.
26. Versalovic, J.; Koeuth, T.; Lupski, R. Distribution of repetitive DNA sequences in *eubacteria* and application to fingerprinting of bacterial genomes. *Nucleic Acids Res.* **1991**, *19*, 6823–6831. [CrossRef] [PubMed]
27. Hunter, P.R.; Gaston, M.A. Numerical index of the discriminatory ability of typing systems: An application of Simpson's index of diversity. *J. Clin. Microbiol.* **1988**, *26*, 2465–2466. [CrossRef]
28. Harrell, F.E., Jr.; Dupont, M.C. *R Package Hmisc*; CRAN; R Foundation for Statistical Computing: Vienna, Austria, 2014. Available online: https://CRAN.R-project.org/package=Hmisc (accessed on 3 February 2020).
29. R Core Team. *R: A Language and Environment for Statistical Computing*; R Foundation for Statistical Computing: Vienna, Austria, 2017. Available online: https://www.R-project.org/ (accessed on 3 February 2020).
30. Taiyun, W.; Viliam, S. R package "corrplot": Visualization of a Correlation Matrix: Version 0.84. 2017. Available online: https://github.com/taiyun/corrplot (accessed on 3 February 2020).
31. Battikh, H.; Chaieb, K.; Bakhrouf, A.; Ammar, E. Antibacterial and antifungal activities of black and green kombucha teas. *J. Food Biochem.* **2013**. [CrossRef]
32. Sievers, M.; Sellmer, S.; Teuber, M. *Acetobacter europaeus* sp. nov., a main component of industrial vinegar fermenters in central Europe. *Syst. Appl. Microbiol.* **1992**, *15*, 386–392. [CrossRef]
33. Ludwig, W.; Klenk, H.P. Overview: A phylogenetic backbone and taxonomic framework for procaryotic systematics. In *Bergey's Manual of Systematic Bacteriology*, The Proteobacteria, Part A, Introductory Essays, 2nd ed.; Brenner, D.J., Krieg, N.R., Staley, J.T., Garrity, G.M., Eds.; Springer: New York, NY, USA, 2005; Volume 2, pp. 49–65.
34. La China, S.; Bezzecchi, A.; Moya, F.; Petroni, G.; Di Gregorio, S.; Gullo, M. Genome sequencing and phylogenetic analysis of K1G4: A new *Komagataeibacter* strain producing bacterial cellulose from different carbon sources. *Biotechnol. Lett.* **2020**, *42*, 807–818. [CrossRef]
35. Jakmuangpak, S.; Prada, T.; Mongkolthanaruk, W.; Harnchana, V.; Pinitsoontorn, S. Engineering bacterial cellulose films by nanocomposite approach and surface modification for biocompatible triboelectric nanogenerator. *ACS Appl. Electron. Mater.* **2020**, *2*, 2498–2506. [CrossRef]
36. Cleenwerck, I.; Vandemeulebroecke, K.; Janssens, D.; Swings, J. Re-examination of the genus *Acetobacter*, with descriptions of *Acetobacter cerevisiae* sp. nov. and *Acetobacter malorum* sp. nov. *Int. J. Syst. Evol. Microbiol.* **2002**, *52*, 1551–1558. [CrossRef]
37. Gevers, D.; Huys, G.; Swings, J. Applicability of rep-PCR fingerprinting for identification of *Lactobacillus* species. *FEMS Microbiol. Lett.* **2001**, *205*, 31–36. [CrossRef]
38. Švec, P.; Vancanneyt, M.; Seman, M.; Snauwaert, C.; Lefebvre, K.; Sedláček, I.; Swings, J. Evaluation of (GTG)$_5$-PCR for identification of *Enterococcus* spp. *FEMS Microbiol. Lett.* **2005**, *247*, 59–63. [CrossRef] [PubMed]

39. De Vuyst, L.; Camu, N.; De Winter, T.; Vandemeulebroecke, K.; Van De Perre, V.; Vancanneyt, M.; De Vos, P.; Cleenwerck, I. Validation of the (GTG)$_5$-rep-PCR fingerprinting technique for rapid classification and identification of acetic acid bacteria, with a focus on isolates from Ghanaian fermented cocoa beans. *Int. J. food Microbiol.* **2007**. [CrossRef]
40. Cleenwerck, I.; De Wachter, M.; González, Á.; De Vuyst, L.; De Vos, P. Differentiation of species of the family *Acetobacteraceae* by AFLP DNA fingerprinting: *Gluconacetobacter kombuchae* is a later heterotypic synonym of *Gluconacetobacter hansenii*. *Int. J. Syst. Evol. Microbiol.* **2009**, *59*, 1771–1786. [CrossRef] [PubMed]
41. Azuma, Y.; Hosoyama, A.; Matsutani, M.; Furuya, N.; Horikawa, H.; Harada, T.; Hirakawa, H.; Kuhara, S.; Matsushita, K.; Fujita, N.; et al. Whole-genome analyses reveal genetic instability of *Acetobacter pasteurianus*. *Nucleic Acids Res.* **2009**, *37*, 5768–5783. [CrossRef]
42. Matsutani, M.; Nishikura, M.; Saichana, N.; Hatano, T.; Masud-Tippayasak, U.; Theergool, G.; Yakushi, T.; Matsushita, K. Adaptive mutation of *Acetobacter pasteurianus* SKU1108 enhances acetic acid fermentation ability at high temperature. *J. Biotechnol.* **2013**, *165*, 109–119. [CrossRef] [PubMed]
43. Pothimon, R.; Gullo, M.; La China, S.; Thompson, A.K.; Krusong, W. Conducting High acetic acid and temperature acetification processes by *Acetobacter pasteurianus* UMCC 2951. *Process Biochem.* **2020**, *98*, 41–50. [CrossRef]

applied sciences

Article

Immobilization of Pectinolytic Enzymes on Nylon 6/6 Carriers

Sana Ben-Othman [1,*] **and Toonika Rinken** [1,2,*]

1   ERA Chair for Food (By-) Products Valorisation Technologies Valortech, Estonian University of Life Sciences, Kreutzwaldi 56/5, 51006 Tartu, Estonia

2   Institute of Chemistry, Faculty of Science and Technology, University of Tartu, Ravila 14a, 50411 Tartu, Estonia

*   Correspondence: sana.benothmanepaloulou@emu.ee (S.B.-O.); toonika.rinken@emu.ee (T.R.)

Abstract: Pectinolytic enzymes are an important tool for sustainable food production, with a wide range of applications in food processing technologies as well as the extraction of bioactive compounds from pectin-rich raw materials. In the present study, we immobilized commercial pectinase preparation onto pellet and thread shaped nylon 6/6 carriers and assessed its stability and reusability. Five commercial pectinase preparations were tested for different pectin de-polymerizing activities (pectinase, polygalacturonase, and pectin lyase activities). Thereafter, Pectinex® Ultra Tropical preparation, exhibiting the highest catalytic activities among the studied preparations ($p < 0.0001$), was immobilized on nylon 6/6 using dimethyl sulfate and glutaraldehyde. The immobilization yield was in accordance with the carrier surface area available for enzyme attachment, and it was 1.25 ± 0.10 U/g on threads, which was over 40 times higher than that on pellets. However, the inactivation of immobilized enzymes was not dependent on the shape of the carrier, indicating that the attachment of the enzymes on the surface of nylon 6/6 carriers was similar. The half-life of enzyme inactivation fast phase at 4 °C was 12.8 days. After 5 weeks, the unused threads retained 63% of their initial activity. Reusability study showed that after 20 successive cycles the remaining activity of the immobilized pectinase was 22%, indicating the good prospects of reusability of the immobilized enzyme preparations for industrial application.

Keywords: pectinase immobilization; nylon 6/6 carrier; pectinolytic activity; reusability; stability

Citation: Ben-Othman, S.; Rinken, T. Immobilization of Pectinolytic Enzymes on Nylon 6/6 Carriers. *Appl. Sci.* **2021**, *11*, 4591. https://doi.org/10.3390/app11104591

Academic Editors: Isidoro Garcia-Garcia, Jesus Simal-Gandara and Maria Gullo

Received: 27 April 2021
Accepted: 14 May 2021
Published: 18 May 2021

Publisher's Note: MDPI stays neutral with regard to jurisdictional claims in published maps and institutional affiliations.

1. Introduction

Pectinolytic enzymes, also known as pectinases, are a heterogeneous group of enzymes catalyzing different steps of the hydrolysis of pectin, a major component of plant cell walls [1]. Based on their mode of action, pectinases are classified as (i) de-esterifying enzymes (pectin esterase), and (ii) de-polymerizing enzymes (hydrolases, lyases). De-polymerizing enzymes are further classified based on their specific action site to endo- and exo-polygalacturonases and rhamnogalacturonases [2], or separated as per the nature of action mechanism, primary substrate, and products [3].

Pectinases have a wide range of applications in different industries including food, textile, and paper industries. Microbial pectinases are one of the most commercially produced enzymes, with 25% share in the global food and beverage enzyme market [4]. Pectinases are considered as a sustainable and environmentally-friendly industrial catalyst as they are produced by fungi (*Aspergillus, Penicillium, Trichoderma, Rhizomucor*, etc.) and bacterial (*Bacillus, Streptomyces, Erwina*) fermentation of different by-products like wheat bran, citrus peels, and sugar beet pulp [2,5]. Commercial pectinase preparations are widely utilized in the extraction and clarification of fruits and vegetables juices [6]. They are also used for the extraction of vegetable oil [7] and the fermentation of coffee beans and tea leaves [2]. Moreover, along with cellulases, pectinases are an important tool for 'green' enzyme-assisted extraction of bioactive compounds from different plant matrices including citrus peel, carrot pomace, tomato, etc. [8,9].

Several studies have highlighted the attractiveness of pectinases' immobilization allowing multiple reuses and increasing the efficiency of production processes, easy separation of the enzyme from the final product [10], but also the development of continuous automated processes particularly relevant in juice production and winemaking [11]. Various techniques have been used for the immobilization of different pectinase preparations including adsorption [12–14], entrapment [15–17], covalent binding to insoluble carriers [18–21], as well as to magnetic nanoparticles [22–25]. Despite the variety of immobilization methods used, the most stable enzyme coupling is achieved using covalent attachment. Commonly, enzymes are coupled via their amino groups, but other functional groups like carboxylic, thiol, sulfhydryl, and other groups can be also used.

Carriers for the immobilization of pectinase and other enzymes which are used in food industry must be non-toxic, biocompatible, and insoluble under reaction conditions, and thermally stable [26]. Functional groups for protein binding onto an insoluble carrier are usually generated through chemical activation of the support surface. In addition, linker molecules (e.g., glutaraldehyde, carbodiimide, etc.) are often used to enhance enzyme flexibility and activity [27]. For the technological production of immobilized enzymes, important criteria are the durability of the carrier, but also its availability and low cost [26]. Nylon 6/6 is an interesting, inexpensive, and convenient carrier for enzyme immobilization. It is inert, non-toxic, mechanically strong, and readily available in different forms (pellets, discs, threads, fabric, etc.) [28,29]. The inertness of nylon 6/6 makes it unable to bind proteins as it is, so it should be activated by partial acid hydrolysis or O-alkylation with dimethyl sulfate (DMS) [30]. Nylon 6/6 activation generates free amino (-NH$_2$) and carboxyl (-COOH) groups on its surface, allowing the covalent binding of enzymes using glutaraldehyde as a crosslinking agent [31].

In the present study, we used nylon 6/6 pellets and threads as carriers for the immobilization of multi-enzyme pectinase commercial preparation used for degrading fibrous plant material and providing higher juice yields and clarity. We optimized the protocols for the attachment of enzyme onto nylon 6/6 and assessed the stability of immobilized pectinase in the course of its storage and during application for multiple successive cycles.

2. Materials and Methods

2.1. Materials

Pectinase preparations Pectinex® Ultra SPL ($\geq$3800 U/mL, Sigma-Aldrich, Steinheim, Germany), Pectinex® BE XXL (13,600 PECTU/g), Pectinex® Ultra Mash (9500 PECTU/g) and Pectinex® Ultra Tropical (5000 PECTU/g) from Novozymes (Bagsvaerd, Denmark), and Panzym® Univers (11,000 PECTU/mL) from Begerow (Langenlonsheim, Germany) were used in this work. Apple pectin, polygalacturonic acid, and galacturonic acid were all purchased from Sigma-Aldrich (Steinheim, Germany). Nylon 6/6 pellets were from Sigma-Aldrich and threads consisting of 60 filaments with a diameter of 25 μm twisted together from Oja, Turkey. All chemical reagents and solvents used were of analytical grade.

2.2. Determination of Pectinolytic Activity

Pectinase activity was measured using apple pectin and polygalacturonase activity was measured using polygalacturonic acid as substrates. Aqueous substrate solution (0.5%, *w/v*) was prepared by adding purified water to polygalacturonic acid or apple pectin, heating the solution under continuous stirring until complete dissolution of compounds, and then cooling the solution to 25 °C. The pH value of the substrate solution was adjusted to 4.8 by adding NaOH solution (1N). Enzyme preparation 40-fold diluted (0.1 mL) or a definite amount of immobilized enzyme was added to 5 mL apple pectin or polygalacturonic acid solution (0.5%, *w/v*, pH 4.8) and incubated 5–30 min in a water bath at 40 °C. All activity measurements were carried out in triplicates.

2.2.1. Galacturonic acid Assay

The amount of galacturonic acid released after the enzymatic reaction was quantified spectrophotometrically using the copper reduction procedure reported by Rajdeo et al. [14]. Briefly, 0.3 mL of 10 mM copper sulfate solution (in 0.2 M CH3COONa buffer containing 2 M NaCl, pH 4.8) was mixed with 0.3 mL of sample in a capped 15-mL tube and incubated at 80 °C for 30 min. After incubation, 2.4 mL of diluted Folin-Ciocalteu reagent (40-fold, final concentration 0.05 N) was added. A blue-colored product immediately formed and the color intensity was determined by measuring the absorbance at 750 nm with a spectrophotometer (UV−1800, Shimadzu, Kyoto, Japan). Galacturonic acid content in the sample was calculated based on galacturonic acid standard curve in concentration range from 30 to 300 nmol.

2.2.2. Iodine Titration Method

The other method used for the measurement of galacturonic acid produced was the standard process involving galacturonic acid oxidation in the presence of I_2 and titration of excess I_2 with thiosulfate [32]. The enzyme reaction was stopped by adding 0.1 M I_2 solution (5 mL) and 1 M Na_2CO_3 solution (1 mL) and the mixture was incubated for 20 min in the dark at room temperature. After adding 2 N sulphuric acid (2 mL), the excess I_2 was titrated with 0.1 M $Na_2S_2O_3$ solution until the solution became colorless. The following equation was used for the calculation of enzyme activity:

$$\text{Activity}\left(\mu\text{mol GA} \times \text{min}^{-1} \times \text{mL}^{-1}\right) = \frac{\left(\text{mL } Na_2S_2O_3 \text{ for Blank} - \text{mL } Na_2S_2O_3 \text{ for Test}\right) \times 1 \times \text{df} \times 100}{2 \times V_{enzyme} \times t_{reaction}} \tag{1}$$

where:

1 = µmole galacturonic acid is oxidized by 1 microequivalent of I_2
100 = microequivalents of S_2O_3 per mL of titrant
df = dilution factor
2 = microequivalents of S_2O_3 oxidized per microequivalent of I_2 reduced
V_{enzyme} = volume (in mL) of enzume used
$t_{reaction}$ = time of incubation (in minutes) for the enzyme reaction

2.2.3. Pectin lyase Activity

Pectin lyase activity was measured using apple pectin as substrate. The activity was determined spectrophotometrically according to the method described by Dal Magro et al. [33] by monitoring the increase in the absorbance at 235 nm due to the formation of unsaturated uronide. An enzyme preparation 40-fold diluted (0.05 mL) was mixed with 1.95 mL pectin solution (0.5%, w/v, pH4.8) and incubated for 1 min at 40 °C. The reaction was stopped by adding 3 mL of 0.5 M HCl and the absorbance at 235 nm was measured using spectrophotometer (UV-1800, Shimadzu, Japan). Unsaturated uronide concentration was calculated according to the molar attenuation coefficient $\varepsilon = 5500$ $M^{-1}.cm^{-1}$ and one unit of pectin lyase activity is defined as one nmol of unsaturated uronide released per minute per mL enzyme preparation.

2.3. Enzyme Immobilization onto Nylon 6/6

Pectinase preparation Pectinex Ultra Tropical was immobilized on different nylon 6/6 carriers (pellets and thread). Nylon thread was cut into 2 m length fragments and coiled around Teflon rings (4 cm outer diameter, 3.5 cm inner diameter).

First, nylon carriers were activated with dimethyl sulfate (DMS) at 60 °C for 1 to 7.5 min, washed two times with ice-cold methanol and several times with ice-cold 0.1 M phosphate buffer (pH 7.0). Then, the carriers were treated for 60 min with glutaraldehyde solution (2.5% or 12.5% in 0.1 M phosphate buffer, pH 7) at room temperature. After washing several times with phosphate buffer, the carriers were incubated with diluted pectinase preparation (20-fold in 0.2 M sodium acetate buffer pH 4.8) for 24 h at 4 °C. After 24 h, nylon carriers were removed from the enzyme solution, washed thoroughly 6 times with ~30 mL 0.2 M sodium acetate buffer (pH 4.8), and stored in the same buffer at 4 °C. For easier handling of threads during the immobilization process and assessment of the activity of immobilized pectinases on threads, we used threads coiled around Teflon rings (Ø 4 cm) (Figure 1).

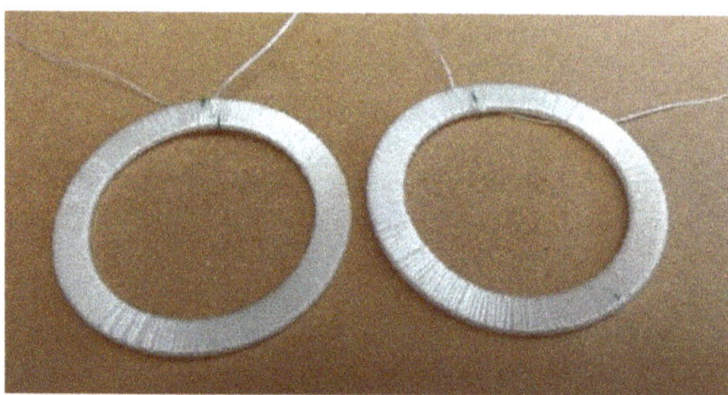

Figure 1. Nylon threads coiled around Teflon rings.

2.4. Statistical Analysis

The statistical analysis was conducted using Prism version 5 (GraphPad Software, San Diego, CA, USA). Data were analyzed by one-way analysis of variance (ANOVA), followed by Tukey's test. Differences at $p < 0.05$ were considered to be significant.

3. Results and Discussion

3.1. Comparison of Different Methods for the Assessment of Pectinase Activity

There are several methods for the determination of pectinase hydrolytic activity. In many studies, it has been assessed by measuring the amount of reducing sugars released during the reaction by the 3,5-dinitrosalicyclic acid (DNS) method [15,33,34]. However, this method is not specific to galacturonic acid as it is sensitive to other reducing sugars like glucose and galactose as well. Other methods for the determination of the produced galacturonic acid are iodometric titration method [32] and spectrophotometric method based on copper-reducing reaction [35]. We evaluated the latter two methods for the determination of pectinase and polygalacturonase activities expressed in international units U (μmol galacturonic acid released per one minute). The results for Pectinex Ultra SPL preparation as an example are shown in Table 1.

Table 1. Pectinase and polygalacturonase activities of Pectinex Ultra SPL preparation determined using titration and spectrophotometric methods.

Method	Pectinase Activity (U */mL Enzyme)	Polygalacturonase Activity (U/mL Enzyme)
Iodometric titration	1333 ± 249 **	2133 ± 94
Spectrophotometric method	1065 ± 94	1732 ± 97

* U is defined as 1 μmol galacturonic acid released per 1 min [36]. ** All values are expressed as mean ± standard deviation (n = 3).

Compared to pectinase activity declared by the producer ($\geq$3800 U/mL), the activity determined was considerably lower ($\geq$3 times) with both methods used. Compared to the titration assay, the spectrophotometric method indicated approximately 20% lower estimations of both pectinase and polygalacturonase activity. As the relative activities are similar, we selected the spectrophotometric method for the determination of the activity of immobilized enzymes in the current study because the assessment procedure was less time-consuming and the experimental errors were smaller. In addition, this method is selective for uronic acids over neutral sugars like glucose, galactose, and rhamnose [35].

3.2. Characterization of Different Commercial Pectinase Preparations

Five commercial preparations of pectinolytic enzymes: Pectinex Ultra SPL (from Sigma-Aldrich), Pectinex BE XXL, Ultra Mash, Ultra Tropical (from Novozymes), and Panzym-Univers (from Begerow) were evaluated regarding their specific pectin depolymerizing activity (Figure 2).

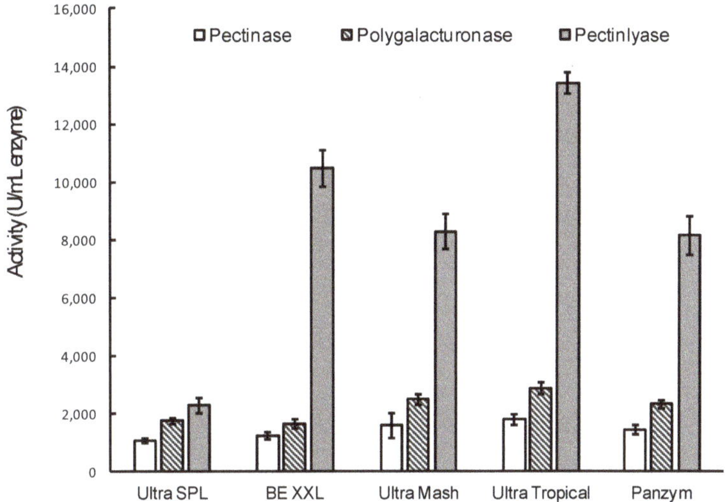

Figure 2. Pectinase, polygalacturonase, and pectin lyase activities of commercial pectinolytic enzyme preparations. All values are expressed as means ± standard deviation (n = 3).

All five commercial pectinase preparations used are recommended for use in fruit mash for increasing juice yield as well as for juice clarification and improved color stability. Using the above-described spectrophotometric method, pectinase activity of the analyzed preparations was found to be between 1065 and 1784 U/mL enzyme (Figure 2, white bars). The determined pectinase activity was considerably lower (3–10 times lower than the activity declared by the producer (see Section 2.1), which ranged between 3800 and 13,600 PECTU/mL. This can be related to the difference in the method and conditions used for activity measurement. For instance, Biz et al. [34] demonstrated that depending on the particular combination of incubation time, pectin concentration, and reaction temperature, the same extract could be reported to have activities that differ by an order of magnitude. All preparations exhibited a polygalacturonase activity higher than their pectinase activities. The detected polygalacturonase activities measured towards polygalacturonic acid ranged from 1633 to 2861 U/mL enzyme (Figure 2, dashed bars). The difference between the two activities can be explained by the nature of the substrate, as pectin has a more complex structure than polygalacturonic acid, which includes rhamnogalacturonan regions with side chains containing rhamnose, galactose, arabinose, and xylose [37]. In addition, polygalacturonic acid is characterized by low content of methyl esters which facilitates its hydrolysis by polygalacturonases.

The analyzed commercial preparations also exhibited pectin lyase activity ranging between 2283 and 13,417 U/mL enzyme (Figure 2, grey bars). Pectin lyases act on pectin with high degree of methylation without the need of pectin de-esterifying enzymes [33]. Pectin lyases catalyze the cleavage of α (1–4) glycosidic bonds by trans-elimination, resulting in a double bond at the non-reducing end of the galacturonic acid.

Statistical analysis showed that pectinase activities of the analyzed preparations were not significantly different (p = 0.1750), whereas, considering polygalacturonase and pectin lyase activities determined, Pectinex® Ultra Tropical preparation exhibited significantly higher enzymatic activities than the remaining four preparations ($p < 0.0001$). Thus, we selected this preparation for further immobilization experiments.

3.3. Optimization of Immobilization Process

Nylon 6/6 is a polyamide of which secondary amide groups can be converted into secondary imidate groups by O-alkylation with DMS. This process introduces a reactive functional group into nylon without any accompanying polymerization, and the formed functional groups react directly with amino groups of enzymes to produce a stable amidine link [38]. However, additional use of glutaraldehyde as a linker results in more stable immobilization of enzymes [28].

Nylon threads were first treated with DMS at 60 °C, which has been previously reported as the optimum temperature for nylon activation [30]. We examined the effect of DMS treatment duration (1–7.5 min) on pectinase immobilization, as in addition to generating active groups, the mechanical strength of nylon 6/6 can decrease after a longer treatment over 5 min. Assuming that the thread with immobilized enzyme is homogenous, the activities of immobilized enzymes can be expressed in meters, which is extremely convenient for technological applications [39]. The effect of the length of DMS treatment on the activity of immobilized on nylon 6/6 threads pectinase is presented in Table 2.

Table 2. The effect of incubation time in DMS on the activity of immobilized pectinase.

DMS Incubation Time (min)	1	2	2.5	5	7.5
Immobilized Pectinase Activity (mU/m thread)	72 ± 2 *	64 ± 1	45 ± 2	43 ± 5	38 ± 4

* All values are expressed as means ± standard deviation (n = 3).

Increasing incubation time in DMS resulted in a decreased immobilized pectinase activity. In fact, over O-alkylation of nylon carrier can result in intra-cross-linkage of the amino groups with glutaraldehyde hindering the attachment of the enzyme [30]. For further enzyme immobilization procedure, we incubated nylon carriers in DMS for 2 min at 60 °C, as this treatment time resulted also in more stable immobilized enzymes (as measured after five working cycles). Two different concentrations of glutaraldehyde, 2.5% [40] and 12.5% [28], were used for enzyme immobilization. (Table 3). We did not use higher glutaraldehyde concentrations, as, in high rate immobilization processes, glutaraldehyde promotes crosslinking between enzyme molecules [41].

Table 3. The effect of glutaraldehyde concentration on pectinase immobilization.

Glutaraldehyde Concentration (%)	2.5	12.5
Immobilized Pectinase Activity (mU/m thread)	56 ± 1 *	64 ± 1
Residual activity after 5 cycles (mU/m thread)	14 ± 1	37 ± 2

* All values are expressed as means ± standard deviation (n = 3).

Treatment with 12.5% glutaraldehyde solution resulted in higher immobilized pectinase activity and also residual pectinase activity after five cycles was 2.5 times higher compared to the threads treated with 2.5% glutaraldehyde solution. So, for further immobilization experiments, nylon carriers were treated with 12.5% glutaraldehyde solution for 1 h at room temperature.

After the immobilization process, the carriers should be washed carefully for several times with buffer using an orbital shaker (150 rpm) for 5 to 15 min each time to remove all non-covalently attached enzymes.

3.4. Nylon 6/6 Carriers

Nylon 6/6 is one of the most common polymers for textile and plastic industries and it is available in different forms including pellets, discs, threads, nets, fabric, etc. that have been previously used also for enzyme immobilization [28–30,40]. In the current work, we compared the immobilization of pectinase enzyme preparation onto two different nylon 6/6 carriers—threads and pellets (Figure 3). The surface area per 1 g thread- and pellet-shaped carrier was calculated considering the micrometer scale carrier dimensions, and was found to be 0.5760 m^2/g and 0.00702 m^2/g, respectively. This difference for ~80 times indicated that the higher enzyme loads per amount of carrier can be expected using nylon 6/6 threads.

(a) **(b)**

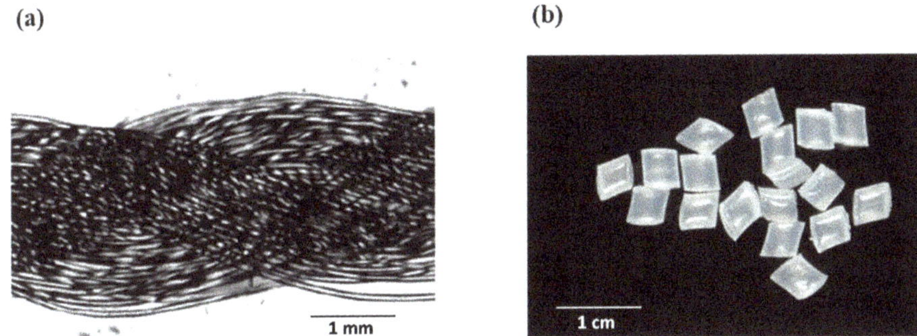

Figure 3. Nylon 6/6 threads ((**a**), shot with Olympus CX31) and pellets ((**b**), shot with UV transilluminator, Cleaver Scientific Ltd., Warwickshire, UK).

3.5. Pectinase Immobilization Yield

Pectinex Ultra Tropical preparation was immobilized onto nylon 6/6 threads and pellets using the selected protocol. Carriers were first soaked into DMS at 60 °C for 2 min, after that, washed with ice-cold methanol and buffer, and kept in 12.5% glutaraldehyde solution 1 h at room temperature. Finally, activated nylon carriers were incubated in 20-fold diluted enzyme solution (89 U/mL) for 24 h at 4 °C. After careful washing to remove unbound enzyme, we measured the pectinase activity of immobilized enzymes and the immobilization yield was expressed as pectinase activity per g of carrier. Results are presented in Table 4.

Table 4. Pectinase immobilization yield on different nylon 6/6 carriers.

Nylon-6,6 Carrier	Beads	Coiled Thread
Immobilization yield (mU/g)	27 ± 2 *	1248 ± 97

* All values are expressed as means ± standard deviation (n = 3).

As expected from the comparison of the surface area of thread and pellets, the immobilization yield of pectinase per 1 g carrier was over 40 times higher for threads than pellets, although the pectinase activity per 1 mm^2 surface area was ~2 times higher for pellets (262 and 140 $\mu U/mm^2$ for pellets and threads, respectively).

Up to now, few studies have investigated the immobilization of pectinases on nylon carriers. Shukla et al. [40] immobilized purified polygalacturonase from *Aspergillus niger* (MTCC 3323) on nylon-6 beads and reported an immobilization yield of 17.54 U/g. This immobilization yield is higher than the measured yield obtained in the present study. This apparent lower yield can be related to the use of the commercial enzyme preparation, which contains different enzymes that are potentially immobilized onto the carrier, although only pectinase activity was assessed.

3.6. Stability and Reusability of the Immobilized Pectinase

Stability and reusability are some of the major features of immobilized enzymes to characterize their industrial applicability. We studied the inactivation of the immobilized enzymes during storage and throughout the active use of both the threads and pellets (Figure 4).

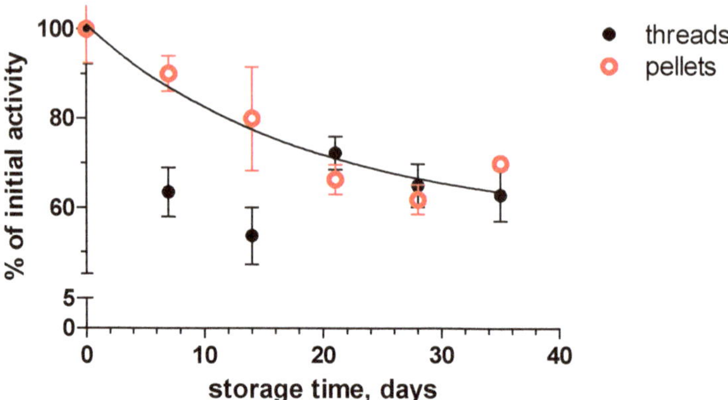

Figure 4. The storage stability of immobilized pectinase in 0.2 M sodium acetate buffer (pH 4.8) at 4 °C.

The activity decreased in an exponential mode and the relative inactivation was not dependent on the shape of the carrier, indicating that the attachment of the enzymes on the surface of nylon 6/6 carriers was similar. The inactivation process was characterized with a 2-exponential decay model, as there is a stabilization phase after the first quick decay of the activity. The F test revealed that there was no statistically relevant difference in the rate of inactivation process of immobilized pectinase ($p = 0.9791$) on nylon 6/6 threads or pellets. The half-life of the enzyme fast phase inactivation in 0.2 M sodium acetate buffer (pH 4.8) at 4 °C was 12.8 days and after 5 weeks, the remaining pectinase activity of the unused threads was 63% of the initial activity. Pectinase immobilized on nylon 6/6 threads and pellets exhibits storage stability comparable to other covalent immobilization carriers like sodium alginate/graphene oxide beads via amide bonds, which retained 52% of initial pectinase activity after 30 days in 0.25 M citrate buffer (pH 4.0) at 4 °C [20]. Meanwhile, pectinase immobilized onto chitosan-grafted magnetic nanoparticles has exhibited a better storage stability, retaining 74% of initial activity after 75 days in 5 mM sodium acetate buffer (pH 4.5) at 4 °C [42].

To determine the reusability of immobilized pectinase onto nylon 6/6 threads and pellets, enzyme activity was measured at 40 °C for 30 min for 20 successive cycles (Figure 5). After each cycle, substrate solution was removed and nylon 6/6 carriers were washed with 0.2 M sodium acetate buffer (pH 4.8) for 10 min before adding new substrate solution. Additionally, in this case, the enzyme inactivation obeyed exponential decay and there was no statistically significant difference in inactivation of immobilized pectinase onto nylon 6/6 threads or pellets, but the activity dropped to 40% after five cycles. After 20 successive cycles, the remaining activity of the immobilized pectinase was 22.0 ± 1.6%, indicating the good reusability of the immobilized enzyme. An earlier study by Shukla et al. [40] reported that polygalacturonase immobilized onto nylon-6 particles retained 50% of its initial activity for three cycles and only 17% after seven cycles. Recent studies on covalent immobilization of pectinase onto magnetic nanoparticle carriers reported better reusability potential where immobilized enzymes retained 50 to 60% of initial pectinase activity after 15 cycles [25,42].

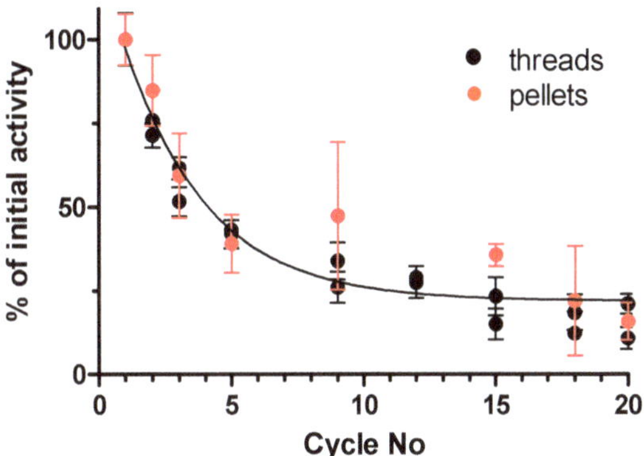

Figure 5. The reusability of immobilized pectinase onto nylon 6/6 threads and beads during 20 successive cycles.

4. Conclusions

In the present work, commercial pectinase preparation was immobilized onto DMS/glutaraldehyde-activated nylon 6/6 carriers. The obtained results showed that pectinase immobilization yield per g of carrier was more than 40 times higher in nylon 6/6 thread than pellets in accordance with a larger surface area per gram of thread compared to pellets. However, the stability and reusability studies showed that the immobilized enzyme inactivation was independent from the shape of the nylon carrier, indicating similar attachment of the enzyme to the carriers. Immobilized pectinase was relatively stable during storage, retaining more than 60% of its initial activity after 35 days at 4 °C. In addition, the immobilized pectinase exhibited good reusability, retaining 40% of its initial activity after five successive cycles and more than 20% after twenty successive cycles. The obtained results indicate the high potential of immobilized pectinase on nylon 6/6 pectinase preparation for industrial application, as it retains sufficient activity after numerous working cycles and the nylon 6/6 carrier is extremely suitable for operating in food production processes due to its inertness, sturdiness, and availability. However, further studies of practical utilization are required to assess the effect of potentially available enzyme inhibitors present in plant material.

Author Contributions: Conceptualization, T.R. and S.B.-O.; methodology, S.B.-O. and T.R.; investigation, S.B.-O. and T.R.; writing—original draft preparation, S.B.-O. and T.R.; writing—review and editing, S.B.-O. and T.R. All authors have read and agreed to the published version of the manuscript.

Funding: This research is funded by VALORTECH project, which has received funding from the European Union's Horizon 2020 research and innovation program under grant agreement No. 810630. Funding received from Mobilitas Pluss ERA-Chair support (Grant no. MOBEC006 ERA Chair for Food (By-) Products Valorisation Technologies of the Estonian University of Life Sciences) is also acknowledged.

Institutional Review Board Statement: Not applicable.

Informed Consent Statement: Not applicable.

Conflicts of Interest: The authors declare no conflict of interest.

References

1. Jayani, R.S.; Saxena, S.; Gupta, R. Microbial pectinolytic enzymes: A review. *Process Biochem.* **2005**, *40*, 2931–2944. [CrossRef]
2. Garg, G.; Singh, A.; Kaur, A.; Singh, R.; Kaur, J.; Mahajan, R. Microbial pectinases: An ecofriendly tool of nature for industries. *3 Biotech* **2016**, *6*, 1–13. [CrossRef] [PubMed]
3. Satapathy, S.; Rout, J.R.; Kerry, R.G.; Thatoi, H.; Sahoo, S.L. Biochemical Prospects of Various Microbial Pectinase and Pectin: An Approachable Concept in Pharmaceutical Bioprocessing. *Front. Nutr.* **2020**, *7*, 1–17. [CrossRef]

4.   Amin, F.; Bhatti, H.N.; Bilal, M. Recent advances in the production strategies of microbial pectinases—A review. *Int. J. Biol. Macromol.* **2019**, *122*, 1017–1026. [CrossRef]

5.   John, J.; Kaimal, K.K.S.; Smith, M.L.; Rahman, P.K.S.M.; Chellam, P.V. Advances in upstream and downstream strategies of pectinase bioprocessing: A review. *Int. J. Biol. Macromol.* **2020**, *162*, 1086–1099. [CrossRef]

6.   Sharma, H.P.; Patel, H. Sugandha Enzymatic added extraction and clarification of fruit juices—A review. *Crit. Rev. Food Sci. Nutr.* **2017**, *57*, 1215–1227. [CrossRef] [PubMed]

7.   Ortiz, G.E.; Ponce-Mora, M.C.; Noseda, D.G.; Cazabat, G.; Saravalli, C.; López, M.C.; Gil, G.P.; Blasco, M.; Albertó, E.O. Pectinase production by Aspergillus giganteus in solid-state fermentation: Optimization, scale-up, biochemical characterization and its application in olive-oil extraction. *J. Ind. Microbiol. Biotechnol.* **2017**, *44*, 197–211. [CrossRef]

8.   Puri, M.; Sharma, D.; Barrow, C.J. Enzyme-assisted extraction of bioactives from plants. *Trends Biotechnol.* **2012**, *30*, 37–44. [CrossRef]

9.   Marathe, S.J.; Jadhav, S.B.; Bankar, S.B.; Kumari Dubey, K.; Singhal, R.S. Improvements in the extraction of bioactive compounds by enzymes. *Curr. Opin. Food Sci.* **2019**, *25*, 62–72. [CrossRef]

10.   Dal Magro, L.; Hertz, P.F.; Fernandez-Lafuente, R.; Klein, M.P.; Rodrigues, R.C. Preparation and characterization of a Combi-CLEAs from pectinases and cellulases: A potential biocatalyst for grape juice clarification. *RSC Adv.* **2016**, *6*, 27242–27251. [CrossRef]

11.   Ottone, C.; Romero, O.; Aburto, C.; Illanes, A.; Wilson, L. Biocatalysis in the winemaking industry: Challenges and opportunities for immobilized enzymes. *Compr. Rev. Food Sci. Food Saf.* **2020**, *19*, 595–621. [CrossRef] [PubMed]

12.   Ramirez, H.L.; Briones, A.I.; Úbeda, J.; Arevalo, M. Immobilization of pectinase by adsorption on an alginate-coated chitin support. *Biotecnol. Apl.* **2013**, *30*, 101–104.

13.   Chauhan, S.; Vohra, A.; Lakhanpal, A.; Gupta, R. Immobilization of Commercial Pectinase (Polygalacturonase) on Celite and Its Application in Juice Clarification. *J. Food Process. Preserv.* **2015**, *39*, 2135–2141. [CrossRef]

14.   Rajdeo, K.; Harini, T.; Lavanya, K.; Fadnavis, N.W. Immobilization of pectinase on reusable polymer support for clarification of apple juice. *Food Bioprod. Process.* **2016**, *99*, 12–19. [CrossRef]

15.   Cerreti, M.; Markošová, K.; Esti, M.; Rosenberg, M.; Rebroš, M. Immobilisation of pectinases into PVA gel for fruit juice application. *Int. J. Food Sci. Technol.* **2017**, *52*, 531–539. [CrossRef]

16.   de Oliveira, R.L.; Dias, J.L.; da Silva, O.S.; Porto, T.S. Immobilization of pectinase from Aspergillus aculeatus in alginate beads and clarification of apple and umbu juices in a packed bed reactor. *Food Bioprod. Process.* **2018**, *109*, 9–18. [CrossRef]

17.   Martín, M.C.; López, O.V.; Ciolino, A.E.; Morata, V.I.; Villar, M.A.; Ninago, M.D. Immobilization of enological pectinase in calcium alginate hydrogels: A potential biocatalyst for winemaking. *Biocatal. Agric. Biotechnol.* **2019**, *18*, 101091. [CrossRef]

18.   Mohammadi, M.; Khakbaz Heshmati, M.; Sarabandi, K.; Fathi, M.; Lim, L.T.; Hamishehkar, H. Activated alginate-montmorillonite beads as an efficient carrier for pectinase immobilization. *Int. J. Biol. Macromol.* **2019**, *137*, 253–260. [CrossRef]

19.   Abdel Wahab, W.A.; Karam, E.A.; Hassan, M.E.; Kansoh, A.L.; Esawya, M.A.; Awad, G.E.A. Optimization of pectinase immobilization on grafted alginate-agar gel beads by 24 full factorial CCD and thermodynamic profiling for evaluating of operational covalent immobilization. *Int. J. Biol. Macromol.* **2018**, *113*, 159–170. [CrossRef]

20.   Dai, X.Y.; Kong, L.M.; Wang, X.L.; Zhu, Q.; Chen, K.; Zhou, T. Preparation, characterization and catalytic behavior of pectinase covalently immobilized onto sodium alginate/graphene oxide composite beads. *Food Chem.* **2018**, *253*, 185–193. [CrossRef]

21.   Yang, S.-Q.; Dai, X.-Y.; Wei, X.-Y.; Zhu, Q.; Zhou, T. Co-immobilization of pectinase and glucoamylase onto sodium aliginate/graphene oxide composite beads and its application in the preparation of pumpkin–hawthorn juice. *J. Food Biochem.* **2019**, *43*, e122741. [CrossRef]

22.   Sojitra, U.V.; Nadar, S.S.; Rathod, V.K. A magnetic tri-enzyme nanobiocatalyst for fruit juice clarification. *Food Chem.* **2016**, *213*, 296–305. [CrossRef] [PubMed]

23.   Ladole, M.R.; Nair, R.R.; Bhutada, Y.D.; Amritkar, V.D.; Pandit, A.B. Synergistic effect of ultrasonication and co-immobilized enzymes on tomato peels for lycopene extraction. *Ultrason. Sonochem.* **2018**, *48*, 453–462. [CrossRef] [PubMed]

24.   Nadar, S.S.; Rathod, V.K. A co-immobilization of pectinase and cellulase onto magnetic nanoparticles for antioxidant extraction from waste fruit peels. *Biocatal. Agric. Biotechnol.* **2019**, *17*, 470–479. [CrossRef]

25.   Dal Magro, L.; de Moura, K.S.; Backes, B.E.; de Menezes, E.W.; Benvenutti, E.V.; Nicolodi, S.; Klein, M.P.; Fernandez-Lafuente, R.; Rodrigues, R.C. Immobilization of pectinase on chitosan-magnetic particles: Influence of particle preparation protocol on enzyme properties for fruit juice clarification. *Biotechnol. Rep.* **2019**, *24*, e00373. [CrossRef]

26.   Yushkova, E.D.; Nazarova, E.A.; Matyuhina, A.V.; Noskova, A.O.; Shavronskaya, D.O.; Vinogradov, V.V.; Skvortsova, N.N.; Krivoshapkina, E.F. Application of Immobilized Enzymes in Food Industry. *J. Agric. Food Chem.* **2019**, *67*, 11553–11567. [CrossRef]

27.   Wu, X.; Fraser, K.; Zha, J.; Dordick, J.S. Flexible Peptide Linkers Enhance the Antimicrobial Activity of Surface-Immobilized Bacteriolytic Enzymes. *ACS Appl. Mater. Interfaces* **2018**, *10*, 36746–36756. [CrossRef]

28.   Kivirand, K.; Rinken, T. Preparation and characterization of cadaverine sensitive nylon threads. *Sens. Lett.* **2009**, *7*, 580–585. [CrossRef]

29.   Damle, M.; Badhe, P.; Mahajan, G.; RV, A. Immobilisation of marine pectinase on nylon 6,6. *J. Text. Eng. Fash. Technol.* **2018**, *4*, 181–187. [CrossRef]

30.   Nan, C.; Zhang, Y.; Zhang, G.; Dong, C.; Shuang, S.; Choi, M.M.F. Activation of nylon net and its application to a biosensor for determination of glucose in human serum. *Enzym. Microb. Technol.* **2009**, *44*, 249–253. [CrossRef]

31. Pahujani, S.; Kanwar, S.S.; Chauhan, G.; Gupta, R. Glutaraldehyde activation of polymer Nylon-6 for lipase immobilization: Enzyme characteristics and stability. *Bioresour. Technol.* **2008**, *99*, 2566–2570. [CrossRef] [PubMed]
32. Enzymatic Assay of Pectinase | Sigma-Aldrich. Available online: https://www.sigmaaldrich.com/technical-documents/protocols/biology/enzymatic-assay-of-pectinase.html (accessed on 25 March 2021).
33. Dal Magro, L.; Goetze, D.; Ribeiro, C.T.; Paludo, N.; Rodrigues, E.; Hertz, P.F.; Klein, M.P.; Rodrigues, R.C. Identification of Bioactive Compounds From Vitis labrusca L. Variety Concord Grape Juice Treated With Commercial Enzymes: Improved Yield and Quality Parameters. *Food Bioprocess Technol.* **2016**, *9*, 365–377. [CrossRef]
34. Biz, A.; Farias, F.C.; Motter, F.A.; De Paula, D.H.; Richard, P.; Krieger, N.; Mitchell, D.A. Pectinase activity determination: An early deceleration in the release of reducing sugars throws a spanner in the works! *PLoS ONE* **2014**, *9*, e109529. [CrossRef] [PubMed]
35. Anthon, G.E.; Barrett, D.M. Combined enzymatic and colorimetric method for determining the uronic acid and methylester content of pectin: Application to tomato products. *Food Chem.* **2008**, *110*, 239–247. [CrossRef] [PubMed]
36. IUPAC. *Compendium of Chemical Terminology*, 2nd ed.; The "Gold Book"; Blackwell Science: Oxford, UK, 1997; Volume 1.
37. Tapre, A.R.; Jain, R.K. Pectinases: Enzymes for fruit processing industry. *Int. Food Res. J.* **2014**, *21*, 447–453.
38. Morris, D.L.; Campbell, J.; Hornby, W.E. A chemistry for the immobilization of enzymes on nylon. The preparation of nylon tube supported hexokinase and glucose 6 phosphate dehydrogenase and the use of the co immobilized enzymes in the automated determination of glucose. *Biochem. J.* **1975**, *147*, 593–603. [CrossRef]
39. Rinken, T.; Järv, J.; Rinken, A. Production of biosensors with exchangeable enzyme-containing threads. *Anal. Chem.* **2007**, *79*, 6042–6044. [CrossRef]
40. Shukla, S.; Saxena, S.; Thakur, J.; Gupta, R. Immobilization of polygalacturonase from Aspergilus niger onto glutaraldehyde activated Nylon-6 and its application in apple juice clarification. *Acta Aliment.* **2010**, *39*, 277–292. [CrossRef]
41. Betancor, L.; López-Gallego, F.; Alonso-Morales, N.; Dellamora, G.; Mateo, C.; Fernandez-Lafuente, R.; Guisan, J.M. Glutaraldehyde in Protein Immobilization. In *Immobilization of Enzymes and Cells*; Guisan, J.M., Ed.; Humana Press, 2006; pp. 57–64. Available online: https://link.springer.com/protocol/10.1007/978-1-59745-053-9_5#citeas (accessed on 25 March 2021). [CrossRef]
42. Soozanipour, A.; Taheri-Kafrani, A.; Barkhori, M.; Nasrollahzadeh, M. Preparation of a stable and robust nanobiocatalyst by efficiently immobilizing of pectinase onto cyanuric chloride-functionalized chitosan grafted magnetic nanoparticles. *J. Colloid Interface Sci.* **2019**, *536*, 261–270. [CrossRef]

Article

Effects of Different LED Light Recipes and NPK Fertilizers on Basil Cultivation for Automated and Integrated Horticulture Methods

Silvia Barbi [1,*], Francesco Barbieri [2], Alessandro Bertacchini [2,3], Luisa Barbieri [3,4] and Monia Montorsi [2,3]

[1] Interdepartmental Research Center for Industrial Research and Technology Transfer in the Field of Integrated Technologies for Sustainable Research, Efficient Energy Conversion, Energy Efficiency of Buildings, Lighting and Home Automation–EN&TECH, University of Modena and Reggio Emilia, Via Amendola 2, 42122 Reggio Emilia, Italy

[2] Department of Science and Methods for Engineering, University of Modena and Reggio Emilia, Via Amendola 2, 42122 Reggio Emilia, Italy; francesco.barbieri1@unimore.it (F.B.); alessandro.bertacchini@unimore.it (A.B.); monia.montorsi@unimore.it (M.M.)

[3] Interdepartmental Center for Applied Research and Services in Advanced Mechanics and Motoring, INTERMECH-Mo.Re., University of Modena and Reggio Emilia, Via P. Vivarelli 10/1, 41125 Modena, Italy; luisa.barbieri@unimore.it

[4] Department of Engineering "Enzo Ferrari", University of Modena and Reggio Emilia, Via Vivarelli 10/1, 41125 Modena, Italy

* Correspondence: silvia.barbi@unimore.it

Citation: Barbi, S.; Barbieri, F.; Bertacchini, A.; Barbieri, L.; Montorsi, M. Effects of Different LED Light Recipes and NPK Fertilizers on Basil Cultivation for Automated and Integrated Horticulture Methods. *Appl. Sci.* **2021**, *11*, 2497. https://doi.org/10.3390/app11062497

Academic Editors: Isidoro Garcia-Garcia, Jesus Simal-Gandara and Maria Gullo

Received: 23 February 2021
Accepted: 9 March 2021
Published: 11 March 2021

Publisher's Note: MDPI stays neutral with regard to jurisdictional claims in published maps and institutional affiliations.

Featured Application: The specific application of this work is related to basil cultivation in indoor horticulture and is devoted to the investigation of specific light recipes and fertilizers addition to promote its germination and growth in a controlled environment.

Abstract: This study aims to optimize the conditions for "Genovese" basil (*Ocimum Basilicum*) germination and growth in an indoor environment suitable for horticulture through a synergic effect of light and fertilizers addition. In fact, several studies determined that specific light conditions are capable of enhancing basil growth, but this effect is highly dependent on the environmental conditions. In this study, the effect of different light sources was determined employing a soil with a negligible amount of fertilizer, demonstrating substantial improvement when light-emitting diode (LED) lights (hyper red and deep blue in different combinations) were applied with respect to daylight (Plants height: +30%, Total fresh mass: +50%). Thereafter, a design of experiment approach has been implemented to calculate the specific combination of LED lights and fertilizer useful to optimize the basil growth. A controlled-release fertilizer based on nitrogen, phosphorus, and potassium (NPK) derived from agro-residues was compared with a soil enriched in macronutrients. The results demonstrate significant improvements for the growth parameters with the employment of the controlled-release NPK with respect to enriched soil combined with a ratio of hyper red and deep blue LED light equal to 1:3 (Total fresh mass: +100%, Leaves number: +20%).

Keywords: basil; design of experiments; valorization strategies

1. Introduction

In recent years, the interest toward the applicability of light-emitting diode (LED) lights for indoor cultivation has significantly grown, with the aim of creating a more affordable and ecological production for plants compared to fluorescent lighting and high-pressure sodium (HPS) lamps [1,2]. In fact, artificial light supply presents additional costs compared to cultivation under natural daylight due to energy demand, but it is mandatory in an indoor environment and wherever natural light is not sufficient. For example, these systems are necessary at northern latitudes, where a strong variation of daylight is observed during the year, to address the current challenges that the greenhouse industry is facing (i.e., energy, environment, and market) [3]. Moreover, it is beneficial to increase the product yield and quality of the final product because growing plants in controlled environment allows drastically reducing the use of pesticides. Finally, controlled-environment agriculture is a key subject of the worldwide food optimization system, due to the upcoming population growth and climate changes scenarios [4]. In this context, LED lights are expected to replace conventional light sources due to their longer functional life, lower operating temperatures, and lower energy consumption, thereafter decreasing the operational costs of indoor urban farming. In addition, compared with conventional light sources, LED technology offers an easier and cheap way for light spectrum manipulation, and this is beneficial for crop growth regulation [5,6].

Indeed, LED lights provide tunable wavelengths to be matched to plant photoreceptors, including phytochrome and cryptochrome, in order to have optimal plant growth in terms of morphology (e.g., height, leaf area, thickness, and stem length) and quality (e.g., metabolites) [7–11]. Therefore, this provides a new opportunity to manipulate the quality and quantity of vegetable products for markets and meet the demands of retailers in a future perspective involving artificial intelligence control over a real-time autonomous horticultural system [12]. Artificial intelligence techniques ensure minimal human manipulation by defining very tailored target parameters, ensuring an increased efficiency of the system from the environmental and economic point of view. This vision seeks the full digital control of the light quantity, in order to optimize the crop productivity, which may contribute to an optimized production of food [3,13]. For example, it could be possible not only to control artificial light based on specific natural light conditions but to account for potential physiological changes in the crop [13]. As a result, research is strongly focusing on the application of different LED lights on different plants, and several types of LED-based lamps became available for commercial plant production [14]. Taking into account the more recent research on this topic, generally, blue light is necessary for the morphologically healthy growth of plants mainly involving a transpiration mechanism and stem elongation [15–17]. In contrast, red light plays a major role in plant photosynthesis [18,19]. In fact, it has been demonstrated that plants' chlorophyll A and B efficiently absorb blue and red lights wavelengths, respectively [20]. It has been confirmed that the optimal ratio between blue and red light is of great relevance in determining plant yield, as it has a strong effect on the conversion efficiency of natural photosynthesis, which is peculiar for each vegetable [12,21]. In addition, research on this topic is generally conducted using commercial available light that is not varied during the experimental procedure, and principal results indicate that the complexity of the plant physiology requires a multi-objective optimization approach [12].

In this context, "Genovese" basil (*Ocimum Basilicum*) has been previously studied employing LED lights as the only light source and compared with daylight or fluorescent lamps as control, but no clear trend emerged on the capability for certain light recipes to enhance basil germination or growth. Specifically, the use of red and blue together seems to be preferable; however, the optimal proportions may vary. In fact, depending on several factors, the employment of blue light must be preponderant over red or vice versa, with the possible addition of green or white LED light [22–25]. These variations are mainly due to thedifferent environmental conditions applied during the tests, such as the light surrounding the experiments' area (e.g., into the dark or into a greenhouse) or time of exposure (varying from 21 to 64 days) or daily photoperiod (from 12 to 18 h). In any case, the use of red LED light alone is not recommended, because it leads to a too long time of growth [21]. The different age of collection and analysis of the growth parameters may be one of the other factors leading to the different influence of light. Thus, if the results of some studies lead to the hypothesis that in the first weeks of plant life, blue light is more important than red, other studies refuse it [22]. In addition, in general, the influence of LED lights on the germination phase is ignored.

In the present research, sustainable conditions for "Genovese" basil (*Ocimum Basilicum*) indoor horticulture were considered as well, as it is one of the plant species most cultivated in horticulture due to its high nutritional value and cultivation density [26]. As a first innovative aspect, with respect to previous literature, the germination phase was considered in this study. In addition, specific LED light recipes of hyper red and deep blue were investigated, also in combination with different soils and the addition of a specific controlled-release fertilizer based on nitrogen, phosphorus, and potassium macronutrients (NPK). Moreover, in a circular economy perspective, this controlled-release NPK fertilizer was derived from agri-food, industrial, and post-consumer activities as previously described in the literature [27,28]. In the first part of the experimental procedure, different substrates with a negligible quantity of fertilizers were tested to identify the most promising one capable to clearly highlight the effect of light source only on the plant germination and growth. In the second part, the fertilizer effect was considered in combination with LED lights recipes, employing a design of experiments (DoE) approach leading to a reduction of the number of experiments needed to derive statistically reliable correlation among data [29].

2. Materials and Methods

2.1. Materials

For this study, basil (*Ocimum Basilicum*) seeds belonging to the variety called "Genovese" (Producer: Magnani Sementi) were employed as reference. In the first part of the experimental procedure, related to soils with a negligible quantity of fertilizers, three substrates were considered: Irish peat (Vigorplant Italia S.r.l., Fombio, Italy), Vegetal soil (Dal Zotto S.r.l., Fombio, Italy), and Organic potting soil for aromatic plants (OBI Smart Technologies GmbH, Wermelskirchen, Germany). In the second part of the study, with the aim of considering the fertilizer effect, a soil with considerable quantity of fertilizers was considered: Floradur B pot coarse universal potting soil (Floragard Vertriebs GmbH, Oldenburg, Germany). Detailed specifications of these substrates are reported in Table 1.

Table 1. Substrates properties as indicated by the different manufacturers.

Property	Irish Peat	Vegetal Soil	Organic Soil	Floradur B
Apparent density (kg/m^3) (fresh matter)	200	300	320	120
Electrical conductivity (mS/cm)	0.0058	0.0053	0.0062	0.3800
pH	3.5	6.5	6.8	6.1
Salt content (kg/m^3)	-	-	1.5	1.2
N (kg/m^3)	0.70	-	0.05–0.30	0.21
P$_2$O$_5$ (kg/m^3)	-	-	0.08–0.30	0.12
K$_2$O (kg/m^3)	-	-	0.08–0.40	0.26

With the aim to evaluate the combined effect of a very specific type of controlled release fertilizer and different light sources, in the second part of this study, a NPK lightweight aggregate from agro-residues, wastes, and by-products resulted from previous research [27,28] was employed and compared with the employment of only soil enriched in fertilizers (Floradur B). According to the literature, the NPK controlled-release fertilizer demonstrated a relevant content of nutrients such as nitrogen (8%), potassium (8%), and phosphorous (5%) [27].

As an alternative to daylight, different LED modules specifically designed for horticulture applications were used. All the considered modules (from Intelligent Led Solutions, [30]) are equipped with 12 Oslon®SSL ThinGan LEDs (UX:3) and driven by a constant 370 mA current driver. The modules exploit a specific lens from LEDiL [31] to reduce the radiation angle of the LEDs from the original 80 deg to about 30 deg to obtain a more constant irradiation for all the plants. The considered modules use hyper red (HR) and deep blue (DB) LEDs with 660 nm and 451 nm wavelength at the emission peak, and full width at half maximum (FWHM) values of 25 nm and 20 nm, respectively [32,33]. Three different light recipes have been obtained exploiting modules with different ratios in terms of number of LEDs per type; see Table 2.

Table 2. Hyper red–deep blue light-emitting diodes (LEDs) ratio for the considered light recipes.

LEDs Ratio (Light Recipe Code)	Number of HR LEDs	Number of DB LEDs
3HR:1DB	9	3
1HR:1DB	6	6
1HR:3DB	3	9

All the three considered LED modules have been experimentally characterized by means of a low-cost spectral sensor (AS7341 from Ams, [34]) to obtain a qualitative comparison of the light spectrums provided. The results are shown in Figure 1.

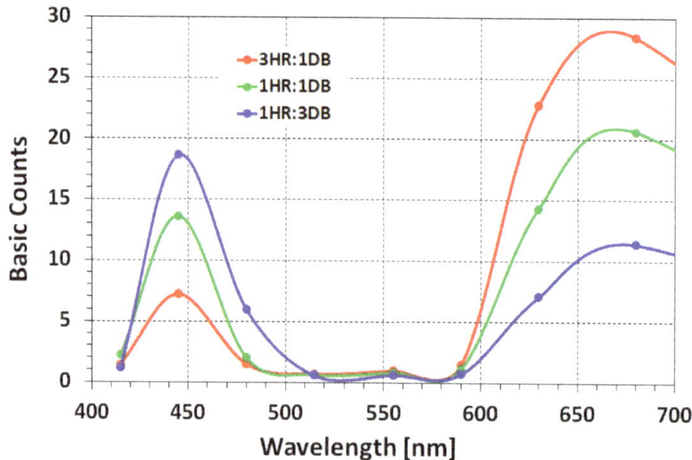

Figure 1. Comparison of the qualitative light spectrums for the considered modules: 3HR:1DB, 1HR:1DB, and 1HR:3DB, obtained using the low-cost AS7341 spectral sensor. Each curve is obtained by the interpolation of the dimensionless output values of the eight optical channels of the sensor with filters centered at wavelengths of 415, 445, 480, 515, 555, 590, 630, and 680 nm, respectively (marker dots). DB: deep blue, HR: hyper red.

More in detail, the used sensor has a sensing element comprised of a matrix of photodiodes used in combination with embedded selective filters to obtain an 11-channel spectrometer. The filters are embedded using nano-optic deposited interference filter technology, and its package provides a built-in aperture to control the light entering the sensor array. Eight of the 11 available channels are reserved to the visible spectrum with filters centered at wavelengths of 415, 445, 480, 515, 555, 590, 630, and 680 nm, respectively. In addition, it has an optical channel dedicated to measure near-infrared light (NIR), an optical channel with a photodiode without filter (called "clear"), and a dedicated channel to detect 50 or 60 Hz ambient light flicker. Light-to-frequency converters embedded into the sensor are used to convert the analog optical channel's output of the photodiodes to dimensionless digital output values of the sensors indicated as basic counts.

Concerning the growth of plants one of the most important parameters is the Photosynthetic Photon Flux (PPF) emitted by the light source used (i.e., the LED modules in our case). PPF refers only to the photons with wavelengths in the range 400–700 nm, because they are the only photons that contribute to the photosynthesis of the plants. Consequently, qualitative light spectrums shown in Figure 1 have been obtained by the interpolation of only the sensor output data provided by the 8 optical channels covering the visible spectrum.

It is important to note that the photosynthetic photon flux (PPF) composition does not reflect the number of LEDs per type ratio of the light recipes. The reason is that the overall photon flux (PF), in μmol/s, depends on the radiant flux (RF), in W, and is proportional to the light wavelength, λ, (1). Consequently, for a given RF, the number of photons generated by an HR LED is higher than the one of a DB LED. Vice versa, since the energy per photon, E_{ph}, is inversely proportional to λ, (2), DB LEDs have a higher PF than HR LEDs for a given number of photons.

$$PF = RF \cdot \lambda \cdot 0.00836 \tag{1}$$

$$E_{ph} = h \cdot (c/\lambda) \tag{2}$$

where the coefficient $0.00836 = 1/(h \cdot c \cdot N)$, h is Planck's constant, c is the speed of light, and N is the number of Avogadro.

In addition, both RF and PF depend on the current flowing through the LEDs. For all the considered modules, all the embedded LEDs were connected in series; therefore, they were driven by the same current and this means that HR and DB LEDs provide a different contribution to the total PPF, as shown in Table 3. The data shown in Table 3 were calculated combining the measurements carried out with the low-cost spectral sensor, direct measurements of power supply voltage and current of the LED modules carried out with digital multimeters, and the data reported in the datasheets of the modules [30] and the datasheet of the LEDs embedded into the modules [32,33]. Temperature and relative humidity were recorded using wireless sensors with accuracy equal to ± 1 °C (resolution: 0.1 °C) for temperature and $\pm 5\%$ (resolution: 1%) for humidity.

Table 3. Photosynthetic photon flux (PPF) composition.

Light Recipe Code	PPF HR LEDs [μmol/s]	PPF DB LEDs [μmol/s]	Total PPF [μmol/s]	%PPF HR	%PPF DB
3HR:1DB	18.13	6.79	24.92	72.76	27.24
1HR:1DB	12.08	13.57	25.65	47.10	52.90
1HR:3DB	6.04	20.36	26.40	22.88	77.12

2.2. Germination and Growth Trials without Fertilizer Addition

The aim of these trials was to determine the compatibility of basil with three substrates without fertilizer (Irish peat, Vegetable soil, and Organic soil) and the effect of the LED light source employment. Approximately 90 basil seeds per unit were laid in cylindrical nonwoven containers with a volume of 165 cm^3 with 6 repetitions for each soil for a total of 18 units. Trials lasted 20 days, at the end of which the number of plants grown, their mass, and average height were evaluated on randomly sampled plants, 10 for each substrate type. Taking into consideration the most promising soil, 6 repetitions were performed for each light source: daylight, 3:1, and 1:3 HR:DB ratios separately. The photoperiod was set to 15 h/day for 30 days. Humidity in the culture room was maintained between 60 and 80% through irrigation with distilled water daily as required during the experimental period. Artificial light treatments were applied using a growth structure (Figure 2) divided into different sectors. Each sector was separated from external sources of light through fixed wood panels as walls. Plants were placed on the bottom in a fixed position, while LED modules were placed on an adjustable system to ensure a constant height equal to 40 cm and a constant illumination of the top canopy of basil in an area equal to 400 cm^2.

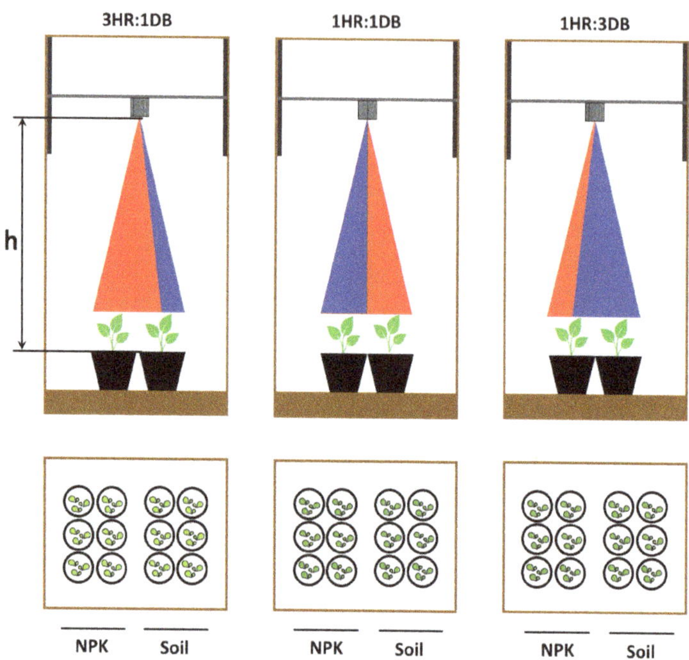

Figure 2. Simplified sketch of the experimental setup: front view (**top**) and top view (**bottom**).

2.3. Design of Experiments (DoE)

A rational approach, codified by the design of experiments (DoE), was employed to obtain the highest amount of information using the minimum number of experiments, saving time and costs [29]. In the second part of the present study, the synergic effect of the amount of the LED light and controlled release fertilizer was investigated on the germination and growth performances. Two factors were considered: NPK as categoric (2 levels: YES/NO) and HR:DB ratio as discrete numeric (3 levels: 1:1, 3:1, and 1:3). The other variables occurring in the process and not specifically considered in this study, such as temperature, humidity, and growth media, were kept constant during all the tests, according to the procedure as explained in Sections 2.1, 2.2 and 2.4. The Design Expert 12.0 (Stat-Ease) code was used both to set up the experimental plan and to analyze the results. Due to the limited number of factors, a full factorial design was selected. A total of 36 experiments were performed including three repetitions for pure error estimation; thereafter, 12 different combinations among NPK and LED lights were observed (Table 4). Central points, considered as the arithmetic average of the factors' levels, were included to investigate the presence of curvature in the data analysis. All the experiments (runs) were carried out randomly to avoid the presence of systematic errors, following the method reported in Section 2.4.

Analysis of variance (ANOVA) was used to point out the cause–effect relationship between controlled-release NPK fertilizer and LED lights on the germination and growing performance. The main assumptions of the ANOVA are that each input factor is independent from each other, normally distributed, and that the variation of the response can be decomposed into different components to evaluate the effect of each factor, their interactions, and experimental error (or unexplained residual) [29]. Through the F-test, variation among all the samples, usually due to process difference or factor changes, is estimated as larger or not than the variation within samples obtained in same experimental conditions. The *p*-value is the statistical parameter used to evaluate the significance of the model and of each factor and represents the probability that the considered model or factor is significant

(p-value < 0.05) or not [35]. The quality of the fit in terms of regression analysis and the predictive power of the model were evaluated by using the R^2 and Pred-R^2, respectively. R^2 is the proportion of the variance in the dependent variables that is predictable from the independent variables, and Pred-R^2 is analogous but associated with predicted values [36].

Table 4. Experimental plan results.

Run	HR:DB	NPK	Height (cm)	NoL	TFM (g)	TDM (g)	LAI (%)	SLA (cm^2/g)
1	3:1	yes	14.63	9.0	4.641	0.353	7.56	566.70
2	3:1	yes	13.33	8.4	3.931	0.293	8.17	601.84
3	3:1	no	12.69	8.0	2.346	0.217	4.56	463.71
4	3:1	no	14.41	8.0	2.772	0.259	4.20	448.71
5	1:1	yes	16.96	11.0	8.124	0.563	6.48	601.78
6	1:1	yes	12.09	8.0	4.512	0.328	7.06	542.05
7	1:1	no	12.17	7.0	2.772	0.285	4.12	422.19
8	1:1	no	12.11	7.6	2.187	0.210	4.23	439.59
9	1:3	yes	14.94	9.3	5.518	0.373	6.60	679.11
10	1:3	yes	15.09	10.7	6.128	0.407	7.42	637.12
11	1:3	no	11.84	8.0	2.452	0.208	4.82	523.50
12	1:3	no	11.33	7.6	2.174	0.192	4.23	479.51

2.4. Growth Experiments with Fertilizer Addition through DoE

Basil seeds were buried in Floradur B soil in plastic pots with a volume of 600 cm^3 and surface area of 78.5 cm^2. Following the experimental plan in Table 4, 36 buried plants were employed, and in some of them, NPK aggregates were included in group of 10. This quantity was determined by considering the 2018 Integrated Production Regulations of the Emilia Romagna Region (Italy) [37] for the production of basil. According to it, under standard soil conditions, the quantity of allowed nutrients such as nitrogen phosphorus and potassium (NPK) are 100–70–80 kg/ha, respectively. In particular, considering the nitrogen amount in the coating, the amount of nitrogen provided is equal to the maximum allowed according to the above-mentioned guidelines and previous NPK characterization, assuming a final release of 50% of the nitrogen contained in the coating [27].

Plants were exposed to the action of the LED modules at a constant distance of 40 cm to ensure homogeneous illumination in an area of 20 × 20 cm. The distance remained constant throughout the experiment, raising the modules as the seedlings grew. The photoperiod was set to 15 h/day for 30 days. The temperature and relative humidity of the samples were monitored during the experiment using one wireless sensor for every set of samples under the same LED module as described in Section 2.2.

2.5. Characterization

At the conclusion of each trial, the following properties were measured for each plant: Total Fresh Mass (TFM), Total Dry Mass (TDM), Height, and Number of Leaves (NoL). For each leaf, fresh mass and area were measured. By adding up all the masses and areas of the same plant, the Total Leaves Mass (TLM) and Total Leaves Area (TLA) were calculated for each plant. Average Leaf Mass (ALM) and Average Leaf Area (ALA) were calculated as well, (1) and (2) formulae:

$$ALM = TLM/NoL \tag{3}$$

$$ALA = TLA/NoL \tag{4}$$

For masses, a laboratory balance (G&G GmbH, model PLC200B-C) with sensitivity ±0.001 g was employed, and for heights, a digital caliper (Borletti CDJB15-20 series) with a resolution of 0.01 mm and accuracy of ±0.02 mm was used. For dry mass measurement, plants and leaves were dried at 80 °C for 24 h. Leaf area measurement was performed by scanning each leaf at 400 dpi on graph paper and measuring using ImageJ software (version 1.52, NIH, Bethesda, MD, USA). Prior to scanning, leaves were cut at certain points to extend their full area on the paper and to better assess their area. Following these measurements, the LAI (Leaf Area Index), given by the ratio of Average Leaf Area to the area of the pot in which the plants had grown, and the SLA (Specific Leaf Area) index, given by the ratio of Average Leaf Area to Average Leaf Dry Mass, were also calculated [22,38].

3. Results and Discussion

3.1. Germination and Growth Trials without Fertilizer Addition

As can be seen in Table 1, Irish peat is distinguishable for its low pH value with respect to the other soils, but the density and content of fertilizer (negligible) are comparable with the others soils without fertilizer. Organic soil compared with Vegetal soil shows a slight content of fertilizer and thereafter a moderate increase in electrical conductivity. Floradur B, on the other hand, is distinguishable due to the relevant concentrations of the three main plant nutrients: nitrogen, phosphorus, and potassium. Concerning conductivity, for all the soils, the limit generally applied to fertilizer compounds necessary to be sold and employed was respected (<2 mS/cm) [39]. At the final harvest, after 30 days under daylight, plants grown in the different the three potting soils without relevant content of fertilizer showed different performances to basil germination and growth. For the first, it must be noted that the employment of Vegetal soil leads to the germination of only one plant, that is a very restrained value when compared to Irish Peat and Organic soil, which were capable of induce the germination of 61% and 75% of the buried seeds, respectively. This result is probably due to the fact that Vegetal soil is the only substrate with a complete absence of nutrients and salts, which is necessary for the plant germination. Thereafter, due to the high incompatibility of Vegetal soil with basil, this substrate was withdrawn from further investigation and characterization. Figure 3 shows the comparison among two principal growth properties related to Irish peat and Organic soil employment: the average height and TFM of the plants. These results highlight that although Irish peat showed excellent compatibility with basil, Organic soil is the most promising soil in terms of TFM and average height of the plants in reference to a quite similar number of sprouted plants. This result is compatible with the fact that Organic soil contains a small amount of nutrients capable of acting as fertilizers and a pH nearer to values more compatible with basil cultivation [40].

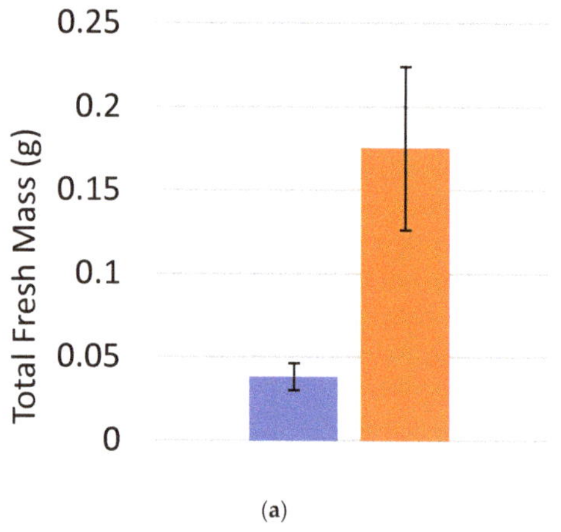
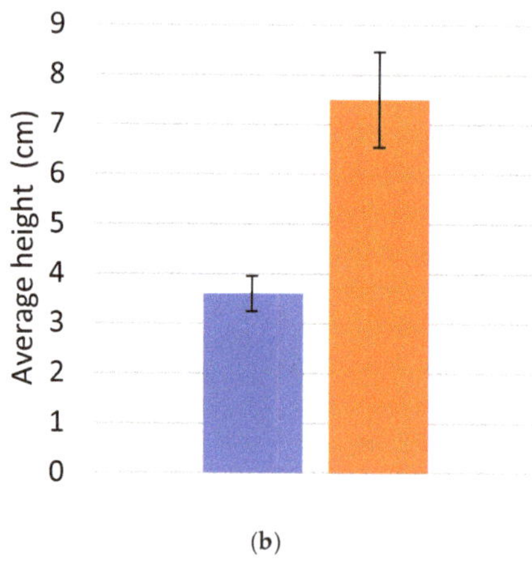

Figure 3. Comparison of Irish peat (in blue) and Organic soil (in orange) considering: (**a**) Total Fresh Mass and (**b**) Average height. Error bars indicate the standard deviation on the average values.

Taking into account the employment of only Organic soil, it has been demonstrated that the light spectrum strongly affects plant growth and development; in fact, basil plants grown under LED lights modules generally show better performances than the plants exposed to daylight, as shown in Table 5 and Figure 4. This is particularly true for parameters such as Height, TFM, TLM, and ALM. On the other hand, the results obtained for the parameters NoL and TDM do not highlight any specific trend. Regarding the comparison between plants grown under two different LED lights combinations, no statistically significant differences emerged. In fact, although plants grown under the 3HR:1DB module showed generally higher mean values of the growth parameters than 1HR:3DB, the standard deviation ranges suggest a similar behavior of the basil plants growth among the two combinations employed. These results are also suggested by a visual comparison of the plants obtained through the three enlightenment methods (Figure 4). For the SLA index, the highest value is guaranteed by 1HR:3DB employment, and this is particularly important because this index depends on the growth environment of the plants and describes their degree of adaptation to it from a morphological point of view. For the LAI parameter, no difference among the two LED module employments can be noted, even if a substantial increment of this index is observed by moving from daylight to LED enlightenment. This parameter is used to determine the size of the interface between the plant and atmosphere, thus the exchange of mass and energy between them to model the photosynthesis mechanism. In addition, it must be noted that a control group of plants grown without any type of light has been tested, obtaining a not significant number of plants without leaves. Thereafter, for this control group, only the plants' height was measured and equal to 2.03 ± 0.62. These results suggest that the LED light combinations investigated in this study play a valuable role as a different source of enlightenment from daylight to improve the basil growth performance in condition with a negligible quantity of nutrients beneficial for cultivation.

Table 5. Growth performance for basil plants under different lights.

	Daylight	**3HR:1DB**	**1HR:3DB**
Height (cm)	6.60 ± 1.09	6.11 ± 1.03	9.28 ± 0.96
LoN	7.0 ± 1.4	6.0 ± 0.0	5.6 ± 0.5
TFM (g)	2.895 ± 0.630	5.788 ± 0.392	5.145 ± 0.488
TDM (g)	0.388 ± 0.201	0.589 ± 0.038	0.465 ± 0.072
TLM (g)	2.330 ± 0.496	4.914 ± 0.349	4.308 ± 0.503
ALM (g)	0.357 ± 0.037	0.819 ± 0.058	0.769 ± 0.054
TLA (cm^2)	72.73 ± 10.99	110.48 ± 10.33	106.12 ± 11.11
ALA (cm^2)	9.70 ± 0.71	18.41 ± 1.72	18.98 ± 1.83
SLA (cm^2/g)	209 ± 56	224 ± 18	282 ± 37
LAI (%)	2.7 ± 1.1	5.1 ± 0.5	4.9 ± 0.5

Figure 4. Basil plants at the end of the growth test performed under different lights and with the Organic soil as substrate.

3.2. Growth Experiments with Fertilizer Addition through DoE

With the aims of finding an optimal LED lights combination, balancing the number of hyper red and deep blue LED lights, that is capable of enhancing all the growth parameters, and investigating the possible effects in combination with fertilizer, a new experimental set-up was managed through an ad hoc DoE experimental plan. Table 4 shows the complete experimental plan and the obtained results, in which the average values of each response among three repetitions of the same experiment are indicated.

As previously stated, with the aim to evaluate only fertilizer and LED lights effects, all the other parameters were kept as constant; thereafter, temperature and humidity were controlled during all the experiments to avoid environmental conditioning. Temperature and humidity were kept almost constant near 30 °C and 60% respectively, with restrained variation around 5% due to the employment of an indoor environment not perfectly conditioned. From a general observation of the results, germination occurred for almost all the basil seeds; in fact, only two of them did not germinate, and thereafter, the number is not statistically relevant. In addition, it must be noted that all the cultivation conditions applied according to this experiment plan were proved to be appropriate for basil, as it was confirmed by the absence of morphological and developmental abnormalities during plant growth. Observing only the data in Table 4, a general comparison can be stated with the results obtained without fertilizer, as shown in Table 5. From this comparison, it is clear that the fertilizer addition, by soil or by controlled-release fertilizer addition, is capable of improving the basil growth performances. In particular, it can be observed that plant height is always greater with fertilizer as well as the NoL and SLA parameters. Figures 5–7 show examples of the plants obtained at the end of the experimentation, respectively those grown under 3HR:1DB, 1HR:1DB, and 1HR:3DB, thereafter moving from a higher to lower proportion of hyper red LED light in relation to deep blue. As can be seen observing Figures 5–7 and the data in Table 4, the introduction of NPK controlled-release fertilizer has a positive effect on the plants' growth by increasing a considerable number of plants parameters (such as Height, TFM, TDM, LAI, and SLA), and in general, this effect is not influenced by the type of LED light. Nevertheless, from Figure 7, it appears clear that a more distinguishable difference among plants with and without NPK is appreciable, suggesting a possible synergic effect between NPK addition and 1HR:3DB LED lights combination that is capable of possibly promoting plant height and TFM. For these reasons, it is essential to study the ANOVA analysis to mathematically identify the possible correlation among NPK and LED lights combination as a function of the different measured parameters.

Figure 5. Basil plants obtained under 3HR:1DB LED lights.

Figure 6. Basil plants obtained under 1HR:1DB LED lights.

Figure 7. Basil plants obtained under 1HR:3DB LED lights.

The normal distribution of the residuals, as well as their homogeneity, has been analyzed (data not reported) for each response, confirming that the mathematical models derived can be used to explore the region of interest. The resulting models allowed describing the relationships between growth condition and the measured responses. The ANOVA results have been presented in Tables 6 and 7, where the quantification of the significance of factors and their interactions were reported as well as the fitting quality parameters as well as the effect sizes for each significant model. An inverse square root transformation was required to normalize the data and codify the hierarchy of the factors for some responses. The model correlating the factors (in single or interaction) to the panel data evaluation is significant, as confirmed by the p-value < 0.0001 for all the responses, which means that the probability of the data variation due to unknown factors is statistically irrelevant. Moreover, it is relevant that the curvature is not significant, and therefore, the central points can be treated as additional data in the regression model, augmenting the design plan. R^2 and Pred-R^2 (Table 6) confirm the sufficient fit of the data and a quite fair predictive power for four responses, having $R^2 > 0.45$, with a particular high fitting for SLA parameters with $R^2 = 0.75$. Considering all the models' equations in Table 6 (where the coefficients of the variables in the model are reported), it is clear that NPK controlled-release fertilizer plays the main role in the performance of the responses, and this is consistent with the previous hypothesis, but a synergic effect has been shown and calculated for the SLA responses, which also corresponds to the model with higher fitting quality (higher R^2).

Table 6. ANOVA results.

Response	Transformation	R^2	Pred-R^2	Models' Mathematical Expression	
				NPK = NO	NPK = YES
Height	NONE	0.13	0.10	-	
NoL	NONE	0.47	0.41	= 8.000	= 9.666
TFM	Inverse Square Root	0.58	0.53	= 2.032	= 1.565
TDM	NONE	0.28	0.19	-	
LAI	Inverse Square Root	0.55	0.49	= 0.543	= 0.413
SLA	Inverse Square Root	0.75	0.68	$= 0.045 + 3.7 \times 10^{-5} \times$ HR:DB	$= 0.039 + 3.7 \times 10^{-5} \times$ HR:DB

Table 7. Effect sizes for significant responses.

Response	Intercept	NPK	HR:DB
NoL	8.83	0.8333	-
TFM	1.80	−0.2340	-
LAI	0.47	−0.0615	-
SLA	0.04	−0.0032	0.0009

Figure 8 shows interaction graphs representing graphically each calculated model. A quite similar behavior can be detected for NoL, TFM, and LAI (Figure 8a–c), where an increasing of these responses is observed by introducing the NPK fertilizer independently of the LED lights combination. In contrast, the SLA parameter (Figure 8d) is clearly influenced by a synergic effect of NPK and LED lights: a restrained quantity of hyper red (1HR:3DB) is favorable to the SLA increase, independently of the presence of NPK, even if the highest value of SLA is obtained employing also controlled release NPK. The overall result is that a high quantity of hyper red LED light should be avoided to promote the basil growth. This result is in agreement with previous literature related to other plants' growth in an indoor and black environment such as lettuce, tomato, and flowers, where the employment of a high quantity of red LED light leads to reduced dry weight, plant height, and leaf area [22,23,41,42].

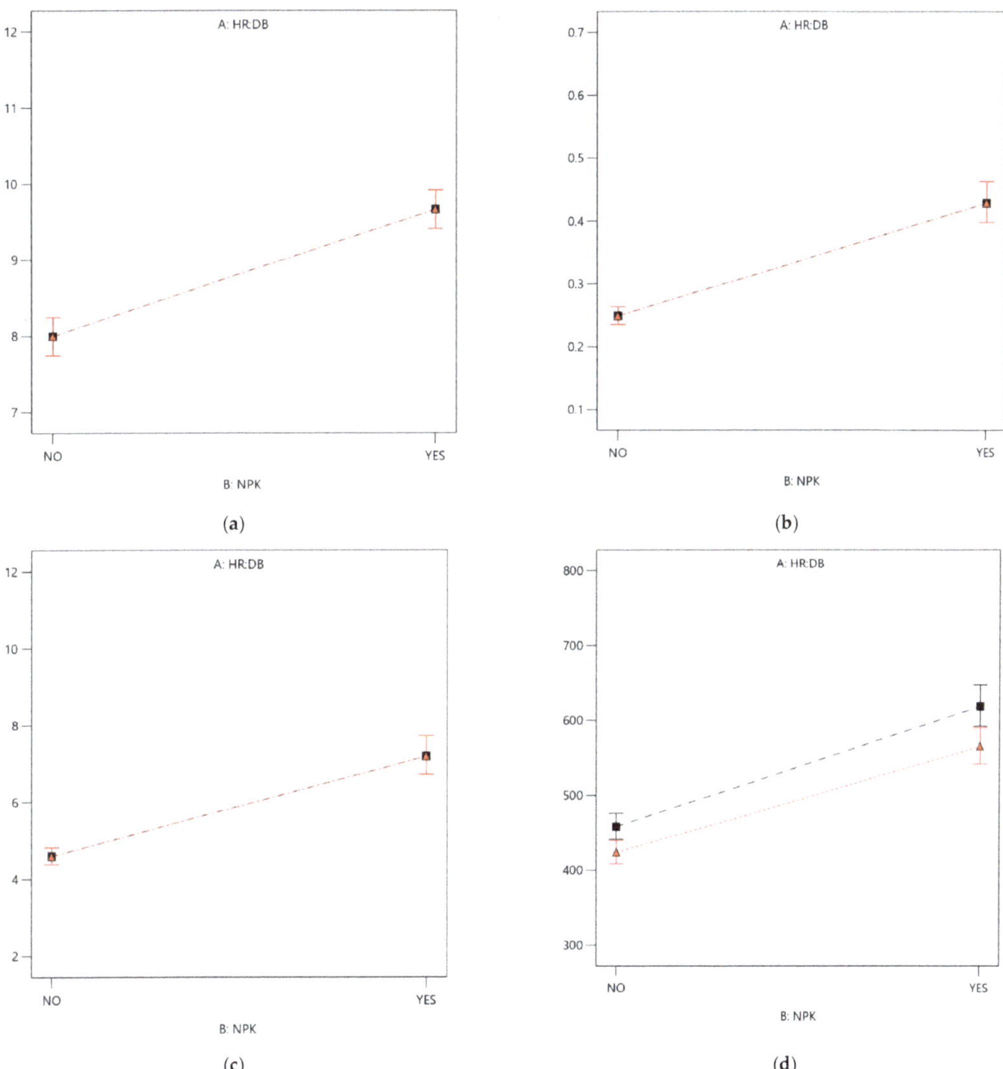

Figure 8. Interaction plots of different responses: in black color, the interaction lines related to 1HR:3DB; in red color, the interaction lines related to 3HR:1DB: (**a**) NoL; (**b**) TFM; (**c**) LAI; (**d**) SLA. Error bars indicate the standard deviation on the average values. LAI: Leaf Area Index, NoL: Number of Leaves, SLA: Specific Leaf Area, TFM: Total Fresh Mass.

4. Conclusions

In this study, it has been demonstrated that the implementation of various combination of LED lights is generally favorable to basil (*Ocimum Basilicum*) germination and growth with respect to daylight (e.g., Height: +30%, Total fresh mass: +50%), although no differences among hyper red and deep blue LED lights combination can be noticed when a soil without any fertilizer is employed as growing media. Applying a design of experiments approach, mathematically reliable information has been derived concerning a possible synergic effect with a soil enriched in macronutrients and with an NPK controlled-release fertilizer derived from agro-residues in a circular economy perspective. It has been demonstrated that the use of NPK controlled-release fertilizer provides better performance, for the cumulative values for individual pots, than the use of fertilized soil alone. In addition, for the SLA parameter, the highest value is obtained by employing controlled-release NPK fertilizer coupled with a specific LED light combination (one hyper red: three deep blue). It should be noted that the present study analyzed the influence of different factors on basil growth considering only one moment of collection without considering the variation of the light recipe in the different phases of life of the same. Thereafter, for a future perspective, the variation of the light recipe during the different phases of life of the basil plant should be investigated. Moreover, by taking some leaves or part of them and storing them in liquid nitrogen, it would be possible to perform biochemical measurements, such as determination of the content of phenolic compounds, flavonoids, antioxidant capacity, nitrate content, chlorophyll content, and carotenoids. Finally, it must be noted that basil is an edible aromatic herb that is extensively employed in the Mediterranean diet; therefore, panel test judging the possible different tastes should be proposed as a future perspective.

Author Contributions: Conceptualization, M.M. and A.B.; software, S.B.; investigation, F.B. and A.B.; resources, L.B. and M.M.; writing—original draft preparation, S.B.; writing—review and editing, F.B. and A.B.; supervision, M.M. and L.B.; funding acquisition, M.M. and L.B. All authors have read and agreed to the published version of the manuscript.

Funding: This research was funded by the Università di Modena e Reggio Emilia under the FAR program supported by Fondazione Cassa di Risparmio di Modena, project title "GREW (Garden from Recycling & Wastes)—New integrated system for house and vertical gardens cultures by synergic application of innovative fertilizer and LED lighting: A circular economy strategy giving to waste materials a new second life".

Institutional Review Board Statement: Not applicable.

Informed Consent Statement: Not applicable.

Data Availability Statement: The data presented in this study are available on request from the corresponding author. The data are not publicly available due to founding regulation.

Acknowledgments: The authors thank Giovanni Verzellesi (University of Modena and Reggio Emilia) for the support and the fruitful discussions.

Conflicts of Interest: The authors declare no conflict of interest.

References

1. Sipos, L.; Boros, I.F.; Csambalik, L.; Székely, G.; Jung, A.; Balázs, L. Horticultural lighting system optimalization: A review. *Sci. Hortic.* **2020**, *273*, 109631. [CrossRef]
2. Folta, K.M.; Childers, K.S. Light as a growth regulator: Controlling plant biology with narrow-bandwidth solid-state lighting systems. *HortScience* **2008**, *43*, 1957–1964. [CrossRef]
3. Pinho, P.; Hytönen, T.; Rantanen, M.; Elomaa, P.; Halonen, L. Dynamic control of supplemental lighting intensity in a greenhouse environment. *Light. Res. Technol.* **2013**, *45*, 295–304. [CrossRef]
4. FAO Statistics Division. Available online: http://www.fao.org/documents/card/en/c/cb3411en (accessed on 10 March 2021).
5. Goto, E. Plant production in a closed plant factory with artificial lighting. *Acta Hortic.* **2012**, *956*, 37–49. [CrossRef]
6. Hernández, R.; Kubota, C. Tomato seedling growth and morphological responses to supplement LED lighting red:blue ratios under varied daily solar light integrals. *Acta Hortic.* **2012**, *956*, 187–194. [CrossRef]
7. Folta, K.M.; Carvalho, S.D. Photoreceptors and control of horticultural plant traits. *HortScience* **2015**, *50*, 1274–1280. [CrossRef]
8. Heijde, M.; Ulm, R. UV-B photoreceptor-mediated signalling in plants. *Trends Plant Sci.* **2012**, *17*, 230–237. [CrossRef] [PubMed]
9. Dueck, T.A.; Janse, J.; Eveleens, B.A.; Kempkes, F.L.K.; Marcelis, L.F.M. Growth of tomatoes under hybrid led and HPS lighting. *Acta Hortic.* **2012**, *952*, 335–342. [CrossRef]
10. Yeh, N.; Chung, J.P. High-brightness LEDs-Energy efficient lighting sources and their potential in indoor plant cultivation. *Renew. Sustain. Energy Rev.* **2009**, *13*, 2175–2180. [CrossRef]
11. Currey, C.J.; Lopez, R.G. Cuttings of Impatiens, Pelargonium, and Petunia propagated under light-emitting diodes and high-pressure sodium lamps have comparable growth, morphology, gas exchange, and post-transplant performance. *HortScience* **2013**, *48*, 428–434. [CrossRef]
12. Durmus, D. Real-Time Sensing and Control of Integrative Horticultural Lighting Systems. *J. Multidiscip. Sci. J.* **2020**, *3*, 20. [CrossRef]
13. van Iersel, M.W. Optimizing LED Lighting in Controlled Environment Agriculture. In *Light Emitting Diodes for Agriculture: Smart Lighting*; Dutta Gupta, S., Ed.; Springer: Singapore, 2017; pp. 59–80, ISBN 9789811058073.

14. Zhang, X.; Bian, Z.; Yuan, X.; Chen, X.; Lu, C. A review on the effects of light-emitting diode (LED) light on the nutrients of sprouts and microgreens. *Trends Food Sci. Technol.* **2020**, *99*, 203–216. [CrossRef]
15. Yanagi, T.; Okamoto, K.; Takita, S. Effects of blue, red and blue/red lights of two different PPF levels on growth and morphogenesis of lettuce plants. *Acta Hortic.* **1996**, *440*, 117–122. [CrossRef]
16. Schwartz, A.; Zeiger, E. Metabolic energy for stomatal opening. Roles of photophosphorylation and oxidative phosphorylation. *Planta* **1984**, *161*, 129–136. [CrossRef]
17. Cosgrove, D.J.; Green, P.B. Rapid Suppression of Growth by Blue Light. *Plant Physiol.* **1981**, *68*, 1447–1453. [CrossRef] [PubMed]
18. Nhut, D.T.; Takamura, T.; Watanabe, H.; Okamoto, K.; Tanaka, M. Responses of strawberry plantlets cultured in vitro under superbright red and blue light-emitting diodes (LEDs). *Plant Cell. Tissue Organ Cult.* **2003**, *73*, 43–52. [CrossRef]
19. Iacona, C.; Muleo, R. Light quality affects in vitro adventitious rooting and ex vitro performance of cherry rootstock Colt. *Sci. Hortic.* **2010**, *125*, 630–636. [CrossRef]
20. Chory, J. Light signal transduction: An infinite spectrum of possibilities. *Plant J.* **2010**, *61*, 982–991. [CrossRef] [PubMed]
21. Tarakanov, I.; Yakovleva, O.; Konovalova, I.; Paliutina, G.; Anisimov, A. Light-emitting diodes: On the way to combinatorial lighting technologies for basic research and crop production. *Acta Hortic.* **2012**, *956*, 171–178. [CrossRef]
22. Piovene, C.; Orsini, F.; Bosi, S.; Sanoubar, R.; Bregola, V.; Dinelli, G.; Gianquinto, G. Optimal red: Blue ratio in led lighting for nutraceutical indoor horticulture. *Sci. Hortic.* **2015**, *193*, 202–208. [CrossRef]
23. Lobiuc, A.; Vasilache, V.; Pintilie, O.; Stoleru, T.; Burducea, M.; Oroian, M.; Zamfirache, M.M. Blue and red LED illumination improves growth and bioactive compounds contents in acyanic and cyanic ocimum Basilicum L. Microgreens. *Molecules* **2017**, *22*, 2111. [CrossRef] [PubMed]
24. Lin, K.H.; Huang, M.Y.; Hsu, M.H. Morphological and physiological response in green and purple basil plants (Ocimum basilicum) under different proportions of red, green, and blue LED lightings. *Sci. Hortic.* **2021**, *275*, 109677. [CrossRef]
25. Carvalho, S.D.; Schwieterman, M.L.; Abrahan, C.E.; Colquhoun, T.A.; Folta, K.M. Light quality dependent changes in morphology, antioxidant capacity, and volatile production in sweet basil (Ocimum basilicum). *Front. Plant Sci.* **2016**, *7*, 1–14. [CrossRef]
26. Avgoustaki, D.D.; Li, J.; Xydis, G. Basil plants grown under intermittent light stress in a small-scale indoor environment: Introducing energy demand reduction intelligent technologies. *Food Control* **2020**, *118*, 107389. [CrossRef]
27. Barbi, S.; Barbieri, F.; Andreola, F.; Lancellotti, I.; Barbieri, L.; Montorsi, M. Preliminary study on sustainable NPK slow-release fertilizers based on byproducts and leftovers: A design-of-experiment approach. *ACS Omega* **2020**, *5*, 27154–27163. [CrossRef]
28. Barbi, S.; Barbieri, F.; Andreola, F.; Lancellotti, I.; García, C.M.; Palomino, T.C.; Montorsi, M.; Barbieri, L. Design and characterization of controlled release PK fertilizers from agro-residues. *EEMJ* **2020**, *19*, 1669–1676.
29. Montgomery, D.C. Design and Analysis of Experiments, 8th ed.John Wiley & Sons: Hoboken, NJ, USA, 2012; Volume 2, ISBN 9781118146927.
30. Intelligent Led Solutions Petunia Led Modules. Available online: https://i-led.co.uk/PDFs/Kits/12Multi-OslonSSL-PetuniaColourV3.pdf (accessed on 28 January 2021).
31. LEDiL C12528 PETUNIA Lens. Available online: https://www.ledil.com/product-card/?product=C12528_PETUNIA (accessed on 10 February 2021).
32. OSRAM LH CP7P 660 nm Hyper Red LED. Available online: https://www.osram.com/ecat/OSLON®//SSL80LHCP7P/com/en/class_pim_web_catalog_103489/prd_pim_device_2402508/ (accessed on 6 March 2021).
33. OSRAM LD CQ7P 451 nm Deep Blue LED. Available online: https://www.osram.com/ecat/OSLON®//SSL80LDCQ7P/com/en/class_pim_web_catalog_103489/prd_pim_device_2402502/ (accessed on 6 March 2021).
34. Ag, A. AS7341 Spectral. Available online: https://ams.com/as7341 (accessed on 10 February 2021).
35. Eriksson, L.; Johansson, E.; Kettaneh-Wold, N.; WikstrÄom, C.; Wold, S. *Design of Experiments: Principles and Applications*; Umetrics Academy: Umeå, Sweden, 2008; ISBN 10:9197373044.
36. Morris, P.; John, P.W.M. Statistical Design and Analysis of Experiments. *Math. Gaz.* **1999**, *83*, 189. [CrossRef]
37. *Region Emilia-Romagna, Disciplinari di Produzione Integrata Norme Tecniche di Coltura*; 2018; pp. 1–7. Available online: EMR_M10.1.1_2016_Racc_Col_Ort.pdf (accessed on 10 February 2021).
38. Stagnari, F.; Di Mattia, C.; Galieni, A.; Santarelli, V.; D'Egidio, S.; Pagnani, G.; Pisante, M. Light quantity and quality supplies sharply affect growth, morphological, physiological and quality traits of basil. *Ind. Crops Prod.* **2018**, *122*, 277–289. [CrossRef]
39. Italian, R. Legislative Decree n. 75/2010 Concerning Fertilizers. Gazz. Uff. Ser. Gen. n.218 del 17-09-2013. 2010. Available online: biostimulants.weebly.com (accessed on 10 February 2021).
40. Frerichs, C.; Daum, D.; Koch, R. Influence of nitrogen form and concentration on yield and quality of pot grown basil. *Acta Hortic.* **2019**, *1242*, 209–216. [CrossRef]
41. Mortensen, L.M.; Strømme, E. Effects of light quality on some greenhouse crops. *Sci. Hortic.* **1987**, *33*, 27–36. [CrossRef]
42. Mortensen, L.M. Effects of temperature and light quality on growth and flowering of Begonia × hiemalis Fotsch. and Campanula isophylla Moretti. *Sci. Hortic.* **1990**, *44*, 309–314. [CrossRef]

applied sciences

Statistical Optimization of a Hyper Red, Deep Blue, and White LEDs Light Combination for Controlled Basil Horticulture

Silvia Barbi [1,*], Francesco Barbieri [2], Alessandro Bertacchini [2,3] and Monia Montorsi [2,3]

[1] Interdepartmental Research Center for Industrial Research and Technology Transfer in the Field of Integrated Technologies for Sustainable Research, Efficient Energy Conversion, Energy Efficiency of Buildings, Lighting and Home Automation–EN&TECH, University of Modena and Reggio Emilia, Via Amendola 2, 42122 Reggio Emilia, Italy

[2] Department of Science and Methods for Engineering, University of Modena and Reggio Emilia, Via Amendola 2, 42122 Reggio Emilia, Italy; francesco.barbieri1@unimore.it (F.B.); alessandro.bertacchini@unimore.it (A.B.); monia.montorsi@unimore.it (M.M.)

[3] Interdepartmental Center for Applied Research and Services in Advanced Mechanics and Motoring, INTERMECH-Mo.Re., University of Modena and Reggio Emilia, Via P. Vivarelli 10/1, 41125 Modena, Italy

[*] Correspondence: silvia.barbi@unimore.it

Featured Application: The specific application of this work is related to basil cultivation in indoor horticulture under artificial light. This work is devoted to the investigation of specific light combinations, based on hyper red, deep blue, warm white LEDs, to promote basil germination and growth. The aim is to improve basil's yield and quality by reducing the overall growth cycle at the same time.

Citation: Barbi, S.; Barbieri, F.; Bertacchini, A.; Montorsi, M. Statistical Optimization of a Hyper Red, Deep Blue, and White LEDs Light Combination for Controlled Basil Horticulture. *Appl. Sci.* **2021**, *11*, 9279. https://doi.org/10.3390/app11199279

Academic Editors: Isidoro Garcia-Garcia, Jesus Simal-Gandara and Maria Gullo

Received: 24 August 2021
Accepted: 1 October 2021
Published: 6 October 2021

Publisher's Note: MDPI stays neutral with regard to jurisdictional claims in published maps and institutional affiliations.

Abstract: This study aims to optimize artificial LEDs light conditions, for "Genovese" basil germination and growth in an indoor environment suitable for horticulture. Following a previous study on the synergic effect of LEDs light and a tailored fertilizer, in this study, the effect of white LED in combination with hyper red and deep blue, as well the plants–lights distance, was correlated to 14 growth and germination parameters, such as height, number of plants, etc. A design of experiments approach was implemented, aiming to derive mathematical models with predictive power, employing a restrained number of tests. Results demonstrated that for the germination phase, it is not possible to derive reliable mathematical models because almost the same results were found for all the experiments in terms of a fruitful germination. On the contrary, for the growth phase, the statistical analysis indicates that the distance among plants and lights is the most significant parameter. Nevertheless, correlations with LED light type emerged, indicating that white LEDs should be employed only to enhance specific growth parameters (e.g., to reduce water consumption). The tailored models derived in this study can be exploited to further enhance the desired property of interest in the growth of basil in horticulture.

Keywords: basil; design of experiments; LED

---

1. Introduction

The principal mechanism that drives plants germination and growth is photosynthesis, as light is the primary source of energy and the principal regulatory variable in plant's cycle. Thereafter, light determines the appearance of plants, their growth rate as well as their quality. In a closed environment, plant growth and morphogenesis can be controlled as desired with artificial light, and, for this reason, it has been demonstrated that the annual production capacity of plants per floor area is about 10 times higher than in the greenhouse [1]. Since under natural light source (that include all the visible spectra), plants must constantly respond with biochemical (carbohydrate content and pigment) and physiological (nutrient uptake and photosynthetic rate) adaptations, with artificial

239

light it is possible to meet plants requirements, therefore improving the overall growth efficiency [2,3]. For example, low irradiance leads to a less controlled leaves growth, breaking the optimization mechanism that regulates the leaf area per unit necessary to maximize the light interception [4] in the growth period of the plants.

In addition, an increased efficiency in plant germination and growth, by a precise control over artificial light, can be a relevant opportunity for environments where natural light is naturally absent, such as northern latitude countries or space travel. Artificial light optimization can be beneficial for leafy greens and herbal crops that are an exceptional source of various human health-promoting macronutrients such as polyphenols, vitamins, and essential oils [5]. For this reason, for these plants, the relationship between artificial light and physiology under different light spectra and/or light intensities is attracting the interest of the more recent research [6,7]. Among leafy greens, *Ocimum basilicum* L. (Basil) belonging to the *Lamiaceae* family is a worldwide-spread herbal crop employed as culinary herb and for essential oil extraction, as well as, widely used in medical industry as component of oral health products and to treat several medical complaints such as gastrointestinal disorders, headaches and stomach aches [8,9]. Considering the beneficial effect on human health and nutrition, basil is considered one of the best candidates to study its potential germination and growth optimization through artificial light [10].

As artificial light source for plant cultivation, light-emitting diodes (LEDs) have received considerable attention as they have the capability to produce a considerable light intensity with low radiant heat output for years, compared to fluorescent lamp, therefore with an increased lifespan [11]. Compared to incandescent light's 1000 h and fluorescent light's 8000 h life span, LEDs have a significantly longer life of 100,000 h [12]. In addition, by using well-tailored LEDs-based arrays, it is possible to have a full control of the spectral composition and an adjustment of light intensity for a specific plant or growth period. In terms of power consumption, the beneficial effect related to the use of LEDs as light source is twofold [13–15]. First, it is well known that, for a given light intensity, the power consumption of a generic LED lamp is significantly lower than any other artificial light source. Second, for a given plant, realizing a custom LED light source accordingly with the optimal light combination of the plant, avoids power waste related to light wavelengths that have no effects on the growth of the plant. These considerations are valid not only for a monochromatic LED, but also for white LEDs. It is well known that white LEDs do not produce directly white light. There are two main manufacturing process to realize white LEDs. The first one is a conventional process that exploits RGB-LEDs dies on the same chip. In this case, it is possible to change the color temperature of the produced white light by acting on the current flowing into the LED that in turn vary the light intensities of the single RGB-LEDs. The second one is called fluorescence (GaN-LED) and exploits a blue LED (or UV in some cases) only with a special yellow coating realized with phosphor. In this case, a phosphor coating is deposited on the blue LED die. The color temperature of the obtained white light depends on the dominant wavelength of the blue LED used [15].

According to the more recent research, plants' chlorophyll molecules absorb the red and blue spectral components most efficiently, and for this reason, red and blue spectral components best drive photosynthetic metabolism and are the most relevant for crop growth [16]. In addition, exposure to red and blue light increases the content of phenolic compounds and improve antioxidant activity in various leafy greens with respect to only white light exposure, including if these effects are different among the species [16–18]. Blue light, having short wavelength, has the capability to regulate plant growth at different stage, promoting, when in considerable presence, compactness, and plant density [19–22]. Red light, having longer wavelength, has beneficial effect on stem elongation and leaf area [23,24]. However, green and yellow light components are less compatible to plant receptor, and they are principally reflected or transmitted and thus are not as important in the photosynthetic process. For basil cultivation the main works are focused on phenolic biosynthesis and accumulation but few of them are related to systematic cause–effect investigation on growth and yield performance under artificial lamps [1,7]. In addition, up until now, few studies investigate this effect in a systematic way and comparison among different studies is not trivial due to the different boundary condition employed for basil cultivation. In particular, hyper red (660 nm) light component effect on plant growth is not fully understood yet, as it seems to contribute to increase the plant' yield and quality, but, at the same time, promoting a reverse effect of phytochromes leading to changes in plant morphology, gene expression, and reproductive responses [25–27].

On these bases, the main objective of the present study is to evaluate the artificial illumination spectrum that is capable to induce the most significant biomass improvement of "Genovese" basil (*Ocimum Basilicum*) plants in a controlled environment. Following the results presented in [28] three different light–plants distances and four different LEDs modules have been considered in combination. From a previous study, the effect of different artificial lights was evaluated as negligible, with respect to a fertilizer addition, to promote basil growth [28]. A slight favorable effect was estimated for a combination of LEDs light containing hyper red and deep blue light in a proportion of 1:3 (LEDs modules quantity), but further investigation was necessary [28]. A statistical approach (design of experiments) has been applied to reduce the number of possible combinations of the variables, and, consequently, of the experiments needed to obtain statistically reliable correlation among data [29]. Data inherent to germination, growth and plant's water consumption have been collected to demonstrate and calculate, numerically, specific effects of artificial light conditions on basil cultivation. Thereafter, as innovative aspects of this study, lights spectral components that have not widely investigated have been studied, and in addition, a systematic approach has been applied, with the aim to calculate specific light combination to further enhance a specific property. Finally, to the best of author's knowledge, this is one of the few researches investigating the effect of different artificial light on plant's water consumption.

2. Materials and Methods

2.1. Materials

For this study, Basil (*Ocimum Basilicum*) seeds belonging to the variety called "Genovese" (Producer: Magnani Sementi) were employed as reference. As burying soil, Floradur B pot coarse universal potting soil (Floragard Vertriebs GmbH, Oldenburg, Germany) was employed. Detailed specifications about this substrate can be found in a previous study, as the same soil was employed to promote the comparison among different soils suitable for basil [28]. This soil respects the limits generally applied to fertilizer compounds necessary to be sold and employed [30].

The experiments were conducted by exploiting as artificial light commercial LED modules realized with OSRAM Oslon®SSL ThinGaN LEDs (UX:3) technology, [31]. The composition in terms of number and type of LEDs are summarized in Table 1. The modules were characterized using an Orb Optronix TEC-100 electrical–thermal–optical (ETO) system with integrating sphere.

Table 1. Composition of the LEDs modules used in the experiments to realize the considered light combinations.

Module Type	Module Code [1]	Total Number of LEDs	Total Number of HR [1] LEDs	Total Number of DB [2] LEDs	Total Number of WW [3] LEDs
1	5HR:1DB:6WW	12	5	1	6
2	9HR:3DB	12	9	3	-
3	6HR:6DB	12	6	6	-
4	3HR:9DB	12	3	6	-

[1] HR = hyper red. [2] DB = deep blue. [3] WW = warm white.

In horticulture, from an application point of view, the main parameter to consider for a light source is the photosynthetic photon flux (*PPF*). The overall photon flux (*PF*) in μmol/s of a light source depends on the radiant flux (*RF*) in W, and is proportional to the light wavelength, λ. Only the portion of photons with wavelengths in the range 400–700 nm (i.e., the so-called photosynthetic active region PAR) contributes to photosynthesis. *PPF* is the portion of *PF* due to photons within the PAR. Since the ETO system provided the data concerning the whole modules, the contribution of each LED type was calculated off-line. For the hyper red [32] and deep blue [33] LEDs, the *PPF* contribution of a single LED (i.e., $PPF_{singleLED\_HR}$ and $PPF_{singleLED\_DB}$) were extracted from datasheets using (1) or (2) depending on the specific LED

$$PPF_{singleLED\_HR} \approx PF = RF \cdot \lambda \cdot 0.00836/1000 \tag{1}$$

$$PPF_{singleLED\_DB} \approx PF = RF \cdot \lambda \cdot 0.00836/1000 \tag{2}$$

where, *RF* is the radiant flux in W, λ is the peak wavelength of the LED in nm, the light wavelength, the coefficient $0.00836 = 1/(h \cdot c \cdot N)$ *h* is Planck's constant, *c* is the speed of light and *N* is the constant of Avogadro. *RF* and λ are reported in the datasheet.

For a given monochromatic LED, the error derived by approximating PPF with PF is negligible because outside the PAR range there are no photons emitted.

For warm white LEDs, instead, it is not possible to apply directly (1) or (2) due to the intrinsic nature of the white light. Different from single color light sources, white light sources have not a single dominant wavelength in the spectrum. For this reason, the datasheets of white LEDs usually do not report the *RF*. They report instead the luminous flux (*LF*, in lumen) for a given color temperature (i.e., 3000 K for the considered LEDs). In this case, to obtain the estimated $PPF_{singleLED\_WW}$ contribution of the warm white LEDs a two-steps empiric procedure was used. First, the relative spectral emission of the considered white LEDs [34], reported in the datasheet was included in the freely available spreadsheet tool provided by lighting analysts [35]. This allowed to obtain a luminous-to-photon flux conversion factor $K_{LF\text{-}to\text{-}PF} = 14.64$. Second, the obtained conversion factor that was employed in (3) to obtain the *PPF* of the warm white LED of interest, $PPF_{singleLED\_WW}$.

$$PPF_{singleLED\_WW} = LF \cdot K_{LF\text{-}to\text{-}PF}/1000 \tag{3}$$

The obtained *PPF* values for each type of LED are summarized in Table 2.

Table 2. Main opto-electrical features of the considered single LED as result by combining data reported in datasheets and off-line computations.

LED Type	Peak Emission Wavelength	Spectral Bandwidth at 50%	Correlated Color Temperature	PPF [μmol/s]
HR	660 nm	25 nm	-	2.01 [1]
DB	451 nm	20 nm	-	2.26 [2]
WW	-	-	3000 K	1.47 [3]

[1] $PPF_{singleLED\_HR}$ obtained by (1). [2] $PPF_{singleLED\_DB}$ obtained by (2). [3] $PPF_{singleLED\_WW}$ obtained by (3).

For each considered LED module, the overall *PPF*, PPF_{module_calc} can be easily obtained as shown in (4) combining the $PPF_{singleLED\_HR}$, $PPF_{singleLED\_DB}$ and $PPF_{singleLED\_WW}$ reported in Table 2 and the LED's composition of the modules reported in Table 1.

$$PPF_{module\_calc} = \#LED_{HR} \cdot PPF_{singleLED\_HR} + \#LED_{DB} \cdot PPF_{singleLED\_DB} + \#LED_{WW} \cdot PPF_{singleLED\_WW} \tag{4}$$

where $\#LED_{HR}$, $\#LED_{DB}$, $\#LED_{WW}$ are the number of hyper red, deep blue and warm white LEDs per module, respectively.

The obtained *PPF* compositions for each considered LED module is shown in Table 3. For a given module, there is not a direct relationship between the percentage of PPF and the ratio between number of LEDs of a given type and total number of LED of the module. The reason is that the overall photon flux (*PF*), in μmol/s, depends on the radiant flux (*RF*), in W, and is proportional to the light wavelength, λ. Conversely, the energy per photon, E_{ph}, is inversely proportional to λ.

Table 3. Calculation of the PPF composition for each considered module.

Module's Code	PPF HR [μmol/s]	PPF DB [μmol/s]	PPF WW [μmol/s]	Total PPF_{module_calc} [μmol/s]	%PPF HR	%PPF DB	%PPF White
5HR:1DB:6WW	10.07	2.26	8.84	21.17	47.57	10.69	41.74
9HR:3DB	18.13	6.79	-	24.92	72.76	27.24	-
6HR:6DB	12.08	13.57	-	25.65	47.10	52.90	-
3HR:9DB	6.04	20.36	-	26.40	22.88	77.12	-

For each considered module, the comparison between the *PPF* measured with the ETO integrating sphere, PPF_{ETO}, and the PPF_{module_calc}, is shown in Table 4.

Table 4. Comparison between measured and calculated PPF for each considered LED module.

Module's Code	PPF_{ETO} [μmol/s]	PPF_{module_calc} [μmol/s]	Error [1] [%]
5HR:1DB:6WW	19.48	21.17	7.98
9HR:3DB	23.61	24.91	5.23
6HR:6DB	24.63	25.66	4.00
3HR:9DB	24.19	26.40	8.38

[1] Error = $100 * 1 - (PPF_{ETO} - PPF_{module\_calc})$.

Off-line computations PPF_{module_calc} have a good match with direct measurements PPF_{ETO} with a maximum deviation of about 8%. In the first approximation, this evidence validates the procedures used to calculate $PPF_{singleLED\_HR}$, $PPF_{singleLED\_DB}$ and $PPF_{singleLED\_WW}$, demonstrating that they provide an easy way to estimate the *PPF* of any custom LED light source, basically using only the data provided by the LED manufacturer and always reported in the LED's datasheet. From a practical point of view, this is important because for a given horticulture scenario, it allows to obtain a reliable *PPF* estimation at design time, without need for field tests on the real system. Consequently, it allows obtaining significant savings in terms of development time, costs, instrumentation needed and design software analysis.

To verify the behavior of each LED module under the same working conditions, and to evaluate the effect of the distance among plants and LEDs in terms of PPF, three different distances, d, (60, 70 and 80 cm respectively) were considered. All the considered LED modules were experimental characterized by means of a low-cost spectral sensor (AS7341 from Ams, [36]). The measurements were conducted by reproducing the same test conditions used for the test with ETO integrating sphere, i.e., constant current flowing through the LEDs of 350 mA and temperature of the modules of 42 °C. Temperature and relative humidity of the surrounding air were recorded using wireless sensors with accuracy equal to ± 1 °C (resolution: 0.1 °C) for temperature, and $\pm 5\%$ (resolution: 1%) for humidity. The results are summarized in Figure 1. As expected, sensor (i.e., plants)-LEDs module distance affects the effective fraction of PPF. The smaller the distance, the larger the effective PPF, and this is due to the radiation angle of each single LED. While ETO captures each emitted photon, in real conditions, the energy associated with photons that do not reach the plants is wasted. This effect is mitigated thanks to the use of specific lens from LEDiL, [37], embedded into the LED modules, that reduce the radiation angle to ≈ 30 deg. These test conditions were held fixed for all the experiments described in the following paragraphs.

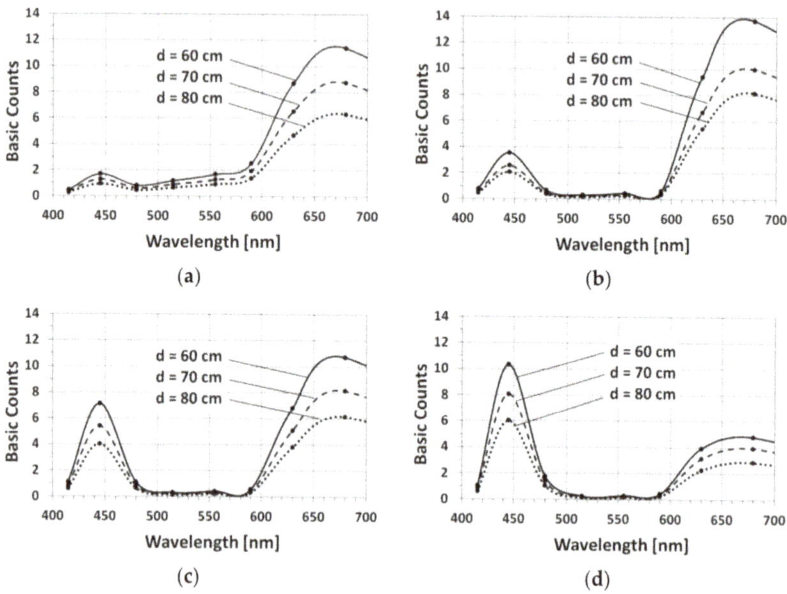

Figure 1. (**a**) Module 5HR:1DB:6WW, (**b**) module 9HR:3DB, (**c**) module 6HR:6DB and (**d**) module 3HR:9DB: light spectrums obtained using the low cost AS7341 spectral sensor. Each curve is obtained by the interpolation of the dimensionless output values of the 8 optical channels of the sensor with filters centered at wavelengths of 415, 445, 480, 515, 555, 590, 630, and 680 nm, in order (marker dots) [37].

2.2. Statistical Methods

A statistical method, known as design of experiments (DoE), was employed to plan the lower possible number of experiments needed to calculate mathematical models, correlating light conditions and germination/growth performances, with predictive power [29]. Three factors were considered as independent variables, as shown in Table 5:

Table 5. Independent variables used for the statistical analysis.

Factor	Type	Levels	Minimum	Zero-Point	Maximum
Distance	Numeric/Discrete	3	60 cm	70 cm	80 cm
HR:DB ratio [1]	Numeric/Discrete	3	1HR:3DB	1HR:1DB	3HR:1DB
White	Categoric/Nominal	2	YES	-	NO

[1] ratio between number of hyper red (HR) and deep blue (DB) LEDs for a given light combination. 1HR:1DB = same number of HR and DB LEDS. 1HR:3DB = number of DB LEDS that is the triple of the number of HR LEDs. 1HR:3DB = number of HR LEDS that is the triple of the number of DB LEDs.

The other variables occurring in the process and not specifically considered in this study, such as humidity and temperature, were kept constant during all the tests, according to the procedure as explained in paragraph 2.3. The Design Expert 13.0 (Stat-Ease) code was used both to plan the experiments and to perform the statistical analysis. To avoid environmental conditioning, all the sample were tested at the same time. Each distance was kept constant for each sample by lifting the light placement by hand, according to plant's increasing high, during growth, that was measured during the experiment with a meter stick having resolution 1 mm.

Due to the high number of possible combinations among factors, a fractional response surface design was selected with I-optimal design type to enhance a lower average prediction variance. A total of 20 experiments were performed, including repetitions of some of them for variance estimation inside a homogenous group of samples, due to intrinsic experimental approximation, with the aim to compare it with the variance among all investigated samples (Table 6).

Table 6. Experimental plan.

	DoE VARIABLES		White	LIGHT COMBINATION [4]			
#Test	d [1] [cm]	HR:DB Ratio [2]	LEDs	COMBINATION'S CODE [3]	# of HR LEDs	# of DB LEDs	# of WW LEDs
1	80	1:03	YES	LR14	8	10	6
2	80	1:03	YES	LR14	8	10	6
3	80	1:01	YES	LR13	11	7	6
4	80	3:01	YES	LR12	14	4	6
5	80	1:03	NO	LR44	6	18	0
6	80	1:01	NO	LR33	12	12	0
7	80	3:01	NO	LR22	18	6	0
8	70	1:03	YES	LR14	8	10	6
9	70	1:01	YES	LR13	11	7	6
10	70	3:01	YES	LR12	14	4	6
11	70	1:03	NO	LR44	6	18	0
12	70	1:01	NO	LR33	12	12	0
13	70	1:01	NO	LR33	12	12	0
14	70	3:01	NO	LR22	18	6	0
15	60	1:03	YES	LR14	8	10	6
16	60	1:01	YES	LR13	11	7	6
17	60	3:01	YES	LR12	14	4	6
18	60	1:03	NO	LR44	6	18	0
19	60	1:01	NO	LR33	12	12	0
20	60	3:01	NO	LR22	18	6	0

[1] LEDs to plants distance. [2] Number of LEDs ratio between hyper red (HR) and deep blue (DB) LEDs. [3] Code obtained by LEDs module's type reported in the first column of Table 1. e.g., LR14 means light combination obtained by combining a type #1 module (5HR:1DB:6WW) with a type #4 module (3HR:9DB). [4] all the light combinations exploit 24 LEDs as result of different combination of 2 LEDs module chosen among the ones reported in Table 3.

Analysis of variance (ANOVA) was employed to investigate single and synergic effects of the artificial light conditions on basil germination and growth. To apply this type of approach, input variables must be independent of each other and normally distributed in the chosen range. In these conditions, each response variation can be divided into different components to evaluate the effect of each factor, their interactions, and experimental error (or unexplained residual) [29]. F-test was applied to estimate if the variation among all the samples, usually due to process difference or factor changes, is larger enough or not than the variation within samples, thereafter, obtained in same experimental conditions. The p value parameter was employed to evaluate the significance of each factor, in single or in interaction, and of the overall model, as it represents the probability that the considered model or factor is significant (p value < 0.05) or not [38]. R^2 and Pred-R^2 parameters were employed to estimate the quality of the fit for the measured dataset, in terms of regression analysis and predictive power of the model, respectively. R^2 is the proportion of the variance in the dependent variables that is predictable from the independent variables and Pred-R^2 is analogous but associated with predicted values [39]. To better highlight the role of the main components on the final considered properties, response contour plots and mathematical equation were derived and discussed. Finally, the desirability function (D) was employed to balance the different responses, considering their peculiar importance (from 1 to 5, where 5 is equal to the highest importance), and objective with respect to the overall purpose of the work.

2.3. Experimental Methods

Five basil seeds per unit were buried in Floradur B pot coarse universal potting soil in plastic containers having a volume of 450 cm^3 each, with 3 repetitions of the 20 pots described in Table 6, for a total of 60 units. During the test, the position of each pot was exchanged with another of the same sample, to avoid any possible slight difference among repetition due to different area illumination under the LEDs light. Trials lasted 35 days, at the end of which the growth and germination data were collected. The experiments were conducted in a room totally isolated from external light sources. Photoperiod was set to 15 h/day (from 00.00 to 15.00) for 30 days. Temperature was kept almost constant near 19 °C, while humidity was always near 60%, with a variation around 5%. These conditions were maintained for all the trials. Artificial light treatments were applied using a growth structure (Figure 2) divided into different areas. Each area was separated from external sources of light through fixed wood panels as walls. As mentioned before, plants were placed on the bottom in a fixed position at three different distances, *d*, from LED module's position: 60, 70, and 80 cm. The three distances were chosen to ensure a homogeneous illumination respectively in areas equal to $31 \times 31, 37 \times 37$, and 43×43 cm. The proportion between these three areas is respectively 1:1.5:2. The fixed distance remained constant throughout the experiment.

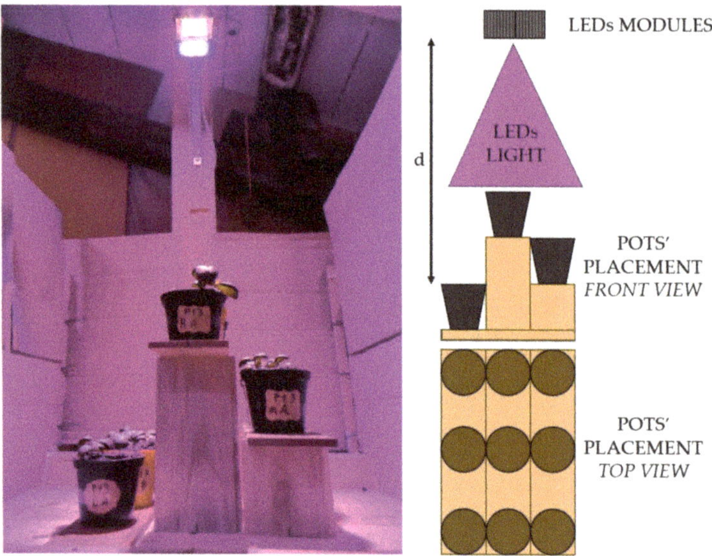

Figure 2. Experimental setup.

Accordingly, with the DoE, six different light combinations were needed. Each combination was obtained matching two LEDs modules among the ones reported in Table 3. The considered light combinations are summarized in Table 7, where the total calculated PPF and the composition in percentage for each type of LED are reported.

Table 7. Summary of LEDs composition for each considered light combination and corresponding Photosynthetic Photon Flux composition.

Light Combination's Code [1]	# of HR LEDs	# of DB LEDs	# of WW LEDs	PPF TOTAL [2] [μmol/s]	%PPF HR	%PPF DB	%PPF White
LR12	14	4	6	46.08	61.19	19.64	19.18
LR13	11	7	6	46.83	47.31	33.82	18.87
LR14	8	10	6	47.57	33.87	47.56	18.58
LR22	18	6	-	49.82	72.76	27.74	0
LR33	12	12	-	51.31	47.1	52.9	0
LR44	6	18	-	52.8	22.88	77.12	0

[1] Notation: LRxy means that the light combination has been obtained combining one module of type x with a module of type y (e.g., LR14 is the light combination obtained combining a LED module of type 1 with a LED module of type 4). [2] Obtained by the PPF's sum of the corresponding single PPFs shown in Table 3.

The same measurements conducted for each single module were repeated for each light combination by exploiting the same AS7341 low-cost spectral sensor at the three distances, *d*, considered in the experiments (i.e., 60, 70 and 80 cm respectively). The obtained results are summarized in Figure 3. In agreement with the theory, for a given light combination and a given distance *d*, the resulting light spectrum is, in first approximation, the sum of the spectrums of the two single LEDs modules used for the light combination. The reasons of the slight differences between theoretical and experimental results are twofold. The first one concerns a minimum misalignment between light source and sensor due to the manual positioning of the sensor itself conducted during the measurements. The second one, is due to a combination of two different aspects. From one side, each LEDs module has a beam angle of about 30 degrees. This means that the divergent angle associated with the modules, would cause the decrease in light intensity, which appears to be higher for samples placed at 70 and 80 cm than for those placed at 60 cm. On the other side, the sensor used has a front adapter with a diffuser that is hold in the right position by a mechanical fixture that could partially shadow the active area of the sensor in case of misalignment.

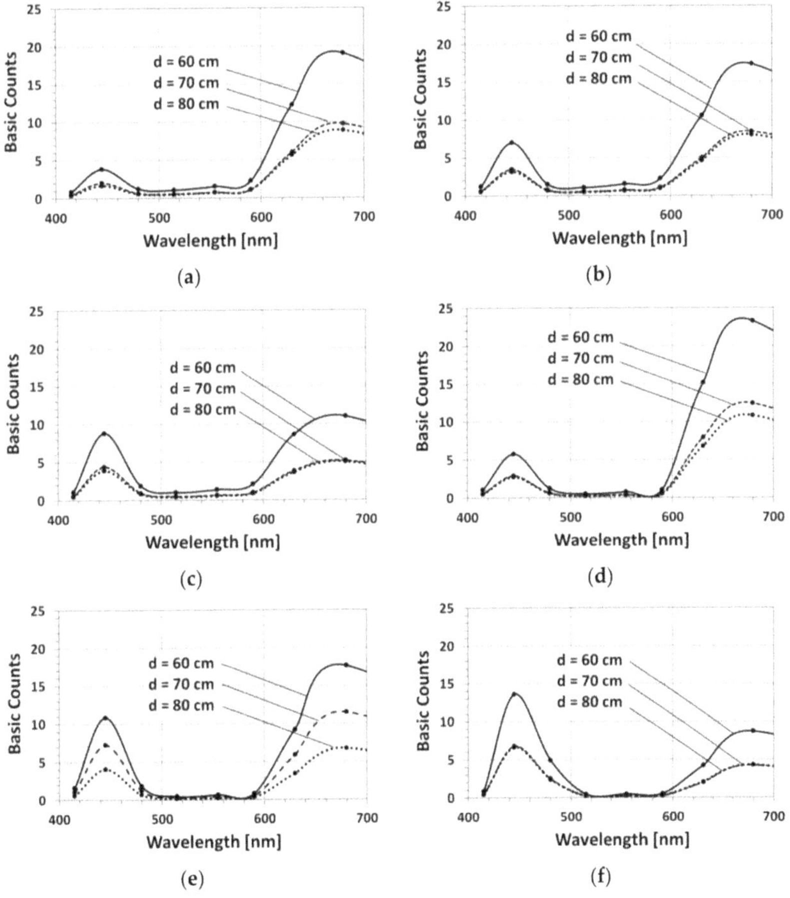

Figure 3. Light spectra of the 6 light combinations considered in the DoE, at the 3 different LEDs to plants distances, *d*. (**a**) LR12, (**b**) LR13, (**c**) LR14, (**d**) LR22, (**e**) LR33, (**f**) LR44.

2.4. Characterizations

A total of 14 responses were investigated to draw reliable considerations about germination and growth of basil. Germination potential of each combination was considered by measuring the Number of Plants and the Days needed for Germination. For the growth phase, at the conclusion of each trial, the following properties were measured for each plant: Wet mass, Dry mass, Height and Number of leaves. For each leaf, wet mass and area were measured. By adding up all the masses and areas of the same plant, the Total leaves mass and Total leaves area were calculated for each plant. Leaf wet mass and Average leaf area were calculated as well, (5), (6) formulae:

$$\text{Leaves wet mass} = \text{Total leaves mass} / \text{Number of leaves} \qquad (5)$$

$$\text{Average leaf area} = \text{Total leaves area} / \text{Number of leaves} \qquad (6)$$

Considering stem properties, average length, diameter, wet mass and dry mass were calculated for each plant. Finally, also the mass of water employed for irrigation during all the experiment was measured by measuring the weight of water employed for each test in all the period of trial. For masses, a laboratory balance (G&G GmbH, model PLC200B-C) with sensitivity ± 0.001 g was employed and for heights a digital caliper (Borletti CDJB15-20 series) with resolution 0.01 mm, accuracy ± 0.02 mm was used. For dry mass measurement, plants and leaves were dried at 80 °C for 24 h. Leaf area measurement was performed by scanning each leave at 400 dpi on graph paper and measuring using ImageJ software (version 1.52, NIH, Bethesda, ML, USA). Prior to scanning, leaves were cut at certain points to extend their full area on the paper and to better assess their area. Following these measurements, the LAI (Leaf Area Index), given by the ratio of Average leaf area to the area of the pot in which the plants had grown, and the SLA (Specific Leaf Area) index, given by the ratio of Average leaf area to Average leaf dry mass, were also calculated [4,16].

3. Results and Discussion

3.1. Preliminary Consideration

From a general and qualitative analysis of the results, germination occurred for almost all the basil seeds and any morphological or developmental abnormalities were observed, thereafter, all the tested light conditions resulted suitable for basil growth, even if with different performances. Figure 4 shows a visual comparison of the plants obtained at the final harvest of six different samples (Table 6), from which qualitative observation can be drawn and discussed. For each sample, all three repetitions were shown to stress the overall reliable repeatability of each experiment, considering that biological systems were investigated. Sample 6, 12 and 19 (Figure 4a–c) have in common the same LEDs combination (LR33) but are different from each other for the light-plant distance that is equal to 80, 70 and 60 respectively. In strong similarity, sample 2, 8 and 15 (Figure 4d–f) are representative of the same light combination (LR14) but with different light intensity, thereafter, moving from greater to smaller LEDs-plants distance. For both the sets, thereafter, independently of the LED combination, an increasing in plants height, number of leaf and leaf area can be observed by reducing placing the plants nearer to the artificial lights. In Figure 4c,f plants' height is clearly over the reference of 12 cm (also considering the pot), whereas in Figure 4b,e plants' height is approximately equal to 12 cm and in Figure 4a,d the overall plants' height is below the value of reference. Nevertheless, regarding the evaluation of the different LEDs combination from a qualitative point of view, is not possible to clearly detect any difference, even if the set of samples employing LR14 combination (Figure 4d–f) seems to better enhance plant growth, with respect to the other set of LEDs. Only observing sample 15 (Figure 4f) it is possible to suppose that this light condition (LR14 at a distance of 60 cm) seems to be the best to promote the basil height and leaf area, among those considered in Figure 4 but further confirmation must be drawn from the statistical analysis. Considering that other variables must be evaluated, such as more different light combination and more growth performance parameters, a statistical approach must be further considered to draw statistically reliable conclusion and to identify mathematical correlations.

Figure 4. *Cont.*

Figure 4. Basil plants at the end of the growth test: (**a**) Sample 6, (**b**) Sample 12, (**c**) Sample 19, (**d**) Sample 2, (**e**) Sample 8, and (**f**) Sample 15. All the sample numbers are referred to in Table 6.

In addition, a control group of plant grown without any type of light has been tested, obtaining a not significant number of plants to derive any specific comparison among data. For the control group, only the plants' height was detectable and was measured equal to 2.03 ± 0.62 mm. This result suggests that LEDs light combinations investigated in this study play a valuable role as a different source of enlightenment from daylight to improve the basil growth performance.

3.2. Anova Analysis

A total of 14 responses were evaluated with statistical methods, and the values considered for each run were the average values of the measurements among three repetitions of the same run (Table 8). As previously stated, with the aim to evaluate only artificial lights effects, all the other parameters were kept as constant, thereafter, temperature and humidity were controlled during all the experiments to avoid environmental conditioning. Temperature and humidity were kept almost constant near 19 °C and 60% respectively, with restrained variation around 5% due to the employment of an indoor environment not perfectly conditioned.

The normal distribution of the residuals, as well as their homogeneity, was analyzed (data not reported) for each response, confirming that each mathematical model derived can be used to explore the region of interest. ANOVA results are presented in Table 9 and Figures 5–10. Models correlating the different lights conditions (in single or interaction) to growth and germination performance are significant as confirmed by the p value < 0.05 for all the responses, indicating that there are any factors that are significant and unknown for the data variation. Moreover, the curvature is not significant, suggesting that the central points can be treated as additional data in the regression model, augmenting the design plan. R^2 and Pred-R^2 (Table 9) confirms the overall good fit of the data, with only two responses with unacceptable quality, *Number of plants* and *Days for germination* having $R^2 < 0.45$. These two responses are mainly related with the germination phase that according to literature is the more critical to model [28]. However, a particular good fitting of the model is shown by *LAI, Dry mass, Stem dry mass* and *Water* responses with values of R^2 equal or above 0.90. The great majority of the resulting models allowed to describe the relationships among light conditions and the measured response. Considering the models' equations (Table 9) not only the *Distance* plays a valuable role to define the growth performance, as already assessed by the preliminary observation, but also the presence of the *White* LED and the proportion among hyper red and deep blue must be considered for most of the response. Thereafter, all the independent variables investigated in this study are relevant to define the growth performance of basil. This result is a new finding with respect to a previous research in which LED lights effect on basil germination and growth were investigated in combination to NPK fertilizer obtained in a circular economy perspective [28].

Table 8. Results.

Sample	High [mm]	Wet Mass [g]	Dry Mass [g]	Number of Leaves	Stem Length [mm]	Stem Wet Mass [g]	Stem Dry Mass [g]	Leaves Wet Mass [g]	Stem Diameter [mm]	Number of Plants [mm]	LAI [%]	SLA [g/cm^2]	Water [g]	Days of Germination
1	64.705	2.0107	0.1977	4	51.577	0.387	0.029	1.61	2.397	4	2.745	335.714	500	14
2	36.555	0.9905	0.0725	3	27.52	0.203	0.014	0.77	1.745	4	1.496	510.03	460	9
3	47.912	1.4652	0.1272	4	35.467	0.2462	0.0165	1.2	2.215	4	2.131	382.482	440	9
4	34.163	0.8396	0.0696	3.33	25.54	0.171	0.013	0.66	1.803	3	1.081	508.694	420	8
5	30.227	0.4682	0.0378	2	21.737	0.111	0.009	0.34	1.372	4	0.825	585.481	400	15
6	35.732	0.8258	0.0665	3	25.297	0.168	0.012	0.72	1.747	4	1.357	497.334	420	8
7	58.768	1.6572	0.1426	4	44.514	0.305	0.018	1.38	2.109	5	3.081	427.778	480	10
8	70.392	2.5177	0.2727	5	59.105	0.478	0.061	2.01	2.405	4	3.388	323.199	520	9
9	24.45	0.4826	0.0383	2	16.913	0.11	0.009	0.36	1.536	3	0.587	527.777	410	11
10	33.03	0.9076	0.0733	3.33	23.03	0.16	0.009	0.74	1.753	3	1.092	451.222	460	11
11	34.76	0.8867	0.0745	2.25	21.1	0.154	0.009	0.73	1.945	4	1.54	470.328	440	12
12	25.557	0.5888	0.0495	2	17.947	0.133	0.01	0.45	1.6	4	0.932	466.9	420	9
13	47.84	1.9922	0.1854	4	32.912	0.295	0.026	1.69	2.564	5	3.18	356.697	480	12
14	38.24	1.1395	0.0817	3	24.73	0.214	0.014	0.91	1.982	4	1.725	500.773	420	17
15	70.368	1.6722	0.1712	4.4	54.75	0.376	0.032	1.28	2.148	5	2.967	400.627	480	10
16	29.103	0.5283	0.0406	2	18.373	0.121	0.012	0.4	1.673	3	0.67	616.623	400	13
17	35.016	0.8626	0.0673	3.33	22.833	0.154	0.011	0.69	1.883	3	1.069	508.54	400	7
18	40.2	1.2056	0.1049	2.8	28.422	0.221	0.0154	0.97	2.056	5	2.387	427.278	440	11
19	56.042	2.2845	0.2596	4	46.81	0.436	0.0516	1.83	2.522	5	3.678	328.818	480	9
20	51.716	1.7271	0.1696	4	36.866	0.27	0.019	1.44	2.35	3	1.85	321.679	440	8

Table 9. ANOVA results.

Response	R^2	Pred-R^2	Equation White = YES	Equation White = NO
Number of plants	0.44	0.32	–	
Days for germination	0.12	0.07	–	
Height	0.76	0.70	$=28.7796 + 15.2203 * Distance$	
Wet mass	0.88	0.83	$=0.6871 + 0.4952 * Distance$	$=0.4957 + 0.8752 * Distance$
Dry mass	0.92	0.82	$=-0.0024 - 0.0093 * Distance + 0.003 * HR{:}DB + 0.0320 * Distance^2$	$=-0.0169 + 0.0346 * Distance + 0.0035 * HR{:}DB + 0.0320 * Distance^2$
LAI	0.93	0.91	$=0.9884 + 0.2941 * Distance + 0.3199 * Distance^2$	$=0.8569 + 0.6771 * Distance + 0.3199 * Distance^2$
SLA	0.81	0.67	$=690.7412 - 82.6704 * Distance - 7.9299 * HR{:}DB + 0.0863 * HR{:}DB^2$	$=653.9108 - 82.6704 * Distance - 7.9299 * HR{:}DB + 0.0863 * HR{:}DB^2$
Number of leaves	0.77	0.63	$=3.3393 + 0.6230 * Distance - 0.0133 * HR{:}DB$	$=2.7819 + 1.1666 * Distance - 0.0135 * HR{:}DB$
Leaves wet mass	0.89	0.84	$=0.5612 + 0.3854 * Distance$	$=0.3784 + 0.7295 * Distance$

Response	R^2	Pred-R^2	Equation White = YES	Equation White = NO
Stem length	0.86	0.76	$=20.8343 + 8.1046 * Distance + 0.0086 * HR{:}DB - 0.1563 * Distance * HR{:}DB\ 6.6094 * Distance^2$	
Stem wet mass	0.87	0.81	$=0.1941 + 0.0280 * Distance - 0.0011 * HR{:}DB + 0.0473 * Distance^2$	
Stem dry mass	0.94	0.88	$=-4.4016 - 0.0272 * Distance - 0.0006 * HR{:}DB - 0.0058 * Distance * HR{:}DB + 0.3453 * Distance^2$	$=-4.5812 + 0.3701 * Distance - 0.0006 * HR{:}DB - 0.0058 * Distance * HR{:}DB + 0.3453 * Distance^2$
Stem diameter	0.83	0.76	$=1.7462 + 0.2290 * Distance$	$=1.5396 + 0.4969 * Distance$
Water	0.90	0.83	$=404.5918 + 39.6258 * Distance + 0.0367 * HR{:}DB$	$=436.0408 + 39.6258 * Distance - 0.5333 * HR{:}DB$

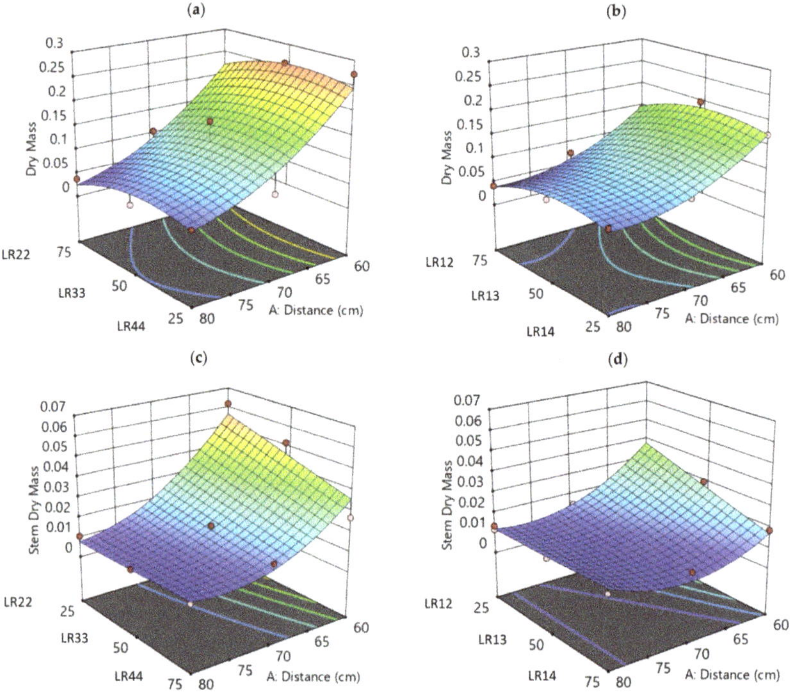

Figure 5. 3D surface contour plots of two different responses: dry mass, without (**a**) and with (**b**) white LEDs; stem dry mass, without (**c**) and with (**d**) White LEDs. Reference to Table 7 for light combination code.

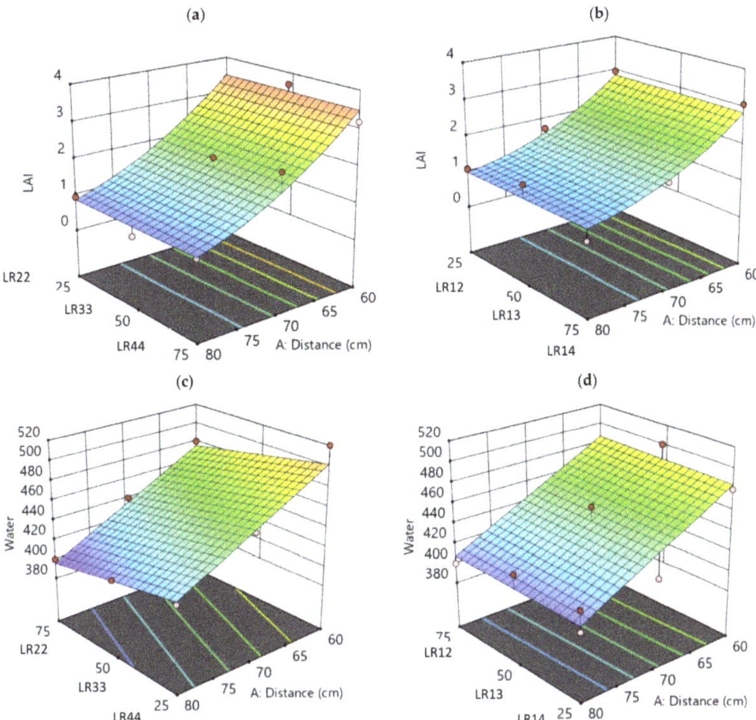

Figure 6. 3D surface contour plots of different responses: LAI, without (**a**) and with (**b**) white LEDs; water, without (**c**) and with (**d**) white LEDs. Reference to Table 7 for light combination code.

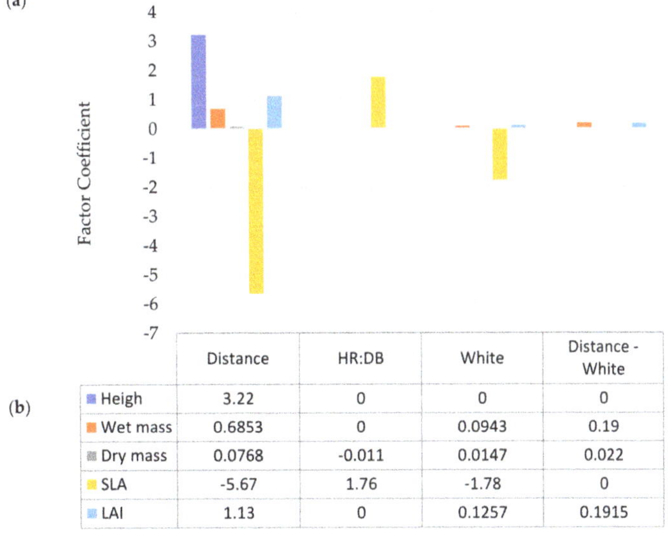

	Distance	HR:DB	White	Distance - White
Heigh	3.22	0	0	0
Wet mass	0.6853	0	0.0943	0.19
Dry mass	0.0768	-0.011	0.0147	0.022
SLA	-5.67	1.76	-1.78	0
LAI	1.13	0	0.1257	0.1915

Figure 7. Effects sizes of the most relevant independent variable on several significant responses related to the overall plant. (**a**) Graphical trend and (**b**) numerical coefficients (error bar = 0.05%, too small to be visible on the graph).

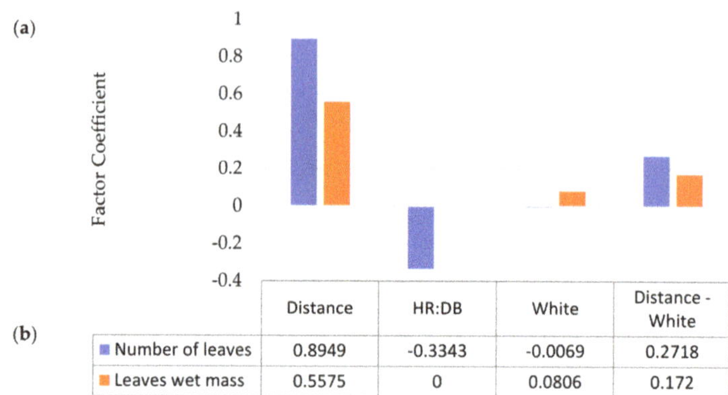

Figure 8. Effects sizes of the most relevant independent variable on several significant responses related to leaf. (**a**) Graphical trend and (**b**) numerical coefficients (error bar = 0.05%, too small to be visible on the graph).

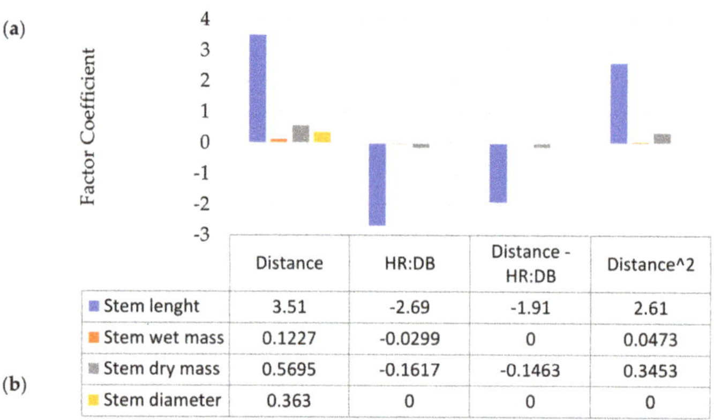

Figure 9. Effects sizes of the most relevant independent variable on several significant responses related to stem. (**a**) Graphical trend and (**b**) numerical coefficients (error bar = 0.05%, too small to be visible on the graph).

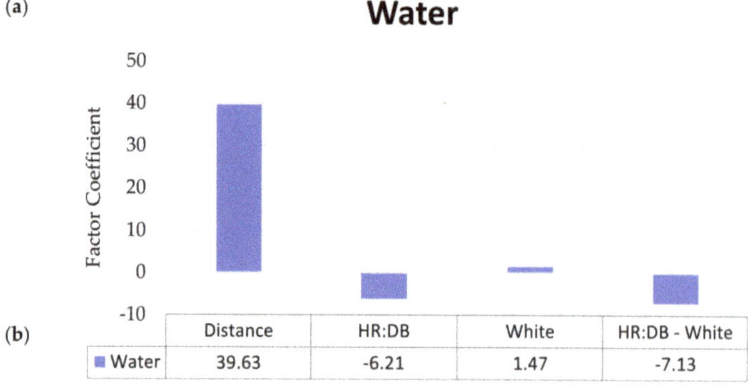

Figure 10. Effects sizes of the most relevant independent variable on the response water. (**a**) Graphical trend and (**b**) numerical coefficients (error bar = 0.05%, too small to be visible on the graph).

In Figures 5 and 6, 3D surface graphs representing graphically the calculated models with $R^2 > 0.9$ are shown. In Figure 5, the two responses related with dry mass (*Dry mass* and *Stem dry mass*) were reported demonstrating for first how much the mathematical model changes by employing (figures on the right) or not (figures on the left) *White* LEDs. Both the responses are favored by the absence of the *White* LED. In addition, it is possible to see clearly that the *Distance* plays the most significant role for both the responses, *Dry mass* and *Stem dry mass*, even if for the first a not negligible effect is due to the proportion among hyper red and deep blue. To promote the dry mass, not only the lowest *Distance* must be selected but also a 1:1 proportion among hyper red and deep blue LEDs that correspond to light combinations LR33 and LR13.

Different behavior can be observed for the responses *LAI* and *Water* (Figure 6). *LAI* (Figure 6a,b) is mainly affected by *Distance* and only a slight effect is observed for the *White* LED. Conditions that optimize this parameter are lower *Distance* and the absence of *White* LED, independently of the ratio among hyper red and deep blue (Light combination LR22, LR33, LR44). Particularly interesting is the evaluation of the *Water* necessary to irrigation over the experimentation (Figure 6c,d) because this one is the only parameter that we desire to minimize to save an important natural resource such as water. As expected, the lower *Distance* increases the water consumption, as that more near the lights, the higher the need of water for plants due to the artificial light heating. The *White* LED alone is not particularly relevant for this response as can be seen by comparing Figures 6c and 6d, because almost the same range of mass of water is necessary; nevertheless, a different shape of the response surface is detectable as the interaction with the other parameter is reliable. Without white LEDs, the light combination LR22 must be preferred to reduce the amount of water to employ (Figure 6c), whereas no particular difference can be noticed among the LEDs combinations LR12, LR13, LR14 if the same distance is applied. This result suggests that the *White LED* addition must be well considered to reduce the water consumption. At the same time, the lower *Distance* that seems to promote that certain growth parameters (Dry mass, LAI and Stem dry mass) should be increased with the aim to minimize the water consumption.

Thereafter, it is interesting to observe quantitatively how much each independent variable, in single or interaction, is capable to affect a specific property, considering each elaborated model. This can be observed from Figures 7–10, where size effects were reported for each significant response, by grouping them for specific basil parts.

Regarding responses related to the overall plant (Figure 7), *Distance* plays absolutely the main role for all the responses; nevertheless, a negative effect is reported for *SLA*. This means that the same *Distance* that is capable to promote some properties (e.g., *Heigh* or *LAI*) is also responsible for a reduction of the *SLA* parameter. As shown from Figure 7, for *SLA* parameter the reduction due to an increasing distance can be recovered by increasing the *HR:DB* ratio, thereafter introducing more hyper-red LEDs with respect to deep blue.

Considering the two responses related to leaf growth (Figure 8) a reliable and positive effect is associated to *Distance* and *White* LEDs employment in single and in interaction. However, the effect of the ratio *HR:DB* is negligible for *Leaves wet mass* and strongly negative for *Number of leaves*, suggesting the need to employ the lights combination LR44 to promote the leaf number, as confirmed by a previous study [28].

Evaluating the responses related to stem properties (Figure 9), again it is possible to detect a strong influence of the *Distance*, as it has a strong effect as single factor and in interaction with *HR:DB*. Even if in this case, a competitive effect emerged related to the response *Stem length*, that is affected in a positive way from *Distance* in single factor but a negative way when its interaction with *HR:DB* is considered, leading to a restrained range of light combination–distance conditions capable to promote this property. A similar situation describes *Stem dry mass* and *Stem wet mass* but with restrained size effect with respect to *Stem length*. Thereafter, conditions that will promote stem growth should be particularly well-tailored to match all these competitive effects.

Finally, the estimated effect size of *Water* (Figure 10) indicates that the variables that should be emphasized to reduce its consumption are *HR:DB*, as it affects Water negatively, and its interaction with *White* LED, for the same reason. However, *Distance* must be severely reduced to decrease the Water consumption as well. This means that the employment of the light combination LR22 or LR12 at a distance equal to 60 cm should be preferable, as argued considering Figure 6c,d.

As shown from these results, it is hard to establish a light condition that can be favorable for all the responses; in particular, is not clear if the overall influence of the white LED is beneficial or detrimental for the basil growth. As suggested in the literature, possible overlapping effect of the white LED with the others LED light that are more absorbable by the plants can have a negative effect on the plants' growth, but this effect depends on the parameters that are investigated [24]. Thereafter, it is necessary to collect all the derived models in one, called desirability function, by balancing them, considering each goal and importance with respect to the overall aim of the research.

The desirability function conditions employed in the present work were shown in Table 10. Number of plants and days for germination responses were discarded from this part of the analysis due to theirs poor R^2 and Pred-R^2, indicating a low fitting and predictive power. The other responses were evaluated through the objectives in Table 10 with the aim to promote the overall plants' best growth with the most possible restrained amount of water.

Table 10. Desirability function conditions.

Responses	Goal	Importance
Height	maximize	4
Wet mass	maximize	5
Dry mass	maximize	4
LAI	maximize	4
SLA	maximize	5
Number of leaves	maximize	4
Leaves wet mass	maximize	3
Stem lenght	maximize	3
Stem wet mass	In range	3
Stem dry mass	In range	3
Stem diameter	maximize	2
Water	minimize	5

As shown in Figure 11, results of the desirability function calculation indicate that the light conditions that better fit the overall objectives indicated in Table 10 are a distance equal to 60 cm and the employment of the LR22 light combination (green area of the graphs). This condition has a desirability value equal to 0.618956 as shown in Figure 11a, and it is the highest one of the overall light conditions investigated in this study. This result agrees with the fact that most of the responses are strongly promoted by a distance equal to 60 cm and the fact that water consumption is reduced avoiding white LEDs light. In addition, this result agrees with previous literature on hyper red effect on basil growth [40,41] However, the blue area of the graphs indicated all the lights combinations that should be avoided, to promote the basil growth. By employing or not white LED, this area changes; without white LED (Figure 11a), the blue area of the graph in mainly due to distance > 75 cm, and by employing white LED (Figure 11b) this area results wider, as the interaction of distances between 65–78 cm and hyper red prevalence in proportion to deep blue (LR 12).

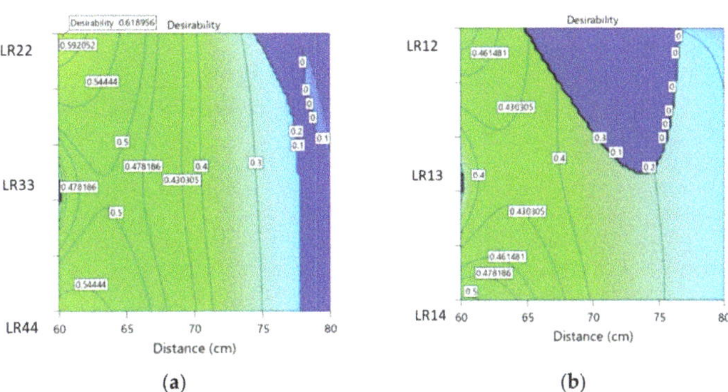

Figure 11. Desirability 2D contour plot without (**a**) and with (**b**) White LED light.

4. Conclusions

In this study, it has been demonstrated that the implementation of various combination of LED lights is generally favorable to basil (*Ocimum Basilicum*) germination and growth. Applying a design of experiments approach, mathematically reliable information has been derived concerning a synergic effect among LED light type and light–plant distance. It has been calculated that by avoiding white LED light, better performance in terms of crop yield enhancement can be reached. Furthermore, LED light combinations, involving hyper red and deep blue light in 3:1 proportion, results the best by considering the desirability function, which includes all the requirements to be satisfied for the overall basil growth. In addition, the distance among plants and light plays a key role; the shorter distance investigated (60 cm) is advisable for the basil growth, even if attention must be paid to water consumption, thereafter a plants-light distance among 65 cm is suggested. The present study has been focused on the influence of different factors on basil growth considering only the final harvest of the plant; but in a future, perspective data at different time of the basil growth can be collected. In this context, the generated models in different times of the plants' growth can be employed in artificial intelligence-driven systems to apply automatically specific light combination or distance in a particular moment of the plants' growth, to further promote the growth and the health of the plants.

Author Contributions: Conceptualization, M.M. and A.B.; software, S.B.; investigation, F.B. and A.B.; resources M.M.; writing—original draft preparation, S.B.; writing—review and editing, F.B. and A.B.; supervision, M.M.; funding acquisition, M.M. All authors have read and agreed to the published version of the manuscript.

Funding: This research received no external funding.

Institutional Review Board Statement: Not applicable.

Informed Consent Statement: Not applicable.

Data Availability Statement: All the data presented and analyzed in this study are available here.

Acknowledgments: The authors thank Giovanni Verzellesi (University of Modena and Reggio Emilia) for the support and the fruitful discussions.

Conflicts of Interest: The authors declare no conflict of interest.

References

1. Frąszczak, B.; Golcz, A.; Zawirska-Wojtasiak, R.; Janowska, B. Growth rate of sweet basil and lemon balm plants grown under fluorescent lamps and led modules. *Acta Sci. Pol. Hortorum Cultus* **2014**, *13*, 3–13.
2. Shahak, Y.; Gussakovsky, E.E.; Cohen, Y.; Lurie, S.; Stern, R.; Kfir, S.; Naor, A.; Atzmon, I.; Doron, I.; Greenblat-Avron, Y. Colornets: A new approach for light manipulation in fruit trees. In Proceedings of the Acta Horticulturae; International Society for Horticultural Science (ISHS), Leuven, Belgium, 11 March 2004; pp. 609–616.

3.   Chang, X.; Alderson, P.G.; Wright, C.J. Solar irradiance level alters the growth of basil (*Ocimum basilicum* L.) and its content of volatile oils. *Environ. Exp. Bot.* **2008**, *63*, 216–223. [CrossRef]

4.   Stagnari, F.; Di Mattia, C.; Galieni, A.; Santarelli, V.; D'Egidio, S.; Pagnani, G.; Pisante, M. Light quantity and quality supplies sharply affect growth, morphological, physiological and quality traits of basil. *Ind. Crops Prod.* **2018**, *122*, 277–289. [CrossRef]

5.   Natesh, H.N.; Abbey, L.; Sk, A. An overview of nutritional and anti nutritional factors in green leafy vegetables. *Horticult. Int. J.* **2017**, *1*, 58–65. [CrossRef]

6.   Stagnari, F.; Galieni, A.; Speca, S.; Pisante, M. Scientia Horticulturae Water stress effects on growth, yield and quality traits of red beet. *Sci. Hortic.* **2014**, *165*, 13–22. [CrossRef]

7.   Bantis, F.; Ouzounis, T.; Radoglou, K. Artificial LED lighting enhances growth characteristics and total phenolic content of Ocimum basilicum, but variably affects transplant success. *Sci. Hortic.* **2016**, *198*, 277–283. [CrossRef]

8.   Hossain, M.A.; Kabir, M.J.; Salehuddin, S.M.; Rahman, S.M.M.; Das, A.K.; Singha, S.K.; Alam, M.K.; Rahman, A. Antibacterial properties of essential oils and methanol extracts of sweet basil *Ocimum basilicum* occurring in Bangladesh. *Pharm. Biol.* **2010**, *48*, 504–511. [CrossRef] [PubMed]

9.   Kathirvel, P.; Ravi, S. Chemical composition of the essential oil from basil (*Ocimum basilicum* Linn.) and its in vitro cytotoxicity against HeLa and HEp-2 human cancer cell lines and NIH 3T3 mouse embryonic fibroblasts. *Nat. Prod. Res.* **2012**, *26*, 1112–1118. [CrossRef]

10. Sipos, L.; Boros, I.F.; Csambalik, L.; Székely, G.; Jung, A.; Balázs, L. Horticultural lighting system optimalization: A review. *Sci. Hortic.* **2020**, *273*, 109631. [CrossRef]

11. Jao, R.C.; Fang, W. An Adjustable Light Source for Photo–Phyto Related Research and Young Plant Production. *Appl. Eng. Agric.* **2003**, *19*, 601. [CrossRef]

12. Yeh, N.; Chung, J.P. High-brightness LEDs-Energy efficient lighting sources and their potential in indoor plant cultivation. *Renew. Sustain. Energy Rev.* **2009**, *13*, 2175–2180. [CrossRef]

13. Morrow, R.C. LED lighting in horticulture. *HortScience* **2008**, *43*, 1947–1950. [CrossRef]

14. Barta, D.J.; Tibbitts, T.W.; Bula, R.J.; Morrow, R.C. Evaluation of light emitting diode characteristics for a space-based plant irradiation source. *Adv. Sp. Res.* **1992**, *12*, 141–149. [CrossRef]

15. Bula, R.J.; Morrow, R.C.; Tibbitts, T.W.; Barta, D.J.; Ignatius, R.W.; Martin, T.S. Light-emitting diodes as a radiation source for plants. *HortScience* **1991**, *26*, 203–205. [CrossRef] [PubMed]

16. Piovene, C.; Orsini, F.; Bosi, S.; Sanoubar, R.; Bregola, V.; Dinelli, G.; Gianquinto, G. Optimal red: Blue ratio in led lighting for nutraceutical indoor horticulture. *Sci. Hortic.* **2015**, *193*, 202–208. [CrossRef]

17. Hasan, M.M.; Bashir, T.; Ghosh, R.; Lee, S.K.; Bae, H. An overview of LEDs' effects on the production of bioactive compounds and crop quality. *Molecules* **2017**, *22*, 1420. [CrossRef] [PubMed]

18. Dou, H.; Niu, G.; Gu, M.; Masabni, J.G. Effects of light quality on growth and phytonutrient accumulation of herbs under controlled environments. *Horticulturae* **2017**, *3*, 36. [CrossRef]

19. Olle, M.; Alsina, I. Influence of wavelength of light on growth, yield and nutritional quality of greenhouse vegetables. *Proc. Latv. Acad. Sci. Sect. B Nat. Exact Appl. Sci.* **2019**, *73*, 1–9. [CrossRef]

20. Huché-Thélier, L.; Crespel, L.; Le Gourrierec, J.; Morel, P.; Sakr, S.; Leduc, N. Light signaling and plant responses to blue and UV radiations-Perspectives for applications in horticulture. *Environ. Exp. Bot.* **2016**, *121*, 22–38. [CrossRef]

21. Wang, J.; Lu, W.; Tong, Y.; Yang, Q. Leaf morphology, photosynthetic performance, chlorophyll fluorescence, stomatal development of lettuce (*Lactuca sativa* L.) exposed to different ratios of red light to blue light. *Front. Plant Sci.* **2016**, *7*, 250. [CrossRef]

22. Miao, Y.; Wang, X.; Gao, L.; Chen, Q.; Qu, M. Blue light is more essential than red light for maintaining the activities of photosystem II and I and photosynthetic electron transport capacity in cucumber leaves. *J. Integr. Agric.* **2016**, *15*, 87–100. [CrossRef]

23. Jishi, T.; Kimura, K.; Matsuda, R.; Fujiwara, K. Effects of temporally shifted irradiation of blue and red LED light on cos lettuce growth and morphology. *Sci. Hortic.* **2016**, *198*, 227–232. [CrossRef]

24. Johkan, M.; Shoji, K.; Goto, F.; Hahida, S.; Yoshihara, T. Effect of green light wavelength and intensity on photomorphogenesis and photosynthesis in *Lactuca sativa*. *Environ. Exp. Bot.* **2012**, *75*, 128–133. [CrossRef]

25. Rahman, M.M.; Vasiliev, M.; Alameh, K. LED Illumination Spectrum Manipulation for Increasing the Yield of Sweet Basil (*Ocimum basilicum* L.). *Plants* **2021**, *10*, 344. [CrossRef]

26. Casal, J.J. Photoreceptor Signaling Networks in Plant Responses to Shade. *Annu. Rev. Plant Biol.* **2013**, *64*, 403–427. [CrossRef]

27. Demotes-Mainard, S.; Péron, T.; Corot, A.; Bertheloot, J.; Le Gourrierec, J.; Pelleschi-Travier, S.; Crespel, L.; Morel, P.; Huché-Thélier, L.; Boumaza, R.; et al. Plant responses to red and far-red lights, applications in horticulture. *Environ. Exp. Bot.* **2016**, *121*, 4–21. [CrossRef]

28. Barbi, S.; Barbieri, F.; Bertacchini, A.; Barbieri, L.; Montorsi, M. Effects of different LED light recipes and NPK fertilizers on basil cultivation for automated and integrated horticulture methods. *Appl. Sci.* **2021**, *11*, 2497. [CrossRef]

29. Montgomery, D.C. *Design and Analysis of Experiments*, 8th ed.; Wiley: Hoboken, NJ, USA, 2012; Volume 2, ISBN 9781118146927.

30. Italian Republic, Legislative Decree n. 75/2010 concerning fertilizers. *Gazz. Uff. Ser. Gen. n.218 del 17-09-2013*, 2010.

31. Intelligent Led Solutions Petunia Led Modules. Available online: https://i-led.co.uk/PDFs/Kits/12Multi-OslonSSL-PetuniaColourV3.pdf (accessed on 1 July 2021).

32. OSRAM LH CP7P 660 nm Hyper Red LED. Available online: https://www.osram.com/ecat/OSLON%C2%AE%20SSL%2080%20LH%20CP7P/com/en/class_pim_web_catalog_103489/prd_pim_device_2402508/ (accessed on 1 July 2021).
33. OSRAM LD CQ7P 451 nm Deep Blue LED. Available online: https://www.osram.com/ecat/OSLON%C2%AE%20SSL%2080%20LD%20CQ7P/com/en/class_pim_web_catalog_103489/prd_pim_device_2402502/ (accessed on 1 July 2021).
34. OSRAM LCW CQDP.EC 3000 K Warm White LED. Available online: https://media.digikey.com/PDF/Data%20Sheets/Osram%20PDFs/LCW_CQDP.EC.pdf (accessed on 1 July 2021).
35. Lighting Analysts Inc. AGi32 Software Documentation. Available online: https://support.agi32.com/support/solutions/articles/22000211885-computing-ppfd-factors-from-spd-data (accessed on 1 July 2021).
36. AMS. AS7341 Spectral Sensor. Available online: https://ams.com/as7341 (accessed on 1 July 2021).
37. LEDiL C12528 PETUNIA Lens. Available online: https://www.ledil.com/product-card/?product=C12528_PETUNIA (accessed on 1 July 2021).
38. Eriksson, L.; Johansson, E.; Kettaneh-Wold, N.; WikstrÄom, C.; Wold, S. *Design of Experiments: Principles and Applications*; Umetrics Academy: Umeǎ, Sweden, 2008; ISBN 10:9197373044.
39. Morris, P.; John, P.W.M. Statistical Design and Analysis of Experiments. *Math. Gaz.* **1999**, *83*, 189. [CrossRef]
40. Lee, M.-J.; Son, K.-H.; Oh, M.-M. Increase in biomass and bioactive compounds in lettuce under various ratios of red to far-red LED light supplemented with blue LED light. *Hortic. Environ. Biotechnol.* **2016**, *57*, 139–147. [CrossRef]
41. Murchie, E.H.; Pinto, M.; Horton, P. Agriculture and the new challenges for photosynthesis research. *New Phytol.* **2008**, *181*, 532–552. [CrossRef] [PubMed]

 *applied sciences*

Article

Differential Response of the Proteins Involved in Amino Acid Metabolism in Two *Saccharomyces cerevisiae* Strains during the Second Fermentation in a Sealed Bottle

María del Carmen González-Jiménez [1], Juan Carlos Mauricio [1,*], Jaime Moreno-García [1], Anna Puig-Pujol [2], Juan Moreno [1] and Teresa García-Martínez [1]

[1] Department of Agricultural Chemistry, Edaphology and Microbiology, Agrifood Campus of International Excellence ceiA3, University of Cordoba, 14014 Cordoba, Spain; b02gojim@uco.es (M.d.C.G.-J.); b62mogaj@uco.es (J.M.-G.); qe1movij@uco.es (J.M.); mi2gamam@uco.es (T.G.-M.)

[2] Department of Enological Research, Institute of Agrifood Research and Technology—Catalan Institute of Vine and Wine (IRTA-INCAVI), Vilafranca del Penedès, 08720 Barcelona, Spain; anna.puig@irta.cat

* Correspondence: mi1gamaj@uco.es; Tel.: +34-957218640

Abstract: The traditional method for sparkling wine making consists of a second fermentation of a base wine followed by ageing in the same bottle that reaches the consumers. Nitrogen metabolism is the second most important process after carbon and takes place during wine fermentation by yeast. Amino acids are the most numerous nitrogen compounds released by this process. They contribute to the organoleptic properties of the wines and, therefore, to their sensory quality. The main objective of this study is to compare the differential proteomic response of amino acid metabolism, specifically their proteins and their interactions in the G1 strain (unconventional yeast) during sparkling wine production *versus* the conventional P29 strain. One of the new trends in winemaking is the improvement of the organoleptic diversity of wine. We propose the use of unconventional yeast that shows desirable characteristics for the industry. For this purpose, these two yeasts were grown at sealed bottle conditions for the second fermentation (*Champenoise* method). No differences were obtained in the middle of fermentation between the yeast strains. The number of proteins identified, and the relationships established, were similar, highlighting lysine metabolism. At the end of the second fermentation, the difference between each strain was remarkable. Hardly any proteins were identified in unconventional *versus* conventional yeast. However, in both strains, the metabolism of sulfur amino acids, methionine, and cysteine obtained a greater number of proteins involved in these processes. The release of these amino acids to the medium would allow the yeast to balance the internal redox potential by reoxidation of NADPH. This study is focused on the search for a more complete knowledge of yeast metabolism, specifically the metabolism of amino acids, which are key compounds during the second fermentation.

Keywords: sparkling wine; protein; interact omics; amino acid metabolism; yeast; GO terms

Citation: González-Jiménez, M.d.C.; Mauricio, J.C.; Moreno-García, J.; Puig-Pujol, A.; Moreno, J.; García-Martínez, T. Differential Response of the Proteins Involved in Amino Acid Metabolism in Two *Saccharomyces cerevisiae* Strains during the Second Fermentation in a Sealed Bottle. *Appl. Sci.* **2021**, *11*, 12165. https://doi.org/10.3390/app112412165

Academic Editor: Maria Kanellaki

Received: 30 September 2021
Accepted: 17 December 2021
Published: 20 December 2021

Publisher's Note: MDPI stays neutral with regard to jurisdictional claims in published maps and institutional affiliations.

1. Introduction

"Méthode champenoise", also known as the traditional method, is a sparkling wine production method whereby wine undergoes a second fermentation process in the bottle to produce CO_2. During the aging of sparkling wines in contact with the yeast lees, the phenomenon of autolysis takes place. Through this process, nitrogen compounds are released that contribute to the organoleptic properties of the wines and therefore to their sensory quality.

Nitrogen is a key nutrient vital to yeast during wine fermentation, and its availability in grape is a key parameter for the progress of wine fermentation, affecting both fermentation kinetics and wine aroma formation. The shortage of these compounds can cause slow or stuck fermentations [1,2] or influence the formation of reduced sulfur compounds, such as hydrogen sulfide [3]. Furthermore, the catabolism of nitrogen sources and the central metabolism of carbon is involved in the synthesis of fermentative aroma precursors in oenological yeasts [4,5].

Saccharomyces cerevisiae can grow on a wide variety of nitrogenous substrates; the metabolism of nitrogenous compounds depends on the yeast strain, its physiological state, and the physicochemical properties of the wine. Nitrogen metabolism is controlled by regulatory mechanisms that depend largely on the nature of the nitrogen sources present in the wine. These mechanisms act at the transcriptional level, as well as in the activity and degradation of enzymes and permeases [6]. The assimilable fraction of nitrogen by yeasts is mainly ammonia and amino acids in similar proportions, but they can also use other nitrogen sources, such as oligopeptides, polypeptides, proteins, amides, biogenic amines, and nucleic acids [7,8]. *S. cerevisiae* incorporates amino acids and small peptides (less than five amino acid residues) through an active transport process using specialized membrane proteins and a pH gradient [6,9]. The amount of these compounds in sparkling wines depends on the grape variety and the aging time [7,8]. To cope with these stressful conditions, which can cause nitrogen shortages, yeasts consume and metabolize these compounds as efficiently as possible during fermentation. The preferred compound for yeast is glutamate, which is stored for *de novo* amino acid biosynthesis. This is particularly relevant when the ethanol content increases, which inhibits amino acid absorption [9]. *S. cerevisiae* uses amino acids differentially; preferred nitrogen sources are alanine, arginine, asparagine, aspartate, glutamate, glutamine, and serine [7].

The amino acids present in sparkling wines have different origins: (i) they can come directly from the grape, without being metabolized by yeast during its growth, (ii) they can excrete at the end of fermentation, (iii) and they can also originate during the autolysis process. Generally, these amino acids are used by yeast for the formation of proteins or as a source of nitrogen by oxidative deamination; alternatively, an amino acid can be broken down and thus liberate nitrogen. During fermentation, yeast incorporates and metabolizes amino acids to grow and produce biomass. In this process, a series of volatile aromatic compounds are produced (esters, higher alcohols, volatile fatty acids, carbonyls, and sulfur compounds), which have a great impact on the organoleptic properties of wines. In addition, the nitrogen released can also be used for the biosynthesis of other nitrogenous cellular constituents, and its carbon structure can be excreted in wine or used as a carbon source for the biosynthesis of other compounds, such as higher alcohols [10,11], which constitute the main group of aromatic compounds in wine [7]. Nitrogen metabolism is very complex because its intermediaries are also shared among other metabolic pathways. Because of this, there is not yet a complete understanding of the impact of amino acids on the formation of different aromatic compounds.

Therefore, it is important to establish a fundamental dataset in order to build solid knowledge. This new study, along with the rest of the work carried out by our research group, seeks to address the lack of information that exists in multi-omics analysis by integrating the proteomics and metabolomics of *S. cerevisiae* during the second fermentation in the production of sparkling wines (cava). For this reason, in previous works, our research group has focused on the study of ester metabolism [12] and the autophagy process [13], as well as the effect of CO_2 overpressure on the aroma [14,15] during the second fermentation in the production of sparkling wines (cava). In addition, the novelty of using an unconventional yeast for the production of this type of wine was introduced. It is a flor yeast that, due to properties such as its high tolerance to ethanol and its flocculation capacity, could be a good candidate for use in the production of sparkling wine. In this way, the diversity, uniqueness, and typicality of sparkling wine yeasts could be improved.

The main objective of this study was to compare the differential proteomic response of amino acid metabolism, specifically their proteins and their interactions in the G1 strain (unconventional yeast) during sparkling wine production *versus* the conventional P29 strain. This relationship never been investigated before in sparkling wines, hence the difficulty of this study and the relevance of its results. This knowledge is intended to serve as a guide for future research and ultimately to improve the quality of the wine.

2. Materials and Methods

2.1. Microorganisms and Fermentation Conditions

The microorganisms used were two strains of *S. cerevisiae*: the conventional P29 strain (CECT11770), a typical yeast in the production of sparkling wines, which was isolated from the designation of origin of the Penedès (DO) Barcelona (Northeast of Spain), and the G1 strain (ATCC: MYA-2451). The G1 strain is a flor yeast isolated from a flor "velum" under the biological aging of "Fino" Sherry wine of the designation of origin (DO) Montilla-Moriles in Córdoba (Southern of Spain).

The yeast strains were incubated for 5 days at 21 °C using a gentle agitation of 100 rpm for their growth in a pasteurized must of Macabeo grape variety, composed of 174.9 g/L of sugar, 18.5 °Bx, 3.6 g/L of total acidity, and 3.43 pH. When an ethanol content of 10.39% (*v/v*) was reached, the yeast cells were introduced into bottles with a standardized commercial base wine (Macabeo: Chardonnay (6:4), 10.21% (*v/v*) of ethanol, 0.3 g/L of sugar, pH 3.29, 5.4 g/L of total acidity, and 0.21 g/L of volatile acidity), 21 g/L sucrose, and 1.5×10^6 cells/mL. The second fermentation was carried out in a thermostatic chamber at 14 °C in bottles with a volume of 750 mL. During the second fermentation, the following samples were taken: at the middle of the second fermentation, MF (3 bar pressure), and at the end of the second fermentation, EF (6.5 bar pressure). In each sample, three bottles were opened for triplicate analysis.

Viable yeast cell counting was carried out using appropriate dilutions with Ringer solution. These were then plated in Sabouraud agar medium for 48 h at 28 °C, and all samples were analyzed five-fold.

2.2. Proteomic Analysis

Cells were collected from each bottle by centrifugation at $4500 \times g$ for 10 min using a centrifuge (Rotina-38, Kirchlengern, Germany), washing the pellet twice with cold sterile distilled water. Subsequently, the washed cell pellets were dissolved in 2 mL extraction buffer (100 mM Tris-HCl pH 8, 0.1 mM EDTA, 2 mM DTT and 1 mM PMSF) and a protease inhibitor cocktail. Lysis of cells was performed using a mechanical technique in a Vibrogen Cell Mill V6 (Edmund Bühler GmbH, Bodelshausen, Germany) using glass beads of 500 μm in diameter. Once the cells were disrupted, the protein extraction was removed. Glass beads as well as cell debris were discarded by centrifugation at $500 \times g$ for 5 min. Protein precipitation was carried out by overnight incubation at -20 °C after the addition of 10% *w/v* of trichloroacetic acid (TCA) and 4 vol of ice-cold acetone to the supernatant. After incubation, samples were centrifuged at $16,000 \times g$ for 30 min, and the protein pellet was then vacuum dried and resuspended in solubilization buffer. Protein concentration was estimated by Bradford (1976) and samples stored at -80 °C until protein analysis [16]. To perform this analysis, 500 μg of total protein for each condition and replica was loaded into an Agilent Technologies OFFGEL 3100 fractionator well tray. Previously, protein samples were solubilized in a Protein OFFGEL fractionation buffer containing urea, thiourea, DTT, glycerol, and ampholytic buffer. The aliquots were distributed in an Agilent Technologies OFFGEL 3100 fractionator (Santa Clara, CA, USA), in a tray with wells. Once the proteins were separated according to their isoelectric point, the fractions were collected from each well and their identification was carried out. For identification, protein fractions were analyzed on an LTQ Orbitrap XL mass spectrometer equipped with an Ultimate 3000 nano LC system at the Central Research Support Service (SCAI) of the University of Córdoba. Proteins had to be digested with trypsin beforehand. Finally, the identified proteins were quantified following the exponentially modified Protein Abundance Index, EmPAI, a method described by Ishihama et al. (2005) [17]. More detailed information is described in articles published by the research group [12].

2.3. Statistical Analysis

After identification, the proteins were filtered to take into account those involved in "amino acid metabolism", according to the Gene Ontology terminology (GO) of the *Saccharomyces* genome database (https://www.yeastgenome.org/, accessed on 15 February 2021), using the GO Slim Mapper tool. The selected processes were: (i) cellular amino acid metabolic process, (ii) amino acid transport, (iii) tRNA aminoacylation for protein translation.

The proteins involved in the "cellular amino acid metabolic process" were classified in different GO terms provided by the SGD database (https://www.yeastgenome.org/, accessed on 15 February 2021). This classification was made using the GO Term Finder tool, provided by the database. The GO terms are descriptive terms that allow relating each gene product with a molecular, cellular, and biological process context, providing a statistical value (p-value). This statistical study was performed with a level of significance (α) of 0.05.

The proteins discussed in this study were strain-specific proteins during the second fermentation in bottle.

2.4. Protein Network Reconstruction

The software STRING version 11 (available online, https://string-db.org/, accessed on 15 February 2021) was used to reconstruct an interaction network for proteins of interest, forming specific protein groups through an MCL (Markov Cluster Algorithm) clustering method. This algorithm accepts a parameter called 'inflation' that is indirectly related to the precision of the clustering. Data were previously normalized through the root square and auto-scaling methods [18].

3. Results and Discussion

The present study was carried out on two strains of *S. cerevisiae*: a conventional strain for the elaboration of this type of wine (P29 strain) and a non-conventional yeast (G1 strain) under real conditions of the second fermentation; the sample times were at the middle of the second fermentation (MF) and at the end of the second fermentation in a sealed bottle (EF).

For this, the total proteins identified in each strain were compared, and those different in each case were selected (from now on named as specific). They were then classified into three processes (amino acid transport, tRNA aminoacylation for protein translation, and cellular metabolism) according to the *Saccharomyces* Genome Database (SGD). Data are shown in Table 1. Finally, the proteins involved in the cellular metabolism process were classified in GO terms according to the SGD database, and the interactions between them were established using STRING software.

Table 1. Biological processes, specific proteins, and ratio of specific to the total proteins related to amino acid metabolism in *S. cerevisiae* P29 and G1 during two sampling times, at the middle of the second fermentation (MF) and at the end of the second fermentation (EF).

	Middle of the Second Fermentation, MF			
	P29 Strain		**G1 Strain**	
Biological Process	**Specific Proteins**	**Specific/Total Proteins**	**Specific Proteins**	**Specific/Total Proteins**
Cellular amino acid metabolic process (GO: 0006520)	Aat1p, Ald2p, Arg81p, Aro9p, Dal81p, Gcv2p, Gdh3p, Ilv6p, Leu9p, Lys1p, Lys4p, Met1p, Mis1p, Put1p, Snz3p, Ths1p, Uga2p, Uga3p, Ura7p, Utr4p, Xbp1p, Ydl168wp, Yml082wp, Ymr084wp	24/103	Aco2p, Adi1p, Aro10p, Bna4p, Cpa2p, Ehd3p, Gcv1p, Gfa1p, Gln4p, Gly1p, His5p, Idp1p, Lap3p, Lys21p, Mae1p, Met12p, Met16p, Mri1p, Nit3p, Pet112p, Ser1p, Sfa1p, Sno3p, Tum1p, Uga1p, Yhr112cp, Yhr208wp, Yll058wp	28/107

Table 1. Cont.

Biological Process	P29 Strain		G1 Strain	
Middle of the Second Fermentation, MF				
	Specific Proteins	**Specific/Total Proteins**	**Specific Proteins**	**Specific/Total Proteins**
tRNA aminoacylation for protein translation (GO: 0006418)	Aim10p, Cdc60p, Ded81p, Dia4p, Dps1p, Msk1p, Slm5p, Ths1p, Wrs1p	9/23	Gln4p, Nam2p	2/16
Amino acid transport (GO: 0006865)	Avt7p, Bap2p, Lst4p, Put4p, Uga4p, Ydl119cp	6/9	Alp1p, Avt4p, Btn2p, Gnp1p, Ssy1p	5/8
End of the second fermentation, EF				
	Specific Proteins	**Specific/Total Proteins**	**Specific Proteins**	**Specific/Total Proteins**
Cellular amino acid metabolic process (GO: 0006520)	Ade3p, Adh3p, Arg1p, Arg7p, Arg8p, Arg80p, Arg82p, Aro1p, Aro2p, Aro3p, Asn1p, Asn2p, Bna1p, Car1p, Car2p, Cys3p, Cys4p, Dtd1p, Gdh1p, Gln1p, Gus1p, His4p, Hom6p, Hpa3p, Idh1p, Ilv6p, Kgd2p, Leu4p, Lpd1p, Lys12p, Lys20p, Mcm1p, Met17p, Mis1p, Mmf1p, Pro2p, Sam2p, Ser1p, Ser3p, Sfa1p, Shm2p, Uga1p, Yhr208wp	43/56	Gcv3p, Gfa1p, Idh2p, Idp1p, Met13p, Met8p, Sno3p, Snz1p, Ybr145wp	9/22
tRNA aminoacylation for protein translation (GO: 0006418)	Arc1p, Grs1p, Gus1p, Ism1p, Msk1p, Ses1p, Vas1p	7/8	Msm1p	1/2
Amino acid transport (GO: 0006865)	Npr2p	1/1		

3.1. Amino Acid Transport

Amino acids are involved in numerous metabolic pathways. *S. cerevisiae* can synthesize them *de novo*; they are also actively imported from the extracellular environment. This is less expensive in terms of energy; the estimated cost of *de novo* amino acid synthesis under aerobic conditions is between 9.5 ATP for glutamate and 75.5 ATP for tryptophan [19]. Yeast export of amino acids enables them to respond to different environments and take advantage of available resources. The amino acids present in the vacuole are histidine, lysine, and arginine, constituting 80 to 90%, while a similar fraction of glutamate and aspartate is present in the cytosol [20–22]. *S. cerevisiae* incorporates amino acids from the medium, making use of proteins known as amino acids permeases by a simport mechanism [23]. The multiplicity and diversity of transporters allows yeast to accumulate amino acids for biosynthesis and catabolism under multiple conditions and in a wide range of external concentrations [24].

Of the total transporters identified in *S. cerevisiae* (41 transporters) [19], 6 transporters (Avt7p, Bap2p, Lst4p, Put4p, Uga4p, Ydl119cp) in MF and 1 (Npr2p) in EF have been obtained in the P29 strain. While in the G1 strain, 5 (Alp1p, Avt4p, Btn2p, Gnp1p, Ssy1p) in MF and 0 in EF have been reported. In view of these results, it would not be expected to find a great difference between the two strains. However, the transporters identified in both types of yeast were different, although all transport the full range of amino acids from the cell exterior to the interior and from the cell interior to the vacuole or mitochondria.

Of the transporters identified in the P29 strain, Avt7p was the one with the highest protein content (0.028 mol%). Data are shown in Table 2. Yeast Avtp proteins can be subdivided into four main groups, defined as Avt1p, Avt2p, Avt3p/4p, and Avt5p/6p/7p [25]. Avt1p intervenes in the active absorption of amino acids in the vacuole, while Avt3p, Avt4p, and Avt6p are involved in the active exit of amino acids from the vacuole into the cytosol. Avt7p is located in the vacuolar membrane of *S. cerevisiae* and participates in the vacuolar absorption of glutamine and proline and in the formation of spores in *S. cerevisiae*, which could be related to the Avt7p-dependent amino acid flux of vacuoles under the conditions of lack of nutrients [26]. On the other hand, the protein that had the highest protein content in the G1 yeast was Btn2p (0.022 mol%); this protein participates in the uptake of arginine [27].

Table 2. Quantification of specific proteins identified in each strain using the EmPAI method (mol%) in *S. cerevisiae* P29 and G1 during two sampling times: at the middle of the second fermentation (MF) and at the end of the second fermentation (EF).

MF						EF					
P29 Strain			G1 Strain			P29 Strain			G1 Strain		
Proteins	mol%	SD	Proteins	mol%	SD	Proteins	mol%	SD	Proteins	mol%	SD
Aat1p	0.024	0.0002	Aco2p	0.028	0.0003	Ade3p	0.007	0.0001	Gcv3p	0.427	0.004
Ald2p	0.010	0.0001	Adi1p	0.048	0.0005	Adh3p	0.078	0.0008	Gfa1p	0.033	0.0008
Arg81p	0.006	0.0001	Aro10p	0.010	0.0001	Arg1p	0.114	0.0011	Idh2p	0.081	0.001
Aro9p	0.028	0.0003	Bna4p	0.010	0.0001	Arg7p	0.055	0.0006	Idp1p	0.0465	0.0005
Dal81p	0.007	0.0001	Cpa2p	0.028	0.0003	Arg8p	0.013	0.0001	Met13p	0.051	0.001
Gcv2p	0.005	0.0001	Ehd3p	0.011	0.0001	Arg80p	0.039	0.0004	Met8p	0.069	0.0007
Gdh3p	0.057	0.0006	Gcv1p	0.013	0.0001	Arg82p	0.0187	0.0002	Sno3p	0.106	0.0011
Ilv6p	0.016	0.0002	Gfa1p	0.015	0.0001	Aro1p	0.004	0.0000	Snz1p	0.076	0.0006
Leu9p	0.048	0.0005	Gln4p	0.005	0.0003	Aro2p	0.150	0.0015			
Lys1p	0.052	0.0005	Gly1p	0.108	0.0011	Aro3p	0.276	0.0028			
Lys4p	0.015	0.003	His5p	0.060	0.0006	Asn1p	0.082	0.0008			
Met1p	0.015	0.0001	Idp1p	0.031	0.0003	Asn2p	0.036	0.0004			
Mis1p	0.005	0.0000	Lap3p	0.046	0.0005	Bna1p	0.039	0.0004			
Put1p	0.011	0.0001	Lys21p	0.138	0.0014	Car1p	0.134	0.0013			
Snz3p	0.016	0.0002	Mae1p	0.016	0.0002	Car2p	0.055	0.0006			
Ths1p	0.011	0.0001	Met12p	0.017	0.0002	Cys3p	0.113	0.0011			
Uga2p	0.011	0.0001	Met16p	0.022	0.0002	Cys4p	0.131	0.0013			
Uga3p	0.010	0.0001	Mri1p	0.014	0.0001	Dtd1p	0.114	0.0011			
Ura7p	0.017	0.0002	Nit3p	0.016	0.0002	Gdh1p	0.139	0.0014			
Utr4p	0.029	0.0003	Pet112p	0.009	0.0005	Gln1p	0.305	0.0031			
Yml082wp	0.008	0.0001	Ser1p	0.043	0.0004	Gus1p	0.041	0.0004			
Ymr084wp	0.019	0.0002	Sno3p	0.023	0.0002	His4p	0.093	0.0009			
Put4p	0.013	0.0001	Tum1p	0.015	0.0002	Hom6p	0.158	0.0016			
Uga4p	0.019	0.0004	Uga1p	0.081	0.0008	Hpa3p	0.116	0.0011			
Lst4p	0.012	0.0001	Yhr112cp	0.014	0.0001	Idh1p	0.079	0.0008			
Bap2p	0.012	0.0001	Yll058wp	0.008	0.0001	Ilv6p	0.068	0.0007			
Avt7p	0.028	0.0006	Yhr208wp	0.161	0.0016	Kgd2p	0.056	0.0006			
Ydl119cp	0.014	0.001	Alp1p	0.016	0.0002	Leu4p	0.041	0.0004			
			Gnp1p	0.009	0.0001	Lpd1p	0.011	0.0001			
			Avt4p	0.009	0.0001	Lys12p	0.326	0.007			
			Btn2p	0.021	0.0002	Lys20p	0.028	0.0003			
			Ssy1p	0.008	0.0001	Mcm1p	0.214	0.0021			
			Odc2p	0.015	0.0002	Met17p	0.054	0.0005			
						Mis1p	0.006	0.0001			
						Mmf1p	0.546	0.009			
						Pro2p	0.040	0.0004			
						Sam2p	0.205	0.003			
						Ser1p	0.017	0.0002			
						Ser3p	0.020	0.0002			
						Sfa1p	0.080	0.0008			
						Shm2p	0.039	0.0004			
						Uga1p	0.013	0.0001			

3.2. tRNA Aminoacylation for Protein Translation

Regarding tRNA aminoacylation for protein translation, more specific proteins were identified in the P29 strain than in the G1 strain: 16 proteins *versus* 3 proteins, respectively (Table 1, Figure 1). The formation of aminoacyl-tRNA (aa-tRNA) is an essential process in protein biosynthesis. Proteins of the aminoacyl-tRNA synthetase family can be classified into two groups, depending on the specificity of amino acids: (i) class I (specific for glutamine, glutamate, arginine, cysteine, methionine, valine, isoleucine, leucine, tyrosine, and tryptophan); (ii) class II (specific for glycine, alanine, proline, serine, threonine, histidine, aspartate, asparagine, lysine, and phenylalanine) [28–31]. At the middle of the second fermentation in the P29 strain, 9 proteins were identified (Aim10p, Cdc60p, Ded81p, Dia4p, Dps1p, Msk1p, Slm5p, Ths1p, and Wrs1p), and at the end of the second fermentation, 7 proteins (Arc1p, Grs1p, Gus1p, Ism1p, Msk1p, Ses1p, and Vas1p) were identified. These proteins are responsible for binding amino acids such as tryptophan (Wrs1p), threonine (Ths1p), asparagine (Ded81p, Slm5p), lysine (Msk1p), aspartate (Dps1p), serine (Dia4p, Ses1p), leucine (Cdc60p), methionine (Arc1p), glutamate (Arc1p, Gus1p), glycine (Grs1p), isoleucine (Ism1p), and valine (Vas1p). These results suggest that yeast may be activating its transcription and translation machinery to try to cope with cell death and autolysis, and they corroborate those previously obtained by our research group [30].

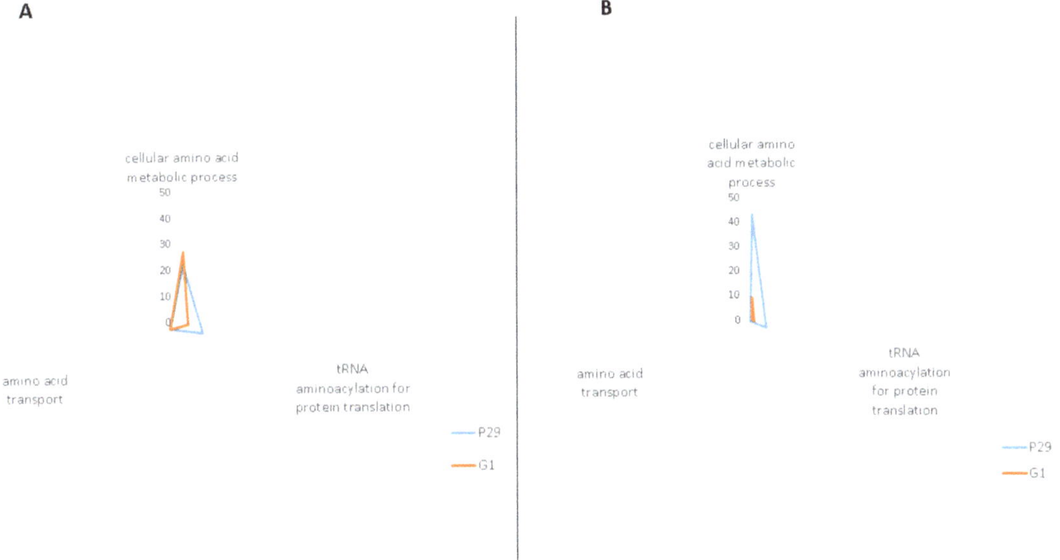

Figure 1. Total number of specific proteins related to amino acid metabolism in *S. cerevisiae* P29 (blue colour) and G1 (orange colour) during two sampling times: at the middle of the second fermentation, MF (**A**), and at the end of the second fermentation, EF (**B**).

On the other hand, the specific proteins identified in the G1 strain were Gln4p and Nam2p, which catalyze the binding of glutamine and leucine, respectively, the nuclear protein Pet112p, required to maintain rho⁺ mitochondrial DNA [31], and Msm1p; this protein is a mitochondrial methionyl-tRNA synthetase. These results seem to indicate that the protein synthesis machinery in flor yeast (strain G1) is not working, though yeast autolysis may be occurring, which would help the survival of the rest of the population in this harsh condition [32].

3.3. Cellular Amino Acid Metabolic Process

In the P29 strain, a total of 67 specific proteins related to the cellular amino acid metabolic processes were obtained (24 in MF; 43 in EF), while in the G1 strain, 37 proteins were identified (28 in MF; 9 in EF). Data are shown in Table 1.

The metabolism of amino acids is affected by the pressure the yeast cells are subjected to in the second fermentation, carried out in the bottle, at both levels of the anabolism and catabolism of amino acids. In order to understand the possible interactions between the differential proteins from each strain involved in amino acid metabolism and to provide a better understanding of the GO terms obtained, a protein–protein interaction network map was created for sampling time (MF, EF) using STRING v11 [18]. Because metabolism is somewhat dynamic, these connections can be conceptualized as networks, and the size and complex organization of these networks present a unique opportunity to obtain a global visualization of the yeast genome during the second fermentation.

3.3.1. Cellular Amino Acid Metabolic Process at the Middle of the Second Fermentation

At the middle of the second fermentation, 24 specific proteins in the P29 strain and 28 specific proteins in the G1 strain were identified. These proteins are represented as nodes in Figure 2A,B, respectively. A total of 28 interactions (number of edges) were observed, with a *p*-value of PPI enrichment $<2.23 \times 10^{-12}$ in the P29 strain; while in the G1 strain, 36 interactions were obtained, with a *p*-value of PPI enrichment $<4.44 \times 10^{-16}$; such enrichment indicates that the proteins are at least partially biologically connected as a group.

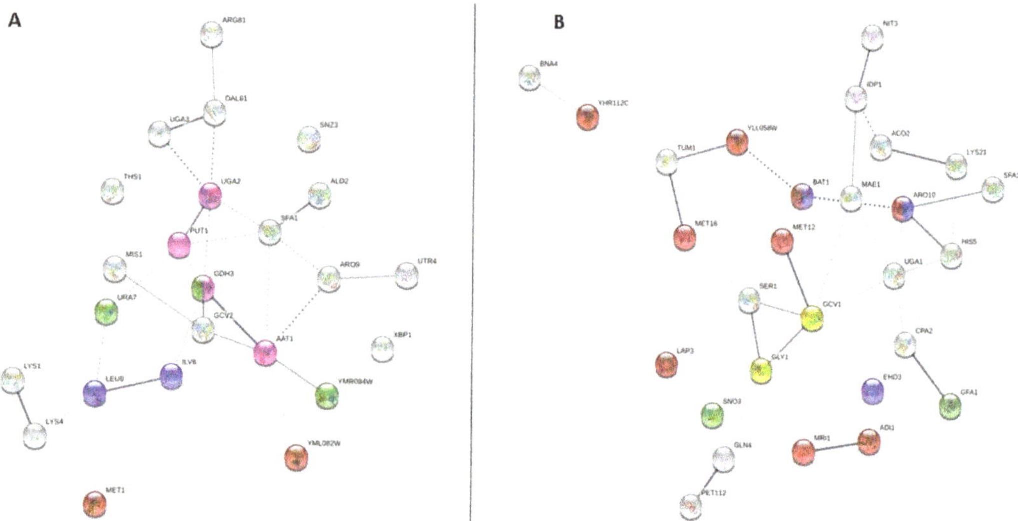

Figure 2. High confidence protein–protein interaction network map built using STRING v11 and based on the proteins of cellular amino acid metabolic process detected in *S. cerevisiae* P29 (**A**) and G1 (**B**) in the middle of the second fermentation. Proteins are shown as nodes, and the existence of interactions between them are represented by lines (the connection between nodes). Line thickness indicates the strength of the different interactions. Nodes with the same color represent specific clusters: sulfur amino acid metabolic process (red nodes), branched-chain amino acid metabolic process (light blue nodes), glutamine metabolic process (light green nodes), glycine metabolic process (yellow nodes), and glutamate metabolic process (light purple nodes).

In the P29 strain, the strongest interactions were observed between proteins belonging to lysine metabolism (white nodes) and the metabolism of branched chain amino acids (dark blue nodes). These amino acids are used for the synthesis of numerous volatile aromatic compounds, such as higher alcohols. Furthermore, interactions were obtained between proteins belonging to different GO terms: Gdh3p-Aat1p, Uga3p-Dal81p, and Put1p-Uga2p. The Snz3p, Xbp1p, and Ths1p proteins showed no interaction with the rest of the proteins (Figure 2A). On the other hand, the strongest interactions present in the G1 strain were between proteins belonging to the metabolism of sulfur amino acids, the metabolism of lysine, and the metabolism of glycine. However, most proteins are involved with the metabolism of sulfur amino acids such as methionine and cysteine, which are necessary for yeast growth [33]. Lap3p, Sno3p, and Ehd3p did not show any interaction but were included in different GO terms (Figure 2B). In both strains, the metabolism of lysine was identified. This could be related to an antioxidant response by yeasts [34].

The main difference between both strains in the middle of the second fermentation is glutamate metabolism in the P29 strain and glycine and tryptophan metabolism in the G1 strain. These data are in agreement with the results obtained by Marsit et al. (2016) [6]. These authors reported an increase in the assimilation of various amino acids, particularly glutamate and glycine. Both amino acids are related to the synthesis of most amino acids. Furthermore, the reversible conversion of glutamate to α-ketoglutarate is an important branch point between carbon and nitrogen metabolism [6]. In addition, glutamate is one of the nitrogen sources that maintain a high growth rate [35]. During the early stages of fermentation, *S. cerevisiae* rapidly accumulates those amino acids necessary for protein synthesis and formation of volatile compounds such as higher alcohols and their respective ethyl and acetic esters [12], while the surplus is stored in the cellular vacuole, because an amino acid deficit causes a decrease in cell biomass [21]. During the fermentation process, bioactive compounds such as melatonin and serotonin may form because of the tryptophan metabolism, in addition to the higher alcohol tryptophol, which are compounds that have an impact on the sensory properties of the wine [22]. In this case, the identified protein belonging to the tryptophan metabolism, Bna4p (0.01 mol%), is a kynurenine 3-monooxygenase. This protein is required for the *de novo* biosynthesis of NAD$^+$ from tryptophan via kynurenine.

3.3.2. Cellular Amino Acid Metabolic Process at the End of the Second Fermentation

At the end of the second fermentation, 43 specific proteins were identified in the P29 strain and 9 specific proteins in the G1 strain. These proteins were represented as nodes in the P29 strain (Figure 3A) and G1 strain (Figure 3B). A total of 234 interactions (number of edges) were observed, with a *p*-value of PPI enrichment $<1.0 \times 10^{-16}$ in the P29 strain, while in the G1 strain, 4 interactions were obtained, with a *p*-value of PPI enrichment <0.00047; such enrichment indicates that the proteins are at least partially biologically connected as a group. In view of the results obtained, the P29 strain increased the number of proteins related to the cellular metabolism of amino acids at the end of the second fermentation, unlike the G1 strain.

In the P29 strain, the strongest interactions were observed between proteins belonging to the arginine metabolic process, the sulfur amino acid metabolic process, and the branched-chain amino acid metabolic process; those that obtained a greater number of proteins involved in their respective processes being the metabolism of arginine and sulfur amino acids (cysteine and methionine), 8 proteins in each process. Sulfur in methionine amino acid can be incorporated into cysteine. Furthermore, methionine has been shown to impact oxidative stress resistance and has the potential to be catabolized into α-ketoglutarate, which can directly enter the central carbon metabolism, providing additional support for why the demand for cell uptake was high. Glutamate, the most abundant supplemented amino acid, showed the highest uptake yield, and although no directly linked catabolic enzymes were significantly up-allocated, glutamate may be utilized metabolically for other transamination reactions because it is the major amino acid for this role [36]. On the other hand, Bna1p and Mmf1p did not interact with any protein. Bna1p is required for the *de novo* biosynthesis of NAD from tryptophan via kynurenine, and Mmf1p is required for the transamination of isoleucine. For the branched-chain amino acids (BCAA) isoleucine, leucine, and valine and the aromatic amino acids (AAA) phenylalanine, tryptophan, and tyrosine, catabolism, via the Ehrlich pathway, is a way for the cell to salvage nitrogen through transamination reactions [37].

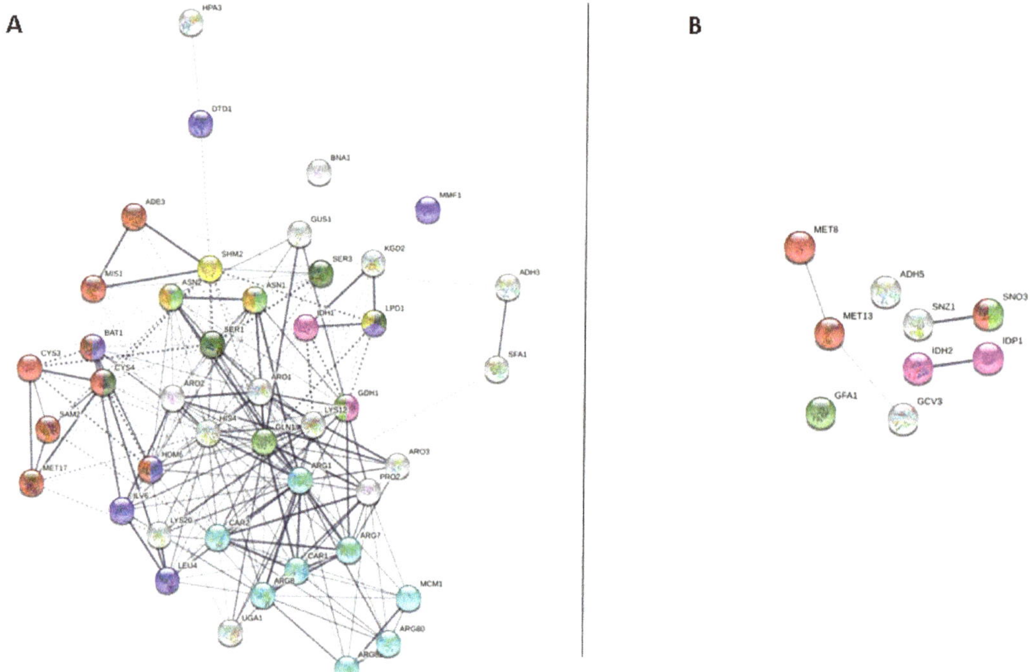

Figure 3. High confidence protein–protein interaction network map built using STRING v11 and based on the proteins of cellular amino acid metabolic process detected in *S. cerevisiae* P29 (**A**) and G1 (**B**) at the end of the second fermentation. Proteins are shown as nodes, and the existence of interactions between them are represented by lines (the connection between nodes). Line thickness indicates the strength of the different interactions. Nodes with the same color represent specific clusters: sulfur amino acid metabolic process (red nodes), branched-chain amino acid metabolic process (light blue nodes), glutamine metabolic process (light green nodes), glycine metabolic process (yellow nodes), glutamate metabolic process (light purple nodes), serine metabolic process (dark green nodes), arginine metabolic process (light blue nodes), and asparagine metabolic process (orange nodes).

In contrast, in the G1 strain, this strong interaction was between proteins involved with the glutamate metabolic process. However, Adh5p and Gfa1p did not interact. Adh5p is an alcohol dehydrogenase isoenzyme V and is involved in ethanol production, while Gfa1p converts fructose-6-phosphate to glucosamine-6-phosphate. This seems to indicate that the yeast is trying to maintain its cell growth and be able to cope with stressful conditions such as overpressure of CO_2.

Despite having a different number of proteins, both strains presented amino acids responsible for a high rate of cell growth [20], such as asparagine (P29 strain) and glutamate (G1 strain). Therefore, maintenance of cell viability should be expected in both strains during the second fermentation. The metabolism of these nitrogen compounds depends on the yeast strain, its physiological state, and the physicochemical properties of the wine and fermentation conditions [38]. This difference obtained with respect to the cellular metabolism of amino acids at the end of the second fermentation between both strains could be because they are different strains of yeast: a typical one for the elaboration of this type of wine (P29 strain) and a flor yeast (G1 strain) subjected to unusual CO_2 overpressure conditions for this type of strain. This suggests that the G1 strain at the end of the second fermentation is stressed, viability is low, and its nitrogen metabolism is strong affected, stopping the biosynthesis of other nitrogenous cellular constituents as well as other compounds formed from its carbon skeleton [10].

The number of identified proteins was higher in the P29 strain compared to the G1 strain, which offered the possibility of establishing a greater number of interactions between these proteins. In both strains, the metabolism of sulfur amino acids, methionine, and cysteine obtained a greater number of proteins involved in these processes: 5 proteins in the P29 strain and 9 proteins in the G1 strain. In strain P29, glutamate metabolism in the middle of the second fermentation and arginine and sulfur amino acids metabolism at the end of the second fermentation stood out; while in the G1 strain, glycine and tryptophan metabolism stood out in the middle of the second fermentation and glutamate metabolism at the end of the second fermentation.

These results represent a first approach in the search for a greater and broader knowledge of the metabolism of yeast during the production of Spanish sparkling wine (cava). However, more research would be necessary, along with a metabolomic study where amino acids are quantified and an assay of enzymatic activities is performed, to reach more solid conclusions. The confirmation of these results by additional tests based on genomic, transcriptomic, and metabolomic activity assays could help achieve a better understanding of the metabolism of amino acids and, therefore, of yeast during the second fermentation in the production of sparkling wines.

Author Contributions: T.G.-M. conceptualized the research, performed experiments, provided resources, and supervised the work; J.M. conceptualized the research, supervised the work, and provided resources; A.P.-P. performed experiments; J.M.-G. participated in the investigation; J.C.M. conceptualized the research, and reviewed and edited the manuscript; M.d.C.G.-J. performed data analysis and wrote the original draft. All authors have read and agreed to the published version of the manuscript.

Funding: This research was granted funding from the "Consejería de Economía, Conocimiento, Empresas y Universidad", Programa Operativo FEDER de Andalucía 2014-2020 (T García-Martínez, J Moreno), Reference 1380480-R.

Institutional Review Board Statement: Not applicable.

Informed Consent Statement: Not applicable.

Data Availability Statement: Not applicable.

Acknowledgments: Kind help of the staff at the Central Research Support Service (SCAI) of the University of Cordoba with the protein analyses is gratefully acknowledged.

Conflicts of Interest: The authors declare no conflict of interest.

References

1. Mauricio, J.C.; Ortega, J.M.; Salmon, J.M. Sugar uptake by three strains of Saccharomyces cerevisiae during alcoholic fermentation at different initial ammoniacal nitrogen concentrations. *Int. Symp. Vitic. Enol.* **1993**, *388*, 197–202. [CrossRef]
2. Moreira, N.; De Pinho, P.G.; Santos, C.; Vasconcelos, I. Relationship between nitrogen content in grapes and volatiles, namely heavy sulphur compounds, in wines. *Food Chem.* **2011**, *126*, 1599–1607. [CrossRef] [PubMed]
3. Giudici, P.; Kunkee, R.E. The effect of nitrogen deficiency and sulfur-containing amino acids on the reduction of sulfate to hydrogen sulfide by wine yeasts. *Am. J. Enol. Viticult.* **1994**, *45*, 107–112.
4. Barbosa, C.; Mendes-Faia, A.; Mendes-Ferreira, A. The nitrogen source impacts major volatile compounds released by Saccharomyces cerevisiae during alcoholic fermentation. *Int. J. Food Microbiol.* **2012**, *160*, 87–93. [CrossRef]
5. Su, Y.; Seguinot, P.; Sanchez, I.; Ortiz-Julien, A.; Heras, J.M.; Querol, A.; Camarasa, C.; Guillamón, J.M. Nitrogen sources preferences of non-Saccharomyces yeasts to sustain growth and fermentation under winemaking conditions. *Food Microbial.* **2020**, *85*, 103287. [CrossRef] [PubMed]
6. Marsit, S.; Sanchez, I.; Galeote, V.; Dequin, S. Horizontally acquired oligopeptide transporters favour adaptation of Saccharomyces cerevisiae wine yeast to oenological environment. *Environ. Microbiol.* **2016**, *18*, 1148–1161. [CrossRef]
7. Henschke, P.A.; Jiranek, V. Yeast-metabolism of nitrogen compounds. In *Wine Microbiology and Biotechnology*; Fleet, G.H., Ed.; Harwood Academic Publishers: Langhorne, PA, USA, 1994; pp. 77–164.

8. Moreno-Arribas, M.V.; Polo, M.C. Amino acids and biogenic amines. In *Wine Chemistry and Biochemistry*; Springer: New York, NY, USA, 2009; pp. 163–189.
9. Kevvai, K.; Kütt, M.L.; Nisamedtinov, I.; Paalme, T. Simultaneous utilization of ammonia, free amino acids and peptides during fermentative growth of Saccharomyces cerevisiae: Simultaneous utilization of ammonia, free amino acids and peptides. *J. Inst. Brew.* **2016**, *122*, 110–115. [CrossRef]
10. Large, P.J. Degradation of organic nitrogen compounds by yeasts. *Yeast* **1986**, *2*, 1–34. [CrossRef]
11. Zhong, X.; Wang, A.; Zhang, Y.; Wu, Z.; Li, B.; Lou, H.; Huang, G.; Wen, H. Reducing higher alcohols by nitrogen compensation during fermentation of Chinese rice wine. *Food Sci. Biotechnol.* **2020**, *29*, 805–816. [CrossRef]
12. González-Jiménez, M.C.; Moreno-García, J.; García-Martínez, T.; Moreno, J.J.; Puig-Pujol, A.; Capdevilla, F.; Mauricio, J.C. Differential analysis of proteins involved in ester metabolism in two Saccharomyces cerevisiae strains during the second fermentation in sparkling wine elaboration. *Microorganisms* **2020**, *8*, 403. [CrossRef]
13. Porras-Agüera, J.A.; Moreno-García, J.; González-Jiménez, M.C.; Mauricio, J.C.; Moreno, J.; García-Martínez, T. Autophagic proteome in two Saccharomyces cerevisiae strains during second fermentation for sparkling wine elaboration. *Microorganisms* **2020**, *8*, 523. [CrossRef] [PubMed]
14. Martínez-García, R.; Roldán-Romero, Y.; Moreno, J.; Puig-Pujol, A.; Mauricio, J.C.; García-Martínez, T. Use of a flor yeast strain for the second fermentation of sparkling wines: Effect of endogenous CO2 over-pressure on the volatilome. *Food Chem.* **2020**, *308*, 125555. [CrossRef] [PubMed]
15. Martínez-García, R.; García-Martínez, T.; Puig-Pujol, A.; Mauricio, J.C.; Moreno, J. Changes in sparkling wine aroma during the second fermentation under CO2 pressure in sealed bottle. *Food Chem.* **2017**, *237*, 1030–1040. [CrossRef] [PubMed]
16. Bradford, M.M. A Rapid and sensitive method for the quantitation of microgram quantities of protein utilizing the principle of protein-dye binding. *Anal. Biochem.* **1976**, *72*, 248–254. [CrossRef]
17. Ishihama, Y.; Oda, Y.; Tabata, T.; Sato, T.; Nagasu, T.; Rappsilber, J.; Mann, M. Exponentially modified protein abundance Index (emPAI) for estimation of absolute protein amount in proteomics by the number of sequenced peptides per protein. *Mol. Cell. Proteom.* **2005**, *4*, 1265–1272. [CrossRef] [PubMed]
18. Szklarczyk, D.; Gable, A.L.; Lyon, D.; Junge, A.; Wyder, S.; Huerta-Cepas, J.; Simonovic, M.; Doncheva, N.T.; Morris, J.H.; Bork, P.; et al. STRING V11: Protein-protein association networks with increased coverage, supporting functional discovery in genome-wide experimental datasets. *Nucleic Acids Res.* **2019**, *47*, D607–D613. [CrossRef]
19. Bianchi, F.; van't Klooster, J.S.; Ruiz, S.J.; Poolman, B. Regulation of amino acid transport in Saccharomyces cerevisiae. *Microbiol. Mol. Biol. Rev.* **2019**, *83*, e00024-19. [CrossRef] [PubMed]
20. Mulero, J.J.; Rosenthal, J.K.; Fox, T.D. PET112, a Saccharomyces cerevisiae nuclear gene required to maintain rho+ mitochondrial DNA. *Curr. Genet.* **1994**, *25*, 299–304. [CrossRef] [PubMed]
21. Vilanova, M.; Ugliano, M.; Varela, C.; Siebert, T.; Pretorius, I.S.; Henschke, P.A. Assimilable nitrogen utilisation and production of volatile and non-volatile compounds in chemically defined medium by Saccharomyces cerevisiae wine yeasts. *Appl. Microbiol. Biotechnol.* **2007**, *77*, 145–157. [CrossRef]
22. Fernández-Cruz, E.; Álvarez-Fernández, M.A.; Valero, E.; Troncoso, A.M.; García-Parrilla, M.C. Melatonin and derived L-tryptophan metabolites produced during alcoholic fermentation by different wine yeast strains. *Food Chem.* **2017**, *217*, 431–437. [CrossRef]
23. Seaston, A.; Inkson, C.; Eddy, A.A. The absorption of protons with specific amino acids and carbohydrates by yeast. *Biochem. J.* **1973**, *134*, 1031–1043. [CrossRef]
24. Chianelli, M.S. Sistemas De Transporte De Aminoácidos Neutros En saccharomyces Cerevisiae, Cepas Silvestres Y Mutantes Transporte-Defectivas. Ph.D. Thesis, Universidad de Buenos Aires (Facultad de Ciencias Exactas y Naturales), Buenos Aires, Argentina, 1998.
25. Russnak, R.; Konczal, D.; McIntire, S.L. A family of yeast proteins mediating bidirectional vacuolar amino acid transport. *J. Biol. Chem.* **2001**, *276*, 23849–23857. [CrossRef]
26. Tone, J.; Yamanaka, A.; Manabe, K.; Murao, N.; Kawano-Kawada, M.; Sekito, T.; Kakinuma, Y. A vacuolar membrane protein Avt7p is involved in transport of amino acid and spore formation in Saccharomyces cerevisiae. *Biosci. Biotech. Bioch.* **2015**, *79*, 190–195. [CrossRef] [PubMed]
27. Kim, Y.; Chattopadhyay, S.; Locke, S.; Pearce, D.A. Interaction among Btn1p, Btn2p, and Ist2p reveals potential interplay among the vacuole, amino acid levels, and ion homeostasis in the yeast Saccharomyces cerevisiae. *Eukaryot. Cell.* **2005**, *4*, 281–288. [CrossRef] [PubMed]
28. Delarue, M. Aminoacyl-tRNA synthetases. *Curr. Opin. Struct. Biol.* **1995**, *5*, 48–55. [CrossRef]
29. Arnez, J.G.; Moras, D. Structural and functional considerations of the aminoacylation reaction. *Trends Biochem. Sci.* **1997**, *22*, 211–216. [CrossRef]
30. González-Jiménez, M.D.C.; García-Martínez, T.; Puig-Pujol, A.; Capdevila, F.; Moreno-García, J.; Moreno, J.; Mauricio, J.C. Biological processes highlighted in Saccharomyces cerevisiae during the sparkling wines elaboration. *Microorganisms* **2020**, *8*, 1216. [CrossRef]
31. Eriani, G.; Delarue, M.; Poch, O.; Gangloff, J.; Moras, D. Partition of tRNA synthetases into two classes based on mutually exclusive sets of sequence motifs. *Nature* **1990**, *347*, 203–206. [CrossRef] [PubMed]

32. Porras-Agüera, J.A.; Moreno-García, J.; Mauricio, J.C.; Moreno, J.; García-Martínez, T. First proteomic approach to identify cell death biomarkers in oenological yeasts during sparkling wine production. *Microorganisms* **2019**, *7*, 542. [CrossRef] [PubMed]
33. Olin-Sandoval, V.; Shu Lim Yu, J.; Miller-Fleming, L.; Tauqeer Alam, M.; Kamrad, S.; Correia-Melo, C.; Haas, R.; Segal, J.; Peña Navarro, D.A.; Herrera-Dominguez, L.; et al. Lysine harvesting is an antioxidant strategy and triggers underground polyamine metabolism. *Nature* **2019**, *572*, 249–253. [CrossRef]
34. Huang, C.W.; Walker, M.E.; Fedrizzi, B.; Gardner, R.C.; Jiranek, V. Hydrogen sulfide and its role in Saccharomyces cerevisiae in a wine-making context. *Yeast Res. FEMS* **2017**, *17*, fox058. [CrossRef] [PubMed]
35. Kessi-Pérez, E.I.; Molinet, J.; Martínez, C. Disentangling the genetic bases of Saccharomyces cerevisiae nitrogen consumption and adaptation to low nitrogen environments in wine fermentation. *Biol. Res.* **2020**, *53*, 1–10. [CrossRef]
36. Eldarov, M.A. Metabolic engineering of wine strains of Saccharomyces cerevisiae. *Genes* **2020**, *11*, 964. [CrossRef] [PubMed]
37. Björkeroth, J.; Campbell, K.; Malina, C.; Yu, R.; Di Bartolomeo, F.; Nielsen, J. Proteome reallocation from amino acid biosynthesis to ribosomes enables yeast to grow faster in rich media. *PNAS* **2020**, *117*, 21804–21812. [CrossRef] [PubMed]
38. Mauricio, J.C.; Millán, C.; Ortega, J.M. Influence of oxygen on the biosynthesis of cellular fatty acids, sterols and phospholipids during alcoholic fermentation by Saccharomyces cerevisiae and Torulaspora delbrueckii. *World J. Microbiol. Biotechnol.* **1998**, *14*, 405–410. [CrossRef]